Analysis of Airborne Particles by Physical Methods

Editor

Hanns Malissa

Institute of Analytical Chemistry and Microchemistry
Technical University of Vienna
Vienna, Austria

Editor-in-Chief
CRC Analytical Chemistry for Environmental Control Uniscience Series

J. W. Robinson

Department of Chemistry
Louisiana State University
Baton Rouge, Louisiana

CRC Press, Inc.
2255 Palm Beach Lakes Blvd · West Palm Beach, Florida 33409

Library of Congress Cataloging in Publication Data

Main entry under title:

Analysis of airborne particles by physical methods.

   Bibliography: p.
   1. Aerosols – Analysis. 2. Dust. I. Malissa,
Hanns. II. Robinson, James W., 1923–
TD890.A48   628.5'3   77-21420
ISBN 0-8493-5275-4

© 1978 by CRC Press, Inc.

International Standard Book Number 0-8493-5275-4

Library of Congress Card Number 77-21420
Printed in the United States

FOREWORD

The assessment of the environmental impact of man's endeavors must be made as rapidly and painlessly as possible. It is not only aesthetically rewarding to retrieve clean lakes, rivers, air, and countrysides; it is possibly vital to man's continued existence. While panic and hysteria serve no useful long-range purpose, but in fact boomerang and disenchant the public after some period of time, nevertheless it is encumbent on man to safeguard his environment. Clearly, the forces involved are quite beyond our comprehension at this time. When we are operating from a position of ignorance as we are today, it is even more important to be sure that the risks we take are minimal.

Discussion on the possible short-term and long-term impacts of various pollutants is valuable but probably endless. The truth ultimately emerges from reliable data interpreted with wisdom and understanding. Analytical chemistry provides the invaluable bridge between speculation and firm data. We can only generate firm data with reliable analytical techniques and skilled scientists.

In an effort to provide means of collecting and dispersing this information to all interested parties, we have invited a number of scientists of stature to produce monographs in their field of expertise. The objective of these monographs is to document analytical procedures and techniques that are useful to the environmentalists.

In general, three groups of people will be interested in using these monographs: industrial engineers and scientists who are monitoring both liquid and gaseous effluents from their plants; government scientists, both federal and state, who are monitoring the industrial effluents; and environmentalists who are trying to assess pollution levels and amass data on long-term health effects and other facets of pollution.

This book, edited by Professor Dr. H. Malissa, is devoted to the broad field of particulate analysis. This monograph has been produced by the joint efforts of a number of acknowledged experts in the field. It will provide invaluable assistance to new laboratories set up by state, federal, industrial, and other research organizations to retrieve pollution data. It will also provide an excellent source of reference materials to laboratories already established and working in the field.

Primarily, the book is directed to the analytical chemist and the control chemist, but these data will concern the engineers who design our industrial facilities, the economists who set the cost of these facilities and their products, the politicians who set our countries policies, and the general public who pays for and hopefully benefits from these endeavors.

J. W. Robinson
Editor-in-Chief
June 1977

EDITOR-IN-CHIEF

Dr. J. W. Robinson obtained his Ph.D. at The University of Birmingham, England in 1952, in Analytical Chemistry. He came to the U.S. in 1955 and worked in industrial research at Exxon Research Corporation and Ethyl Corporation Research in Baton Rouge, Louisiana until 1964, when he joined the faculty at Louisiana State University in Baton Rouge where he is presently Professor of Chemistry.

Dr. Robinson has published over 100 original papers and four books. He is the Editor of the *CRC Handbook of Spectroscopy*, Volumes I and II, and the monthly international member of the Editorial Board of *Spectroscopy Reviews and Analytica Chimica Acta.* member of the Editorial Board of *Spectroscopy Reviews* and *Analytica Chimica Acta.*

In 1975, Dr. Robinson was Chairman of the Gordon Conference on Analytical Chemistry and a Guggenheim Fellow. His research work has been mostly devoted to analytical chemistry and has found application in industrial analysis, but, more particularly, in the research areas of environmental control.

PREFACE

The air we breathe is — as analytical chemistry tells us — a mixture of gaseous, liquid, and solid components. The solid components of the air, usually referred to as airborne particles or dust, can be of natural or anthropogenic origin and vary widely in respect to their concentration in air, their chemical composition, homogeneity, size, shape, and morphology. Such particles can have a direct physiological influence on life; they can act as condensation nuclei or as catalysts for reactions taking place in the atmosphere. Unfortunately, scientific knowledge about airborne particles is still insufficient to provide an accurate assessment of their influence on the environment or their involvement in the chemistry of the atmosphere.

Therefore, it is necessary to perform an accurate and complete characterization of the solid pollutants in air and especially to discuss the analytical data obtained by taking into account the physico-chemical and technological data available concerning the origin, transformation, and the effects of dust. This synoptic approach, which is known as "integrated dust analysis" is the underlying concept of this book.

This monograph presents the analytical possibilities of characterizing airborne particles utilizing mainly physical methods. Initially, there is a description of the situation that confronts the analytical chemist and a definition of the goals of analysis. The different techniques for sampling particulate matter are then presented. After sampling, the analytical task involves obtaining information concerning the average content of the elements in the dust sample or the air. X-Ray fluorescence, atomic emission and absorption spectroscopy, radioactivity measurements, spark source mass spectrometry and neutron activation analysis are discussed regarding their application to this particular task. A primary concern of integrated dust analysis is the identification and determination of the chemical species. Obtaining this information via the analysis of individual particles (which also includes characterization by size, shape, and morphology) with the various techniques of local analysis is discussed as is the use of techniques that directly yield compound specific information like electron spectroscopy, IR-spectroscopy or thermoanalysis. Questions concerning the development and choice of suitable standard reference materials are treated. The final chapter deals with monitoring atmospheric pollution.

Hopefully, this book will serve as an aid to the chemist confronted with the assessment of a pollution situation, to the analyst who must evaluate the different existing techniques for a particular problem, to the environmentalist who desires a deeper understanding of the difficulties associated with obtaining reliable data, and finally, to all those who are interested in the applications of modern analytical chemistry.

H. Malissa

THE EDITOR

Dr. H. Malissa obtained his doctor's degree at the Technical University of Graz, Austria in 1946 in analytical chemistry. He served as a lecturer at the Institute of Biochemical Technology and Food Chemistry at the same University until 1948. After a year as senior research worker at the University of Uppsala, Sweden, he joined the Max Planck Institute for Iron Research in Düsseldorf, Germany. In 1959, Dr. Malissa was asked by the Technical University of Vienna to become full professor and head of the Institute of Analytical Chemistry and Microchemistry.

Dr. Malissa has published three books and approximately 200 papers on subjects of analytical chemistry and microchemistry, with emphasis on environmental analysis, metallurgical analysis, organic reagents, elementary analysis of H, C, N, and S. Included in his studies were physical methods of analysis, such as electron probe microanalysis, atomic absorption spectroscopy and IR-spectroscopy, automation in and with analytical chemistry, and information theory. He is presently on the editorial board of *Angewandte Chemie, Zeitschrift für Analytische Chemie,* and *Analytica Chemica Acta.*

Dr. Malissa was Chairman of the International Symposium on Microchemical Techniques 1970 in Graz, Austria; the 6th Annual Symposium on the Analytical Chemistry of Pollutants 1976, Vienna, Austria; the biennial Symposium on Metallurgical Analysis 1962 to 1976, Vienna, Austria; and a member of organizing committees of numerous others.

Dr. Malissa is also active in scientific societies. Among other activities, he was President of the Austrian Society for Microchemistry and Analytical Chemistry, President of the Analytical Division of the International Union of Pure and Applied Chemistry, and a member of the IUPAC Bureau. Dr. Malissa is presently Chairman of the Working Party on Analytical Chemistry of the Federation of European Chemical Societies.

CONTRIBUTORS

Michael Birkle, Ph.D.
Head, Department for the
    Development of Air and Water
Pollution Measuring Networks
University of Karlsruhe
Karlsruhe, Germany

Karl Buchtela, Ph.D.
Professor
Head, Radiochemical Department
Atomic Institute of the Austrian University
Vienna, Austria

István Cornides, Ph.D.
Scientific Advisor
Director, Mass Spectrometer Laboratory,
Eötvös Loránd University
Budapest, Hungary

Manfred Grasserbauer, Ph.D.
Professor
Head, Working Group on Physical Analysis
Institute of Analytical Chemistry
    and Microchemistry
Technical University of Vienna
Vienna, Austria

Dieter Hochrainer, Ph.D.
Scientist
Institute for Aerobiology of the
    Fraunhofer-Gesellschaft
Schmallenberg-Grafschaft, Germany

Robert Kellner, Ph.D.
Professor
Head, Working Group on Chemical Analysis
    and Infrared Spectrometry
Institute of Analytical Chemistry
    and Microchemistry
Technical University of Vienna
Vienna, Austria

Donald E. Leyden, Ph.D.
Phillipson Professor of Environmental
    and Mining Chemistry
Department of Chemistry
University of Denver
Denver, Colorado

Tibor Meisel, Ph.D.
Associate Professor
Institute for General and Analytical Chemistry
Technical University of Budapest
Budapest, Hungary

Tihomir Novakov, Ph.D.
Group Leader, Energy and Environment Division
Lawrence Berkeley Laboratory
University of California
Berkeley, California

L. Pólos
Research Worker
Institute for General and Analytical Chemistry
Technical University of Budapest
Budapest, Hungary

E. Pungor, Ph.D.
Head, Institute for General and
    Analytical Chemistry
Technical University of Budapest
Budapest, Hungary

Richard C. Ragaini, Ph.D.
Deputy Division Leader for
    Environmental Sciences Division
Lawrence Livermore Laboratory
Livermore, California

Philip A. Russell, M.S.
Research Biologist
Structures Laboratory
Denver Research Institute
University of Denver
Denver, Colorado

C. O. Ruud, Ph.D.
Senior Research Scientist
Manager, Structures Laboratory
Metallurgy and Materials Science Division
Denver Research Institute
University of Denver
Denver, Colorado

ACKNOWLEDGMENTS

I would like to thank all authors of the individual chapters for their contributions and especially Prof. Dr. M. Grasserbauer for his valuable assistance in the preparation of this book. I also owe thanks to the staff at CRC Press, Inc., especially Terri Weintraub and Margaret Saulino for their excellent cooperation.

TABLE OF CONTENTS

Chapter 1
INTRODUCTION

H. Malissa

The profound effects of man-made and natural particulate matter on health, environmental conditions (global and specific), and climate cannot be understood without complete knowledge of the chemical and physical nature (composition and morphology) of such matter. Very recently, Fennelly[1] made an interesting contribution concerning the origin of dust, revealing that the largest amount of total dust originates from nature and not from man. Consequently, there always was and always will be a dust problem. The main task today and in the future will be treating the problems of dust emission, conversion, transportation, and immission. We must bear in mind that the air of our biosphere, which is so essential to all life and welfare, is — in the sense of analytical chemistry — not a homogeneous medium but a mixture of gaseous, liquid, and solid components (see Table 1). The solid airborne particles — varying widely in chemical composition, size, shape, as well as in homogeneity and concentration — are referred to as "dust," regardless of their origin. The physical, chemical, medical, biochemical, and technological effects of dust can rarely be understood by results gained from one analytical method. A synoptic view of the results obtained from as many analytical procedures as possible better serves to elucidate the phenomena of dust and its reactions. The best means for this is Integrated Dust Analysis (IDA).[2] This is primarily the most complete analytical examination of collected solid airborne particles *in situ* — using, for example; light and electron microscopy for the examination of shape and size, X-ray fluorescence (XRF) for elemental analysis, relative conductometry[3] and gas chromatography (GC) for the investigation of decomposition and/or adsorption behavior. Also involved are infrared (IR) techniques and the different microprobe procedures, including Electron Spectroscopy for Chemical Analysis (ESCA) and Auger spectroscopy for the elucidation of chemical bondings of dust components.

We have to evaluate and, if possible, establish a special analytical strategy. Besides appropriate sampling, this strategy will have to contain the following steps:

1. Average analysis of all or individual (toxic) elements
2. Identification and quantitative analysis of individual particles

TABLE 1

Occurrence of Particulate Matter in Air

Sample	Definition
Dust	Solid-particle dispersion of any size
Smoke	Dense condensation of an aerosol with solid or solid and liquid particles
Fog	A cloud of (usually) water droplets of less than 100 μm and a droplet concentration of about 0.01 g/cm^3 at or very near the ground
Rain	Water droplets of 400 μm to 1000 μm in size and of 0.1–3 g/cm^3 in concentration
Mist	Liquid particles aerosol of any size
Haze	Not cloud-forming solid or liquid particle dispersion lower in concentration than fog and causing a slight or moderate diminution of visibility
Smog	Smoke and fog
Smaze	Smoke and haze
Cloud	Any free aerosol system having well-defined boundaries

3. Determination of particle mass (concentration), particle size and particle-size distribution
4. Description of morphology

These analyses should be done simultaneously, if possible, to gain maximum information. Together with the application of stereometric analysis,[4] they form the basis of IDA. Today, the analyst has many methods and instruments at his disposal for carrying out these determinations, and in principle, successful achievement of these analytical activities is surely possible. It is evident that only physical, nondestructive methods of analytical chemistry can contribute significantly to this achievement.

Before describing the different analytical procedures, the main question of "What is Dust" must be considered. The synonym "dust" for solid airborne particles is most common and will be used frequently in this book. By definition, dust is

1. A solid-particle dispersion of any particle size,[5] or according to a more precise definition, within the range of 5.10^{-4} to 5.10^{+2} μm
2. Fine dispersed solids of any kind, shape, structure, and density with a diameter between 0.5 to 500 μm[6]
3. Dispersions of solids in gases, produced by mechanical processes of wind; forming aerosols with fumes and fogs[7]
4. A minute solid particle (or liquid droplet) ranging in size from 5.10^{-3} to about 5.10^{+2} μm[1]

The number of definitions can be extended nearly infinitely without a firm clarification of the "dusty" situation. Herein, the term dust refers to solid airborne particles ranging from 10^{-4} μm (e.g., molecules, clusters of molecules) to 10^{+3} μm (large grains which can be transported by heavy winds). These size limits are arbitrary but indicate the particle size that encompasses seven orders of magnitude. They reflect inhomogeneities of chemical composition (and therefore density) and shape (due to agglomeration), as well as reactions with gases due to absorption and adsorption processes.

The complexity of dust presents the main problem for analysis and even more for the interpretation of analytical data. Figure 1 illustrates the interrelations between particle size, physical behavior, and homogeneity. The behavior of the particles (such as motion, adsorption) is determined by the physical and chemical laws that

are valid for finely divided matter (see Chapter 2). One of the major problems in dust analysis is the question of homogeneity. We must consider homogeneity of elemental concentration (chemical homogeneity) and grain morphology and particle-size distribution.

As in stereometric analysis,[4] the logarithm of the reciprocal of the volume in examination (log 1/V) can be used as a measurement of homogeneity. Within a certain standard deviation, this yields the same results in measurements on samples taken from various sites as repeated measurements on one sample site. The question of homogeneity is directly related to the degree of resolution for the method of examination employed.

For XRF, particles under 100 μm can be considered as "homogeneous," whereas if electron probe microanalysis (EPM) or scanning electron microscopy (SEM) is used, they could be "heterogeneous." Consequently, XRF methods suffice for measurement of homogeneity class 5 (from 24 to 5). If the grain size becomes larger (class 5 to 0), problems will surely occur. The highest homogeneity would be achieved if every atom in the sample was the same as the other atoms (class 24). Artificial homogenization (similar to a dissolution step for atomic absorption) fulfills the requirement for a better statistical elemental distribution for analysis. However, it is characterized by a loss of information about the sample (like bonding conditions).

Beyond the not very satisfying definition of dust and its homogeneity problems, there are still some further aspects to be considered. Particularly for interpretation of analytical results, we must distinguish between primary, secondary, and even tertiary particulate matter. According to Fennelly,[1] primary particulates (usually 1 to 20 μm in size) are those injected directly into the atmosphere by chemical and/or physical processes. Secondary particulate matter (usually 0.005 to several μm in size) is the product of chemical reactions occurring in the atmosphere. It contains sulfates, nitrates, and hydrocarbons originating from the reactions of, e.g., SO_2, NH_3, NO_x, H_2O with primary dust nuclei primarily under the influence of (sunlight) radiation. Frequently, due to the sampling technique, we have still another dust collection which is neither purely primary nor purely secondary but contains conversion products obtained during sampling. If we consider the use

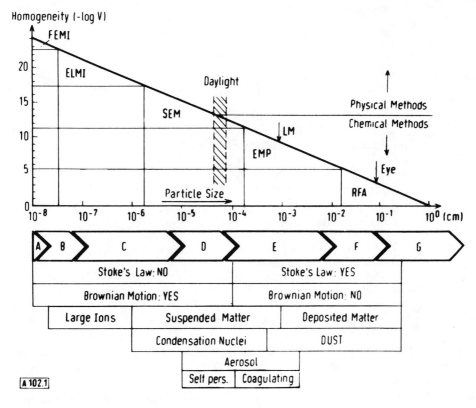

FIGURE 1. Particle size in environmental analysis. LM = light microscope; XRA = X-ray fluorescence analysis; EMP = electron microprobe; SEM = scanning electron microscopy; ELMI = electron microscopy; FEMI = field electron microscopy. A = atoms; B = molecules; C = Aitken particles; D = large particles; E = giant particles; F = fine solids; and G = coarse solids. (From Malissa, H., *Angew. Chem. Int. Ed. Engl.,* 15, 141, 1976. With permission.)

of dust jars for sampling over a longer period of time, the reactions of primary particulate matter with rain water, which contains abundant dissolved acidic ions (sulfuric or nitric compounds), and the agglomeration and chemical conversion possibilities during the concentration step, we can then imagine what may happen and how difficult and dangerous the interpretation of analytical results may be. An impression of the wide variety of types and sizes of primary dust particles can be gained by study of the compilation in the *CRC Handbook of Chemistry and Physics.*[8]

Dust is everywhere, always influencing health and property. The harmfulness of dust does not depend on chemical composition alone but also on particle size and shape. Recent work by Natusch and Wallace[9] illustrates that small particles are generally more dangerous than larger ones. The reason for this lies in the process of breathing and deposition. Figure 2 compares the correlation between particle size, degree, and place of deposition. Usually, the dangerous components, such

as lead, cadmium, selenium, etc., are present as small particles. It was found for example that prolonged exposure to coarse grained MnO_2 was not injurious to health, whereas a finer dust of the same substance (80% of the particles had a diameter <2 μm) caused pneumonitis. Silicosis, a fibrotic disease of lung, is produced after at least 2-year exposure to air containing 10^6 ppm^3 with a diameter of less than 10 μm. Currently, the influence of particle size and shape has not been studied in great detail, but we can assume that angular particles are more dangerous than rounded ones even if the compound itself is not reactive, toxic, or dangerous (SiO_2). This opens up yet another field for cooperative research (analysts, doctors, etc.).

The analytical chemist must remember the essential fact that dust is far from being a homogeneous product — neither from the point of view of chemical composition nor from particle

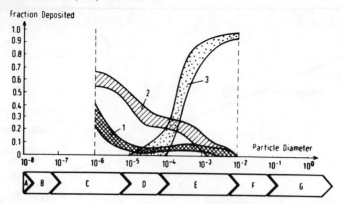

FIGURE 2. Deposition of dust according to particle size. (1) tracheobronchial, (2) pulmonary, (3) nasopharyngeal (cf. also Figure 1). (From Natusch, D. F. S. and Wallace, J. R., *Science,* 186, 695, 1974. Copyright 1974 by the American Association for the Advancement of Science. With permission.)

size and/or shape. In the analytical sense, dust is an agglomerate (coagulate) of a wide variety of substances and particle sizes. The characterization of dust demands special consideration of the problems of homogeneity. For clarification, the methods of physical and chemical analysis including stereometric parameters must be employed. Furthermore, the amount of dust samples available for analysis usually lies in the micro- to milligram range, and this leads to the first unavoidable claim concerning the techniques of analysis to be employed: only micromethods are adequately suitable. If one considers that a series of physiologically important elements is present in dust in pico-, nano-, or microgram quantities in the few milligrams of sample, the second important feature required to the analytical technique becomes apparent: the methods must be highly sensitive, in terms of trace analysis. Consequently, microtrace methods are required for the analysis of airborne particulate matter. Another characteristic of dust is its complex multi-element system. The elements to be determined can be present in a wide range of concentration (from several percents to parts per billion). Hence, the overall analytical problem is simply that the analyst must have at his disposal, if possible, nondestructive, multi-elemental methods that can cope with a concentration range of about eight orders of magnitude and a particle size range of about seven orders of magnitude for the analysis of microsamples.

If the arsenal of modern analytical chemistry is inventoried with regard to the attributes required, we find that physical micromethods are best suited for dust analysis. Only these comply with the requirements of the sample and problems as stated above. The problem cannot, however, be solved on the basis of a simple method, and therefore, the integration of various techniques of physical analysis under the scope of IDA is needed.

Another aspect of our environmental problem is not only the often discussed and described smog situation but also "black and white episodes," recently described by Brosset.[10] Here, the long-range transportation of airborne particles is the major concern, especially when high concentrations of dark particles (soot concentration up to about 40 $\mu g/m^3$) are transmitted to Sweden from the European continent. Mean values of measurements of particle composition during white and black episodes at Onsala Peninsula, Sweden in 1975 showed that continental particles contain carbonaceous material, sulfate, and nitrate (see Table 2). The sulfate formed by catalytic oxidation of SO_2 to H_2SO_4 is subsequently neutralized by NH_3. Such ammonium sulfate crystals are shown in Figure 3 on Nuclepore® membrane filters. During the dry and sunny weather of the summer months, small colorless particles containing different types of ammonium sulfates are responsible for white episodes. Furthermore, the very important role of manganese, iron, and vanadium in fine particles (solid aerosols) in conversion processes of gaseous air components is also a very broad field of environmental research.

In dust analysis, the strategy and tactics of analytical chemistry play an important role. IDA possesses the strategy and tactics to overcome difficulties in analyzing minute quantities of a multicomponent, unique sample to elucidate the problem of dust. In this connection, it is worthwhile to consider the relationship between analytical strategy and tactics and their place in system theory. Strategy is primarily the proper definition of the goal and the principal way to achieve this goal, i.e., the chemical analysis of dust for lead to obtain information about air pollution from motor vehicles. The tactics in this case may be a specific dissolution step for lead compounds or valence band spectrometry.

In this part of operation, the analyst must act like a "strategus" (Greek for general) and think and act in the sense that strategy is the prelude to the battlefield, as tactics are to action on the battlefield itself. This means he must come to an anamnese about the enemy (to collect and analyze information), to survey and align resources (methods), and to define his goals. The primary requirements of system theory are listed as follows:[11]

1. Problem setting, definition of question and goals

2. Alignment of tools, parameters, data and results

3. Modeling, comparison with standard

4. Simulation, correlation of final results with the goal of the problem setting, air standards

No single analytical result will be of great value if the sampling conditions are not suitable for the problem in question. Since sampling of dust is more closely connected with physical principles than anything else, we have to pay full attention to this fact (see Chapter 2). Otherwise, the best use of the physical methods (the tactics) may be worthless and lead to wrong interpretations of results.

Because of the great variety of analytical tools often based on different principles, competent scientists of varied backgrounds wrote the individual chapters of this book, thus providing a solid data base for an analyst concerned with the problems of our environment. The reader concerned with the investigation of airborne particulate matter can select the methods necessary and establish his own strategy to solve his problems.

TABLE 2

Comparison of Particle Composition During White and Black Episodes at Onsala, Sweden, 1975

				Episode type (neq/m^3)				
	SO_4^{2-}	NO_3^-	$SO_4^{2-} + NO_3^-$	NH_4^+	H^+	$NH_4^+ + H^+$	NH_4^+/H^+	Soot ($\mu g/m^3$)
Black	308	33	341	342	7	349	51	23
White	317	<3	~320	233	75	309	3.1	2.3

From Brosset, C., *Ambio*, 5, 157, 1976. With permission.

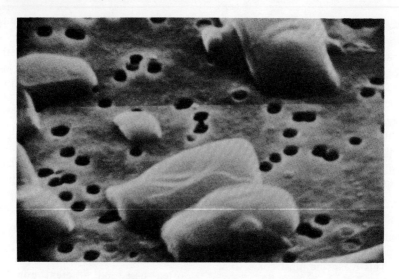

FIGURE 3.   SEM of $(NH_4)_2SO_4$ crystals collected on Nuclepore® membrane. (Magnification × 10,000.) (From Aerosol Filtration, brochure, Nuclepore Corp., Pleasanton, Cal., 1976. With permission.)

REFERENCES

1.   **Fennelly, P. F.,** The origin and influence of airborne particulates, *Am. Sci.,* 64, 46, 1976.
2.   **Malissa, H.,** Integrated dust analysis by physical methods, *Angew. Chem. Int. Ed. Engl.,* 15, 141, 1976.
3.   **Malissa, H., Puxbaum, H., and Pell, E.,** Simultane Kohlenstoff- und Schwefelbestimmung in Stäuben (Simultaneous Determination of Carbon and Sulfur in Dust), *Z. Anal. Chem.,* 282, 109, 1976.
4.   **Malissa, H.,** Stereometrische Analyse mit Hilfe der Elektronenstrahlmikroanalyse (Stereometric Analysis with Electron Probe Microanalysis), *Z. Anal. Chem.,* 273, 449, 1975.
5.   **Bibbero, R. J. and Young, J. P.,** *Systems Approach to Air Pollution Control,* New York, 1974, 31 and 451.
6.   VDI-Richtlinie 2104, *Begriffsbestimmung Reinhaltung der Luft* (Society of German Engineers, Recommendation No. 2104, *Definitions Clean Air*), VDI-Verlag, Düsseldorf, Germany.
7.   German Research Foundation, Report 9 of the Commission for the Study of Health Hazardous Substances, Bonn-Bad Godesberg, Germany, June 28, 1973.
8.   *CRC Handbook of Chemistry and Physics,* Weast, R., Ed., 51st ed., Chemical Rubber Co., Cleveland, 199, 1971.
9.   **Natusch, D. F. S. and Wallace, J. R.,** Urban aerosol toxicity – the influence of particle size, *Science,* 186, 695, 1974.
10.   **Brosset, C.,** Airborne particles: black and white episodes, *Ambio,* 5, 157, 1976.
11.   **Gottschalk, G., Kaiser, R., Malissa, H., Schwarz-Bergkampf, E., Simon, W., Spitzy, H., Werder, R. D., and Zettler, H.,** Systemtheorie in der Analytik (System Theory in Analytical Chemistry), *Z. Anal. Chem.,* 256, 257, 1971.

Chapter 2

PHYSICAL BEHAVIOR AND SAMPLING OF AIRBORNE PARTICLES

D. Hochrainer

TABLE OF CONTENTS

I. INTRODUCTION

Airborne particles form a colloid consisting of a gaseous dispersion medium and dispersed particles, which may be solid or liquid. Liquid particles are always spherical, since the surface tension is much larger than frictional forces. Solid particles show a variety of shapes such as spheres, regular polyhedrons, straight and curved fibers, flakes, and irregular particles. Particles that are approximately the same length in all directions are called isometric.

For spheres, the radius or diameter can be used to characterize the size, and for nonspherical particles, equivalent diameters can be used. If physical properties of the particle, like density or index of refraction, are unknown, equivalent diameters can also be useful for measuring spheres. Aerosols may be classified according to size distribution, e.g., coarse or fine particles, respirable particles and so on or according to physicochemical properties, such as fog, dust, smoke, or smog. The plot of certain properties, for example particle number concentration or mass concentration vs. the particle size, is referred to as size distribution. In the following chapter, the physical properties will be discussed in more detail. Fuchs' book[1] can be used as a reference for problems concerning the mechanics of aerosols.

II. RESISTANCE OF PARTICLES TO THE GASEOUS MEDIUM

Laws for the resistance of a particle to the gaseous medium govern the deposition of particles in almost all precipitators.

A. Spherical Particles

Much effort has been spent in studying the resistance of a particle to the gaseous medium. Most attention has been focused on the resistance of spheres. The proper laws for each dynamic regime have to be selected on the basis of the ratio of the mean free path l to the particle radius (r). We have to distinguish between the free molecular regime, $l \gg r$, the transition regime, $l \approx r$, and the continuum regime, $l \ll r$.

1. Free Molecular Regime

The force F, acting on a particle, is expressed as:

$$F = \frac{6\pi \eta \, r^2 v}{Kl}$$

where η is viscosity of the gas in kg/(sec·m), v is velocity of the particle (m/sec), K is a dimensionless constant between 1.091 and 1.175, and depending on the accommodation coefficient, generally a value of 1.15 can be assumed.[2] l is 6.63 × 10^{-8} m for 1.013 bar and 15°C.

2. Continuum Regime

In this case resistance is governed by Stokes' formula:

$$F = 6\pi\eta r v$$

if the Reynolds number

$$Re = 2r\rho_g v/\eta$$

is much smaller than 1 (ρ_g gas density, kg/m^3). Re gives the ratio between inertial forces and frictional forces. If Re is below 1, the resistance is mainly governed by frictional forces. A sphere, falling in air, reaches an equilibrium between weight and resistance, leading to a terminal velocity. In terms of the particle diameter (d), the equation for the velocity can be written as:

$$v = \frac{d^2 \rho_p g}{18\eta}$$

where ρ_p is particle density (kg/m^3) and g is acceleration due to gravity. For spheres with the density 1000 kg/m^3 falling under the influence of gravity, the Reynolds numbers are below 0.05 if the particles have a radius (r) < 15 μm. In this case, correction due to the Reynolds number is below 1%. For larger Reynolds numbers a correction must be applied. For Re < 6 the equation for the velocity is

$$v = \frac{d^2 \rho_p g}{18\eta (1 + b\, Re^c)}$$

where constant b = 0.13 and constant c = 0.85 can be used.[3] For Reynolds numbers between 0.002 and 400 the following velocity equation can be applied:[4]

$$v = \sqrt{\frac{4\, d\rho_p g}{3\, \rho_g \left(\frac{24}{Re} + \frac{4}{\sqrt{Re}}\right)}}$$

When using either equation, v cannot be calculated explicitly; however, iterative, numerical calculations are possible.

3. Transition Regime

In the transition regime between free molecular flow and continuum flow, a theoretical equation for the drag force cannot be derived. Instead there exist some interpolation formulas. The law for the resistance, including the Cunningham correction[5],[6] is:

$$F = \frac{6\pi\eta r v}{1 + A\frac{l}{r}}$$

where A is a dimensionless constant, which is somewhat dependent on surface properties, but lies near 1.25. For particles below 8 μm radius the Cunningham correction term is larger than 1%. For particles with a radius below 0.2 μm the correction should include an exponential term, given by Knudsen and Weber:[7]

$$F = \frac{6\pi\eta r v}{1 + \frac{l}{r}\left(A + Q \exp\left[-B\frac{r}{l}\right]\right)}$$

where the constant Q is approximately 0.42 and the constant B is about 0.85; both constants are slightly dependent on surface properties. If the particle is very small, r << 1, this equation has the same form as in the free molecular regime. Another interpolation equation is given by Fuchs and Stechkina.[8]

Figure 1 shows the value of the Knudsen-Weber correction and the correction due to the Reynolds number as a function of the particle diameter for particles with a density of 1000 kg/m^3 falling in air in a gravitational field or in a centrifugal field with 1000 or 20,000 m/sec^2. The range, where the uncorrected Stokes' law can be applied, is not even one order of magnitude. Table 1 gives the terminal velocity in gravitational field, the Reynolds number, the Knudsen-Weber correction, the correction due to the Reynolds number, and the mean displacement in 1 sec (see Chapter 3) for particle diameters between 0.01 and 1000μm. The following values have been assumed: ρ_g = 1.225 kg/m^3, ρ_p = 1000 kg/m^3, g = 9.81 m/sec^2, η = 1.79 10^{-5} kg/(sec·m), l = 6.63 × 10^{-8} m, A = 1.25, Q = 0.42, B = 0.87, b = 0.13 and c = 0.85.

B. Nonspherical Particles

For nonspherical particles, shape factors must be applied for the use of Stokes' law. Many particles have orientation-dependent shape factors, which means that more than one shape factor is necessary to characterize them. Usual shape factors are

1. Shape factor: κ' ratio of the drag force on a given particle to the drag force on a sphere with

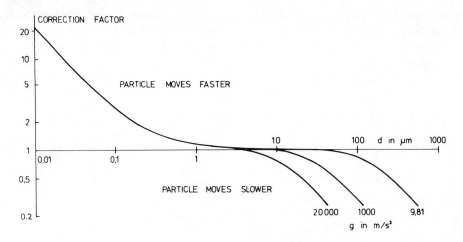

FIGURE 1.   Corrections to Stokes' law.

TABLE 1

Values for the Terminal Velocity, Reynolds Number, Correction Factors for Stokes' Law, and Mean Displacement in 1 sec

d (μm)	v (mm/sec)	log Re (–)	Correction Mean free path (–)	Reynolds number (–)	$\overline{\Delta x}$ (μm)
0.01	0.000 069	–10.32	22.8	1.00	260
0.02	0.000 143	–9.71	11.8	1.00	131
0.05	0.000 39	–8.87	5.1	1.00	67
0.1	0.000 90	–8.21	2.9	1.00	30
0.2	0.002 3	–7.50	1.91	1.00	16.8
0.5	0.010 2	–6.46	1.34	1.00	10.3
1	0.036	–5.61	1.17	1.00	5.9
2	0.131	–4.75	1.08	1.00	4.0
5	0.79	–3.57	1.03	1.00	2.8
10	3.1	–2.67	1.02	1.00	1.74
20	12.2	–1.78	1.01	0.97	1.23
50	72	–0.61	1.00	0.94	
100	250	0.23	1.00	0.81	
200	70	0.98	1.00	0.57	
500	2000	1.84	1.00	0.26	
1000	3900	2.42	1.00	0.13	

the same characteristic length moving in the same medium at the same velocity.

2. Dynamic shape factor: κ ratio of the drag force on a given particle to the drag force on a sphere of the same volume moving in the same medium at the same velocity.

3. Sphericity: ϕ ratio of the surface of a sphere with the volume of the given particle to the surface of the given particle.

The shape factor κ' is used when the volume of the particle is not known, and the coefficient of sphericity is used when the deviation from the sphere is denoted on a strictly geometrical basis. For a sphere this shape, factors are 1, while for other bodies, κ' may be larger or smaller than 1, κ is in most cases larger than 1, and ϕ is always smaller than 1.

Needle-like and flaky particles can be approximated by ellipsoids of revolution, for which formulas for κ have been derived.[9] Other shape factors have been found experimentally.[10–13]

Dahneke[14] has determined slip correction factors for nonspherical bodies.

C. Time and Distance to Reach Terminal Velocity

For particles of the size of 1 μm, the time to reach terminal velocity is on the order of 1 μsec, and the distance that the particle travels to reach this velocity is in the order of 10^{-10} m. Particles of this size and smaller can be considered at almost any time to be at terminal velocity in the gravitational field. A particle of the density of 1000 kg/m^3 and 100 μm in diameter needs, under the influence of gravity, a tenth of a second to reach 99% of the terminal velocity and moves 20 cm, while a particle of 1 nm diameter needs 1.25 sec and moves 3.7 m.

III. BROWNIAN MOTION AND DIFFUSION

Aerosol particles are subject to Brownian motion. The mean square displacement (Δx^2) in a given direction in time (t) is given by Einstein's formula:[15]

$$\overline{\Delta x^2} = \frac{2RTt}{NK_r}$$

where R = gas constant, 8.314×10^{-4} J/(K · mol); T = temperature in degrees Kelvin; N = Avogadro constant, 6.02×10^{23}/mol; K_r = resistance factor; and the diffusion coefficient is

$$D = \frac{R \cdot T}{N \cdot K_r}$$

Since Brownian motion is only of importance for particles where the Reynolds number correction to Stokes' law can be neglected, the resistance factor has the formula:

$$K_r = \frac{F}{v} = \frac{6\pi\eta r}{1 + \frac{1}{r}(A + Q \exp[-Br/l])}$$

Values for the mean displacement in one second are given in Table 1. Comparing the Brownian displacement with the displacement by a physical (e.g., gravitational) force, one has to bear in mind that Brownian displacement increases with $\sqrt{t}$, while other displacement increases directly with time. In addition to the translational Brownian movement, a Brownian rotation of particles exists, but no practical applications are known.

If a particle is suspended in a thermal gradient, Brownian motion exerts a force in the direction from higher to lower temperature on the particle. Particles move toward the cold surface, while the hot surface is surrounded by a dust free space. This thermophoretic velocity depends on material properties of the aerosol particles and on the ratio of the mean free path to the radius. If the radius becomes smaller than the mean free path or if the radius becomes larger than about ten times the mean free path, it is independent of the particle size. Approximative formulas[16] exist for velocity v

for r < 1

$$v = -\frac{1}{3} D \frac{1}{T} \frac{dT}{dx}$$

for r $\gg$ 1

$$v = -\frac{4}{27} D \frac{1}{T} \frac{dT}{dx}$$

where D is the diffusion coefficient of the air (for 1 bar and 0°C 1.85×10^{-5} m^2/sec), T temperature in degrees Kelvin, and $\frac{dT}{dx}$ the temperature gradient. More accurate formulas can be found in Waldmann and Schmidt.[17] This effect is used for the thermal precipitator (see Section IX.G.).

IV. CHARGE OF AEROSOL PARTICLES: ELECTRIC MOBILITY

Many aerosol particles carry electric charges that may result from unipolar or bipolar ions in the environment, from the production of the aerosol, or from radioactivity.

A. Boltzmann Equilibrium

If an aerosol contains sufficient positive and negative ions and the total charge of these ions is zero, after some time the aerosol reaches charge equilibrium with the ions. This is known as Boltzmann equilibrium. The fraction (f) of particles with diameter (d) carrying (n) elementary units of charge (e) is[18,19]

$$f = \frac{1}{\Sigma} \exp\left[-\frac{(ne)^2}{4\pi\epsilon_0 dkT}\right]$$

$$\Sigma = \sum_{-\infty}^{+\infty} \exp\left[-\frac{(ne)^2}{4\pi\epsilon_0 dkT}\right]$$

where ϵ_0 = dielectricity constant of vacuum, 8.85×10^{12} F/m; k = Boltzmann constant, 1.38×10^{-23} J/K; T = temperature in degrees Kelvin; e = 1.6×10^{-19} C; diameter in meters.

B. Diffusion Charging

If the aerosol contains unipolar ions, the particles have unipolar charges. In the absence of an external electric field the mode of charging is called diffusion charging, while in the presence of an external electric field and when diffusion can be neglected, the mode of charging is called field charging.

When diffusion charging occurs, the ions have to move with their thermal energy against the potential of the particle. If the ion comes close enough to the particle, image forces may help to bring the ion in contact with the particle. Thus, for a given particle, the charging mechanism is mainly governed by the ion density of the aerosol and the time span during which ions are present. If the time span is sufficient, the particles almost reach a charge limit, since it becomes highly improbable that an ion with higher thermal energy will approach the particle.

C. Field Charging

For field charging, Brownian motion is neglected, and it is assumed that the ions move along the lines of the electric field through the aerosol cloud. If a field line touches the aerosol particle, an ion moving along this field line will charge this particle. If the particles are sufficiently charged, the field lines are repelled and no further charging can occur. In practice, the assumption of neglecting Brownian motion is usually not valid, and charging with an electric field often involves combined diffusion and field charging.

For the charge reached by a particle under certain conditions, formulae have been verified for certain size ranges and are given in Reference 19. As an order of magnitude estimate, the charge (Q_s) of a sphere of diameter (d) can be calculated by the formula:

$$Q_s = 2\pi\epsilon_0 \ dU$$

where ϵ_0 = dielectricity constant of vacuum, 8.85 $\times$ 10^{-12} F/m, voltage U = 0.1 V can be assumed.

D. Charge from Aerosol Production

Many aerosol particles are charged during their production. Dispersion of powders usually results in high charges where the charge on the particles may reach several tenths of a volt. Low charges, often less than Boltzmann equilibrium, are found in condensation aerosols. Usually, particles become charged during dispersion of liquids. If the

liquids are volatile, the charge remains constant while the diameter shrinks; thus the voltage of the particle increases. Finally, a particle may be charged, if it is radioactive and a charged elementary particle is emitted.

E. Electron and Ion Charge Limit

In addition to the charge limits of diffusion and field charging mentioned above, electron and ion charge limits also exist. If solid particles become very highly charged, the field intensity (E) on the surface of the particle can be sufficiently high to emit electrons (approximately 10^9 V/m) or ions (ca. 2×10^{10} V/m).

The number (n) of elementary units of charge (e) on a particle with diameter (d) is

$$n = \frac{\pi\epsilon_0 \ Ed^2}{e}$$

Liquid drops will disintegrate when the repellent forces from the electric charges are larger than the surface tension (σ). This is known as the Rayleigh limit[20] and the formula is

$$n = \frac{2\pi}{e}\sqrt{2\epsilon_0 \ \sigma d^3}$$

F. Electrical Mobility

The electrical mobility (u) of a particle is defined as the ratio of the terminal velocity, attained by a particle moving under electric forces, to the field intensity and has the dimension m^2/Vsec. It can be calculated from the formula:

$$u = \frac{neC}{3\pi\eta d}$$

where n = number of elementary charge units, e = elementary charge unit, η = viscosity, d = particle diameter, and C = corrections to Stokes' formula. As a function of d, n depends on the charging or discharging mechanisms. The mobility (u) can either increase, decrease, or be independent of diameter (d). Even if a general law relating n as a function of d could be found, there is always the stochastic nature to the charging processes, and for small numbers of n, it is important that n can only take integer values.

V. OPTICAL PROPERTIES

Optical properties of aerosol particles have been studied for many years, and complete solutions exist for bodies of relative simple geometry. But even in this case, the solutions are so complex that they will not be cited here. This chapter will present a circumstantial list of known phenomena. For further studies, the books of Van de Hulst[21] and Kerker[22] can be used.

A. Light Scattering of Spherical Particles
1. Rayleigh Scattering
For particles which are much smaller than the wavelength of light the intensity (I_θ) scattered into direction (θ) (angle between incident beam and scattered beam) in the angle element $(d\theta)$ is

$$I_\theta = I_0 \frac{9\pi^2 V^2}{2R_0^2 \lambda^4} \left(\frac{m^2 - 1}{m^2 + 2}\right)^2 (1 + \cos^2 \theta)\, d\theta$$

where I_0 = intensity of incident beam, V = volume of particle, R_0 = distance from particle to the observer, λ = wavelength of light, and m = index of refraction. This law assumes unpolarized light for the incident beam and has two linear polarized components for scattered light, one with no angular variation and the other with the variation $\cos^2 \theta$.

2. Mie's Theory
For particles having the same dimension as the wavelength of light, the intensity can be calculated using Mie's theory. The formulas are very complicated and need extensive use of computers. A bibliography for published scattering functions is presented in Chapter 3.8 of Kerker's book.[22] When particles that are much smaller than the wavelength of light are growing, forward scattering becomes more pronounced, while in other directions minima and maxima can be close together. When a particle is of approximately the same dimension as the wavelength, the forward scattered light may have a 1000-fold intensity of the light of other directions (Figure 2). By superimposing the intensity functions for different colors as they exist in white light, the light, which is scattered in different directions, may have different colors and is known as the Tyndall spectra.

3. Geometric Optics
For geometric optics, the particle must be much larger than the wavelength of light. A particle that initially has the same dimension as the wavelength and expands to the size where geometric optics are applicable exhibits an increasingly differentiated intensity function, where finally a few intensity maxima become predomin-

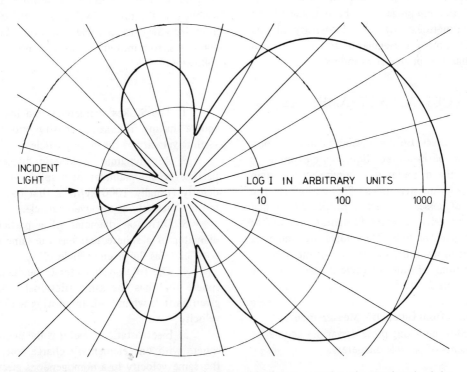

FIGURE 2.   Scattering intensities of a sphere with diameter equal to the wavelength of the light, index of refraction 1.35. Calculation by W. Holländer.

ant, while most of the others vanish. These predominant maxima are commonly known as rainbows. Rayleigh scattering and geometric optics are extreme cases of Mie's theory.

B. Light Absorption

The absorbing material modifies the above-mentioned theories. Small absorption is assumed for Rayleigh's formula, where m is the complex index of refraction. For high absorption, a different formula exists. Absorption may be included in Mie's theory also. Generally, absorption reduces the intensity of the scattered light, especially in the forward direction. Geometric optics is simply understood for strong absorption. Only a reflected intensity from the illuminated hemisphere exists, the highest intensity occurs toward the illuminating lamp, the lowest in the direction of the incident beam.

C.   Light Scattering of Nonspherical Particles

Much less attention has been given to other nonspherical, regular bodies. Complete solutions exist for infinitely long cylinders.[23] Ellipsoids may be treated like spheres, if the longest dimension is not much larger than the wavelength of light.[24] For other regular or irregular bodies, model experiments with the particles and wavelength in the cm-range have been made.[25] All scattering functions show several intensity minima and maxima, but for irregular bodies, they are less distinct than for spheres or cylinders.

VI. EQUIVALENT DIAMETERS

In many cases, only equivalent diameters of a particle can be measured. Equivalent diameter is defined as the diameter of a sphere with certain qualities (e.g., density, index of refraction, electric charge) that possesses a certain property (terminal velocity, intensity of scattered light, etc.) equal to that of a given particle. The equivalent diameters often depend upon the orientation of the particle. For definitions of other equivalent diameters, see VDI-Richtlinie 3491.[26]

A. Diameters from Geometric Measurements

The following listing presents various geometric measurements of particle diameters:

1. Volume equivalent diameter of a particle

is the diameter of a hypothetical sphere that has the same volume as the given particle.

2. Surface equivalent diameter of a particle is the diameter of a sphere that has the same surface as the given particle.

3. Projected diameter of a particle is the diameter of a circle that has the same area as the orthogonal projected particle.

4. Martin's diameter is the length of a chord parallel to a given direction, that divides the projected area into two equal halves.

5. Feret's diameter is the distance between two tangents on both sides of the projected area, that are parallel to a given direction.

B. Equivalent Diameters from Mobility Measurements

For these diameters, only the terminal velocity of the particle relative to the surrounding gas is considered. Except for the equivalent diffusion diameter (described below), Brownian motion is neglected.

The given particle and the equivalent hypothetical sphere are constrained to move in the same environment (gas, temperature, density, pressure, force fields). Mechanical mobility is the ratio of velocity to force, causing the motion. Electrical mobility is the ratio of velocity to electrical field intensity at the particle location. Equivalent diameters from mobility measurements include the following:

1. The Stokes' diameter is the diameter of a sphere having the same density and the same settling velocity as the given particle.

2. Aerodynamic or kinetic diameter of a particle is the diameter of a sphere of specified density (usually 1000 kg/m^3), that has the same terminal velocity as the given particle.

3. Diffusion diameter of a particle is the diameter of a sphere that has the same diffusion coefficient as the given particle.

4. The friction diameter is the diameter of a sphere that has the same frictional force as the given particle when both are moving with the same velocity.

5. Electrostatic diameter is the diameter of a sphere with one elementary charge unit, that has the same velocity in a homogeneous electric field as the given particle.

C. Equivalent Diameters from Extinction and Scattering Measurements

These following definitions are based on the replacement of the particle in an optical instrument by a sphere that gives the same optical signal.

1. Extinction diameter of a particle is the diameter of a sphere with the same optic constants, that withdraws from the incident beam the same portion (by scattering or absorption) as the given particle.

2. Scattering diameter of a particle is the diameter of a sphere with the same optic constants, that scatters into a certain solid angle the same portion of the incident intensity as the given particle.

3. Reduced scattering diameter is the diameter of a sphere with the optical constants of a standard material, that scatters into a certain solid angle the same portion of the incident intensity as the given particle.

VII. CONDENSATION AND EVAPORATION OF PARTICLES

Condensation and evaporation are processes that tend to establish an equilibrium between a liquid particle and a gaseous phase of the particle material. These processes as well as sublimation will be discussed in this chapter. The equilibrium vapor pressure (p) of a small droplet is higher than the equilibrium vapor pressure (p_∞) of a plane surface of the same material at the same temperature. The ratio is given by the Thomsin-Gibbs equation:[27]

$$\frac{p}{p_\infty} = \exp \frac{\sigma \widetilde{V}}{RTd}$$

where σ = surface tension (N/m), $\widetilde{V}$ = molecular volume (m^3/mol); R = gas constant 8.31×10^{-4} J/(K·mol), T = temperature in degrees Kelvin; d = particle diameter in meters. This means that small particles (especially those smaller than 0.1 μm in diameter) evaporate faster than larger ones. Besides this change of vapor pressure, several other effects have to be included, i.e., (a) self-cooling of the particle during evaporation that results in changes in the vapor pressure, gas density, and surface tension; (b) ventilation factors, since the particle is falling during evaporation; (c) insoluble surface layers, which reduce evaporation velocity;

(d) presence of other particles of the same material; and (e) presence of hygroscopic material in the particle when water is evaporated. These contributing factors make it difficult to calculate the rate of evaporation or condensation. As an example of the varying times required for evaporation, water droplets of 1 μm diameter evaporate in milliseconds and droplets of the same size of dioctylphtalat (DOP), which has a vapor pressure of 10^{-10} bar at room temperature, evaporate in days. For all samples in which condensation or evaporation might be important, it should be determined if these effects change the sample.

VIII. COAGULATION

When two aerosol particles collide, the particles usually form a new particle with a mass equal to both of the initial particles. Even solid particles coagulate, since surface forces exceed the inertial forces. The forces which bring particles in contact are usually thermal forces, although additional electrical, magnetic, or gravitational forces may exist.

The number concentration (n_c) of spherical particles that coagulate only with thermal forces can be calculated with Smoluchowski's formula[28] for the continuum regime:

$$n_c = \frac{n_0}{1 + \frac{K_0}{2} n_0 t} = \frac{n_0}{1 + t/t_h}$$

where n_0 = initial number concentration; K_0 = coagulation constant; t = time; and t_h = half-value time of the number concentration. Theoretical values of K_0 for different radii are given in Table 2. The constant K_0 depends upon the radii of the coagulating particles and on the conditions and physical constants of the gas, i.e., temperature, viscosity, and mean free path. Experimental verification of this formula shows little dependence upon particle composition. For example,

TABLE 2

Values for the Coagulation Constant κ_0 for Particles Having Radius r

r (μm)	0.001	0.01	0.1	1.0
κ_0 (cm^3/sec)	803.4	84	12.7	6.6

After Zebel, G., *Aerosol Science*, Davies, C. N., Ed., Academic Press, London, 1966.

the time required to halve the particle number, for $K_0 = 10^{-9}$ cm^3/sec and a concentration of 10^{11} particles/cm,^{3} is 0.02 sec., for 10^6 particles/cm,^{3} 33 min; and for 10^5 particles/cm,^{3} 5.5 hr.

Zebel[29] and Wagner and Kerker[30] discuss coagulation problems of particles of different size, particles of the order or smaller than the mean free path of the gas molecules, nonspherical particles, and the influence of forces between particles or force fields. Bipolar charging increases while unipolar charging reduces the coagulation rate and can be used to stabilize the aerosol.

IX. SAMPLING INSTRUMENTS

A. Principles for Obtaining Samples

Each aerosol measuring device requires an inlet, where the sample is taken from the bulk aerosol. In most cases, this is a round tube. Ideally, the sample collected through this tube is representative of the aerosol. Obtaining samples through the tube is easily accomplished for particles that have an aerodynamic diameter smaller than a few micrometers. However, for larger particles, it can be difficult, since the streamlines of the gas may deviate substantially from the trajectories of the aerosol particles, a result of inertia and gravitation.

The best method of sampling is designated as isokinetic sampling.[1,31] It involves a thin-walled sampling tube having its axes parallel to the wind and the mean gas velocity in the tube equal to the wind velocity. No change of the aerosol concentration will occur, if the wind velocity is much larger than the sedimentation velocity of the particles. The errors with nonisokinetic sampling are shown in Figure 3. Near the entrance of the nozzle, trajectories of large particles deviate from the streamlines of the gas. If the flow through the nozzle is too high, the sampling rate of large particles is too small and vice versa. Approximate calculations of the errors are given in Belyaev and Levin,[32] Zenker,[33] and Rüping.[34] Special nozzles exist, where the static pressure inside and outside of the nozzle can be measured, and the flow can be adjusted until both pressures are equal, thus leading to the same flow velocity inside and outside of the nozzle.[35]

Other cases of sampling are discussed also, sampling from an aerosol with no wind speed, sampling with a hole or slit in a wall, and sampling with a slit in a cylinder and a hole in a sphere.[36] Variable wind speed and direction make sampling

of natural aerosols with large particles difficult. An approximate calculation[32-34] of the possible errors of different wind speeds and velocities is useful. For surveys over larger geographical areas, locations and times of sample collections should be selected by statistical methods[37,38] to obtain the necessary information with a minimum cost (see Chapter 14).

A survey over the applicable size ranges for sampling instruments is given in Figure 4. For analysis purposes, it is necessary to obtain a high concentration of the material on the deposition

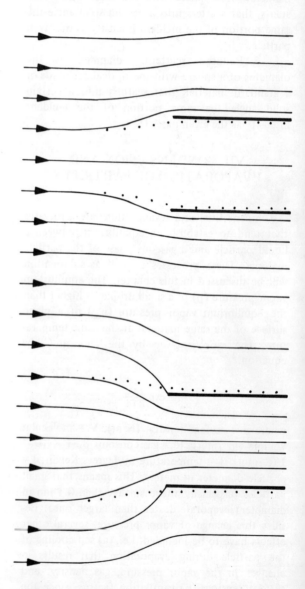

FIGURE 3.   Nonisokinetic sampling. Left: suction flow too high, loss of large particles; right: suction rate too low, excess of large particles.

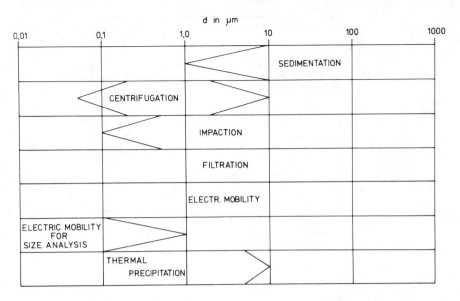

FIGURE 4. Applicable size range for sampling instruments.

surface. The following instruments are listed according to their achievable concentrations under similar conditions (aerosol sampling time), beginning with the highest concentration:

1. Impactors
2. Filters
3. Electric precipitators
4. Centrifuges
5. Instruments based on sedimentation

Descriptions of various sampling instruments can be found in the *Air Sampling Instruments Manual*[39] and in the *Handbook on Aerosols*.[40] In instruments that need a long sampling time, the collected particles may undergo changes. Evaporation and chemical reactions may occur, especially, if the sample is subject to rain water, as in the Bergerhoff collector (described in Section IX.B.).

B. Sampling Instruments Based on Sedimentation
1. Instruments with No Particle Size Separation
The simplest method for collecting dust particles is to use a horizontal tray, into which particles settle. The Berghoff collector uses a glass beaker in a wire cage.[39] An instrument such as this collects mostly large particles, since the falling speed is proportional to $\frac{1}{d^2}$, neglecting the corrections to Stokes' law. It is very useful for surveys, if a correlation between the total amount of dust and

the amount collected in the beaker can be established.

2. Instruments with Cumulative Particle Size Separation
If a laminar aerosol flow through a horizontal channel is maintained, the particles from the aerosol flow settle on the bottom of the channel and form a cumulative size distribution (Figure 5). Particles of a given size form a deposit from the beginning of the channel to a "leading edge". Only particles smaller than this given size will pass this leading edge. A complete calibration exists of the function of the aerodynamic diameter on the deposition length and the area that is covered by particles of a certain size. If this area is not covered uniformly, a density function has to be used as a correction. The frequency of particle number or the particle mass density on the bottom multiplied by the proper areas of certain size particles gives a cumulative number or mass distribution. The frequency distribution is obtained by differentiation. Such cumulative size distribution instruments are frequently used as elutriators for size-selective sampling.[42]

3. Instruments with Differential Particle Size Separation
A horizontal sedimentation channel in which particle free air and a small amount of aerosol are introduced through a slit nozzle at the top of the inlet forms a differential size distribution on the

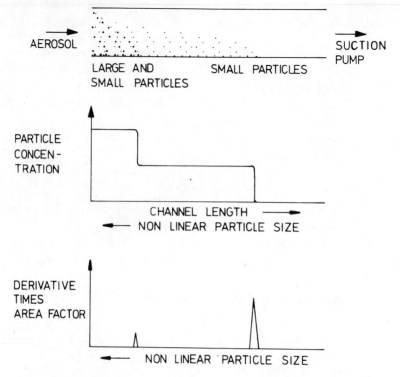

FIGURE 5.   Horizontal channel forming a cumulative size spectrum. The aerosol contains particles of two sizes.

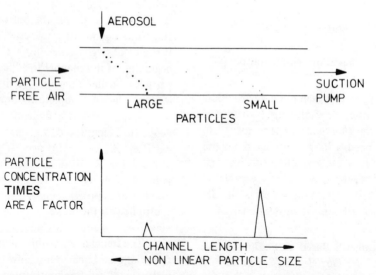

FIGURE 6.   Horizontal channel forming a discrete size spectrum. The aerosol contains particles of two sizes.

bottom plate[43] (Figure 6). Particles of a given size have a defined deposition length. The deposition length at the boundaries is reduced, but it is still well defined. The deposit of particles of a certain size forms a curved line. Usually, only the central portions are used for evaluation. Calibration of the deposition length is a function of the aerodynamic diameter of the deposited particles. Counting the particle frequency or determining the mass distribution, after multiplication with the derivative of the calibration function, determines the number or the mass distribution.

C. Centrifuges

For particles with aerodynamic diameters smaller than approximately 1 μm, it is difficult to achieve a size separation by sedimentation, since the sedimentation velocity is small, and the mean displacement by Brownian motion is similar to or larger than displacement by sedimentation. This limit can be shifted to smaller sizes, if the centrifugal force is used instead of gravitational force to precipitate the particles.

1. Centrifuge with Cumulative Size Distribution

Included in this category is the so-called Goetz aerosol spectrometer.[44,45] It can operate at speeds up to 24,000 r/min and at aerosol flows of several liters per minute. It can be used to deposit particles above ≈ 0.1 μm diameter. It is useful in determining narrow size distributions. For size analysis of a polydisperse aerosol, the advantage of the high flow rate is compensated by the inaccurate differentiating procedure to evaluate the deposit.

2. Centrifuge with Differential Particle Size Separation

Several centrifuges with differential particle size separation have been described in the last decade.[46-49] The Stöber rotor centrifuge has found the widest application.[47] In principle, it is a channel constructed as a spiral around the axes of a rotor (Figure 7). The channel begins tangentially to the axes with particle free air flowing through it. From the inlet in the center of the spiral, a

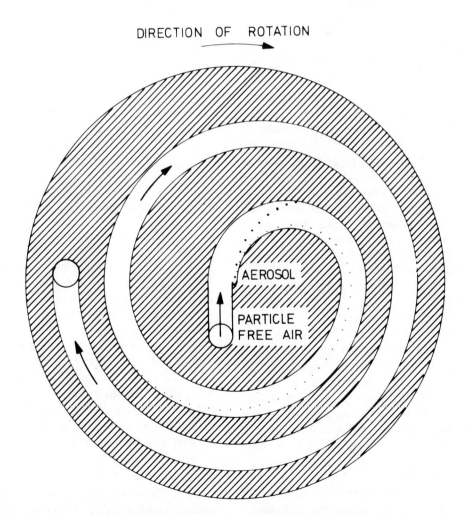

FIGURE 7.   Schematic cross section of the spiral centrifuge (Stöber rotor).

small portion of aerosol is added through a slit nozzle to the particle free air. Particles are deposited on the outer wall of the channel and show decreasing size with increasing distance from the point of deposition. Typical rotor speeds are several thousand revolutions per minute with a total flow rate of several liters per minute and with an aerosol flow rate of 0.1 to 1 l/min. The rotor diameter is approximately 30 cm, the channel length 2 m. Smaller versions of this centrifuge also exist, where the smallest fractions are collected on a filter at the end of the channel. The applicable particle size ranges from below 0.1 to several micrometers aerodynamic diameter. For large particles, above several micrometers in diameter, inlet losses become significant.

D. Impactors

Impactors[50] collect airborne particles by sucking the aerosol through a nozzle with gas speeds on the order of 100 m/sec and depositing the particles on a plate immediately in front of the nozzle. For size separation, cascade impactors, in which several impactor stages are joined in series, are used. The first stages have nozzle dimensions of several millimeters, the last ones are several tenths of a millimeter. An example of an impactor is shown in Figure 8.[51] The flow rate through the instrument is about 1 l/min. Each stage has a sigmoidal characteristic; that is, small particles can pass the stage with almost no deposition, a certain size is collected with a 50% probability, and much larger particles are completely collected. Superimposing the deposition probability for the stages determines, for each stage, a particle size interval, where the particles are deposited. This is demonstrated for a four-stage impactor in Figure 9. The intervals for adjacent stages are always overlapping. Impactors can be used for particles ranging from several tenths of a micrometer to approximately 50 μm. Impactors with many nozzles of the same size at each stage can be used with higher flow rates in order to collect a larger sample.[52,53] The gas flows change their characteristics somewhat from the central nozzles to the jets of the outer nozzles, so that the deposition probability, for each stage as a function of particle size, shows a less steep curve, and, therefore, provides less size resolution than does an impactor with a single nozzle for each stage.

A different concept is used for the spectral impactor.[54] For its operation this impactor needs

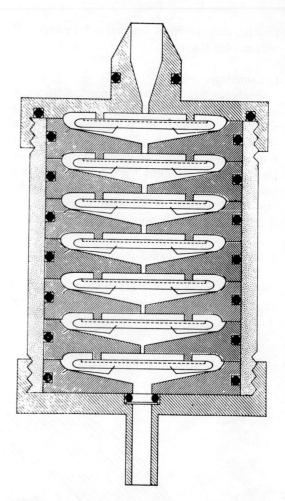

FIGURE 8.   Diagram of the cascade impactor. (From Mercer, T. T., Tillery, M. I., and Newton, G. J., *J. Aerosol Sci.*, 1, 9, 1970. With permission.)

particle free air to operate and has only one stage where larger particles are deposited in the center and smaller ones in the periphery of a circular deposit. Each particle size has a unique radius of deposition. Particles of the same size form a narrow ring as a deposit. The diameter of the deposit is 3 mm, the aerosol flow rate is several cubic centimeters per second, and the flow rate of the particle free air is approximately ten times more. Particle size resolution can reach 5%, and the applicable size range is 0.5 to 5 μm aerodynamic diameter.

E. Filtration

Filtration techniques are used where large particle masses must be collected and size separa-

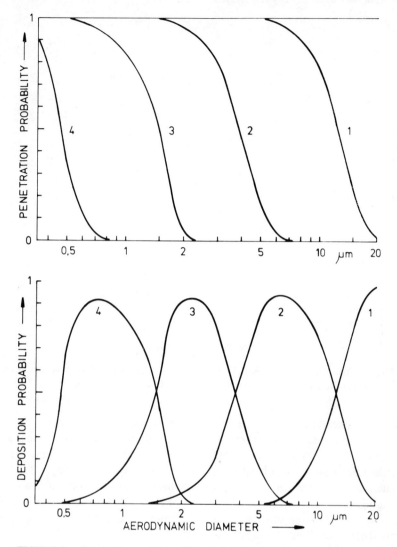

FIGURE 9. Penetration and deposition probabilities of a four-stage impactor.

tion is not required. The mechanisms or forces effective in filtration include:

1. Impaction
2. Diffusion
3. Interception
4. Electrical forces

The idea of a sieve, where particles smaller than the pore size can penetrate and larger ones are retained, cannot be applied to aerosols.[55,56]

Figure 10 illustrates filter efficiency for different mechanisms.[57] Large particles are deposited by impaction, small ones by diffusion. If the trajectory of the center of a particle does not hit the filter, but the particle is large

enough that it touches the filter, it is trapped by interception. Electrical forces usually increase filter efficiency. Between the ranges of deposition by interception, impaction, and diffusion is the so-called filter minimum (see Figure 10.) Fiber filters consist of a mesh of fine fibers. Typical fiber materials are cellulose, synthetic fibers, cloth, cotton, glass, and asbestos. With fiber diameters of about 1 μm and proper air velocities, the efficiency can exceed 99.99% for any size particle.

The classic membrane filter consists of a membrane of cellulose or similar material that has many irregular pores. During the production process, it is possible to control the mean pore size. However, the actual pore sizes may deviate strongly from the mean and the nominal pore size.

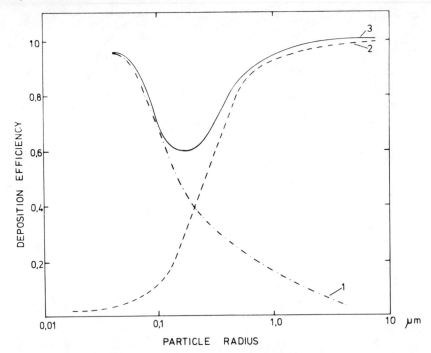

FIGURE 10.   Filter efficiencies (1) diffusion, (2) impaction, (3) total. (From Spurný, K. and Pich, J., *Collect. Czech. Chem. Commun.,* 30, 2276, 1965. With permission.)

Another type of membrane filter is the Nuclepore® filter.[58] It consists of a membrane of polycarbonate, into which parallel holes of uniform size are etched after radioactive irradiation. Particles are usually deposited on the surface, so that after sputtering, direct observation of the deposit in the scanning electron microscope is possible. The resistance to air flow of a membrane filter is generally higher than that of a fiber filter.

Filter theories[59] have been developed, however, they are difficult to apply (except for Nuclepore filters[60]), since the filter structures are very irregular. Deposition efficiencies and resistances can be determined experimentally. Complete information about the deposition efficiency of a given filter would require determination of the efficiency as a function of particle size and flow velocity. Since this would require several monodisperse aerosols or a tedious analysis, it is rather common to make efficiency tests with a few types of aerosols having different size distributions. The following aerosols and materials, among others, are used: dioctylphtalataerosol with a narrow or with a broad size distribution, methylene blue, uranine, radioactive labeled metals, quartz dust, sodium chloride, and atmospheric aerosol. If filter samples are taken for analysis

purposes, it is best to make a preliminary test with two filters in series. By comparing the dust load, it is easily determined if the first one has a sufficiently high efficiency. Besides weighing and chemical analysis of the deposit, the attenuation of β-radiation is used for the determination of the deposited mass.[61-63]

F. Electrical Mobility
1. Sampling with No Particle Mobility Separation

Several instruments exist that use electrical forces to deposit particles, independent of their size. The particles are charged by passing a corona discharge, which is maintained either by a point to plane arrangement[64] or by a wire surrounded with a cylinder.[65,66] The point or wire carries a voltage of about 5 to 10 kV. Air passes through this discharge, either between point and plane or through the cylinder. Aerosol particles charged by the corona discharge move in the electrical field from the point to the plane or from the wire to the cylinder, where they are deposited. The plane or the inner cylinder wall carries a foil or an electron microscope specimen holder that is taken from the instrument after sampling is completed, so that the particles can be analyzed.

2. Sampling with Particle Mobility Separation

For this purpose, the charging and the precipitation elements of the instrument have to be separated. The most developed instrument is the Minnesota electrical aerosol analyzer (Figure 11).[67] The geometric arrangement of the instrument is cylindrical. In the center of the cylinder is a stretched wire that carries a voltage of +5 kV. Ions are produced by corona discharge and fly to or through a concentric grid, which is grounded. Some distance from the grid is a cylinder that carries a voltage of approximately –60 V. Ions penetrate the grid and migrate to the cylinder. The

aerosol flows in an axial direction between grid and cylinder, and ions can accumulate on the particle. The charging mechanism is mainly diffusion charging. In another portion of the instrument, the charged aerosol is fed into a cylinder capacitor. This consists of a central rod, which carries a voltage up to several thousand volts, a concentric flow of particle free air, a concentric flow of aerosol, and the outer electrode. The particles that have higher electric mobility are collected on the central rod, while particles with lower mobility remain airborne. After leaving the capacitor, they are collected on a filter, where the charge of the particles is accumulated and measured by an electrometer. A mobility distribution can be deduced by increasing the voltage on the center electrode stepwise and measuring the charge of the penetrating particles. Since this deduction requires differentiation, it is often inaccurate. This problem can be avoided by using a differential mobility analyzer, where only a fraction of the aerosol within certain mobility limits can pass the capacitor through a ring slit in the central electrode. The measurement of the electric mobility can have an accuracy of about 1%. The results are often interpreted in terms of particle size, in which case the results become questionable.[68]

G. Sampling with Thermal Precipitation

Sampling with thermal precipitators can be applied when small samples are sufficient and the sample is representative. Thermal gradients are maintained between a heated wire,[69] heated ribbon,[70] or heated plate,[71] and a cold plate. The cold plate can carry an electron microscope specimen holder. The temperature differences are usually on the order of 100°C, and the distances between hot and cold parts are between several tenths and a few millimeters. The flow velocity should be sufficiently low so that particles of all sizes are deposited. Thus, the sample collected over the whole plate is representative of the airborne particle size distribution, however, each area element is not necessarily representative.

X. TEST AEROSOLS

Test aerosols are used for calibrating sampling instruments. Table 3 is a compilation of information on frequently used aerosols. Included are the types of aerosol generators, mean particle size,

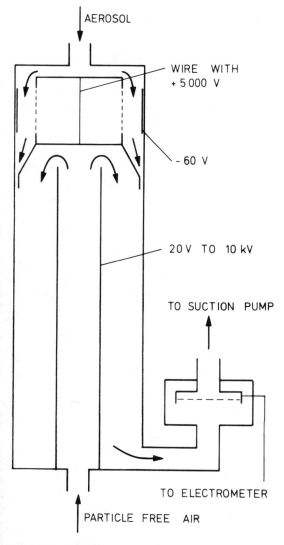

FIGURE 11. Electrical aerosol analyzer. (After Whitby, K. T., *Fine Particles,* Liu, B. Y. H., Ed., Academic Press, New York, 1976.

TABLE 3

Test Aerosol Generators

Aerosol generator	Mean particle size (μm)	Particle concentration (cm^{-3})	Degree of dispersion	Solid or liquid
Latex aerosol generator	0.09–100	10–1000	Monodisperse for particles <2 μm	s
La Mer-type generator	0.03–5	10^6–10^7	Monodisperse	l (s)
Atomizers with nozzles	0.1–5	10^6–10^7	Polydisperse	l and s
Spinning disc generators	0.1–50	10–100	Monodisperse	l and s
Generators with periodic disruption of liquid jets	0.5–50	10–100	Monodisperse	l and s
Generators with dispersion of powders	1–100	100–10^5	Polydisperse	s
PtO aerosol generator	0.005–0.03	10^5–10^8	Polydisperse	s

concentrations, degree of dispersion of the aerosol, and whether the particles are in the liquid or solid state. Monodisperse indicates that all particles are of about the same size with a geometric standard deviation smaller than 1.15 or a coefficient of variation smaller than 14%. For other definitions see References 1 and 26. Monodisperse aerosols are required, if the sampling instrument is calibrated for a single particle size or if a simple relation between particle number concentration and particle mass concentration is required. Calibration with polydisperse aerosols is advantageous, if the size distribution of the aerosol is similar to that of the aerosol to be measured.

The latex aerosol generator atomizes an emulsion of monodisperse latex particles in water. The droplets are dried, finally consisting of only solid latex particles. The degree of monodispersity of the aerosol is less than that of the emulsion due to droplets that contain no latex particle but still leave some residue after evaporation of the water and due to droplets that contain more than one latex particle. The La Mer-type generator uses condensation of oil on condensation nuclei to produce a monodisperse aerosol of high concentra-

tion. The particles show very little electrical charge. Liquids or solutions can be atomized by nozzles. Particles of a solution can dry and leave solid particles. The density of the dry particles may be smaller than that of the bulk material because of pores and holes in the particles. Spinning disc generators can operate with up to 400,000 r/min., liquid being fed to the center of the disc. The size of the produced droplets is determined by the equilibrium between centrifugal force and surface tension. For generators with periodic disruption of liquid jets, the particle size is exactly determined by the flow rate of the liquid and the frequency of disruption. Many techniques exist for the dispersion of powders.[72] For many problems, the so-called fluidized beds are the best solution. The PtO generator works by evaporation of PtO from an electrically heated wire.[73]

A survey on production methods for test aerosols is given in References 40 and 74. Latex aerosol generator, La Mer generator, atomizers with nozzles, spinning disc generators, and generators with periodic disruption of liquid jets are described in an article by Raabe.[75]

REFERENCES

1. **Fuchs, N. A.,** *The Mechanics of Aerosols,* Pergamon Press, Oxford, 1964.
2. **Millikan, R. A.,** Coefficient of slip in gases and the law of reflection of molecules from the surfaces of solids and liquids, *Phys. Rev.,* 21, 217, 1923.
3. **Hänel, G.,** Bemerkungen zur Theorie der Düsen–Impaktoren, *Atmos. Environ.,* 3, 69, 1969.
4. **Klyachko, L. S.,** Equation of the motion of dust particles, *Heat. Vent.,* 5, 27, 1934.
5. **Cunningham, E.,** On the velocity of steady fall of spherical particles through fluid medium, *Proc. R. Soc. London Ser. A,* 83, 357, 1910.
6. **Basset, A. B.,** On the motion of a sphere in a viscous liquid, *Philos. Trans. R. Soc. London,* 179, 43, 1888.
7. **Knudsen, M. and Weber, S.,** Luftwiderstand gegen die langsame Bewegung kleiner Kugeln, *Ann. Phys.,* (Leipzig) 36, 981, 1911.
8. **Fuchs, N. A. and Stechkina, J. B.,** Resistance of a gaseous medium to the motion of a spherical particle of a size comparable to the mean free path of the gas molecules, *Trans. Faraday Soc.,* 58, 1949, 1962.
9. **Oseen, C.,** *Neuere Methoden und Ergebnisse der Hydrodynamik,* Leipzig, 1927.
10. **Kunkel, W. B.,** Magnitude and character of errors produced by shape factors in Stokes' law estimates of particle radius, *J. Appl. Phys.,* 19, 1056, 1948.
11. **Stöber, W., Berner, A., and Blaschke, R.,** The aerodynamic diameter of aggregates of uniform spheres, *J. Colloid Interface Sci.,* 29, 710, 1969.
12. **Stöber, W.,** Dynamic shape factors of nonspherical aerosol particles, in *Assessment of Airborne Particles,* Mercer, T. T., Morrow, P. E., and Stöber, W., Eds., Charles C Thomas, Springfield, Ill., 1972.
13. **Horvath, H.,** Das Fallverhalten nicht-kugelförmiger Teilchen, *Staub Reinhalt. Luft,* 34, 251, 1974.
14. **Dahneke, B. E.,** Slip correction factors for nonspherical bodies, *J. Aerosol Sci.,* 4, 139, 1973.
15. **Einstein, A.,** *Investigations on the Theory of the Brownian Movement,* E. P. Dutton, New York, 1926.
16. **Stetter, G.,** Abscheidung und Fraktionierung von Staub durch Thermodiffusion, *Sitzungsber Oesterr. Akad. Wiss. Math-Naturwiss Kl.,* 169 (Part II), 91, 1960.
17. **Waldmann, L. and Schmitt, K. H.,** Thermophoresis and diffusiophoresis of aerosols, in *Aerosol Science,* Davies, C. N., Ed., Academic Press, London, 1966.
18. **Keefe, D., Nolan, P. I., and Rich, T. A.,** Charge equilibrium in aerosols according to the Boltzmann law, *Proc. R. Ir. Acad. Sect. A,* 60, 27, 1959.
19. **Whitby, K. T. and Liu, B. Y. H.,** The electrical behaviour of aerosols, in *Aerosol Science,* Davies, C. N., Ed., Academic Press, London, 1966.
20. **Rayleigh, L.,** *Philos. Mag.,* 14, 184, 1882.
21. **Van de Hulst, H. C.,** *Light Scattering by Small Particles,* John Wiley & Sons, New York, 1957.
22. **Kerker, M.,** *The Scattering of Light,* Academic Press, New York, 1969.
23. **Farone, W. A., Kerker, M., and Matijević, E.,** Scattering by infinite cylinders at perpendicular incidence, in *Electromagnetic Waves,* Kerker, M., Ed., Pergamon Press, Oxford, 1963.
24. **Roth, C., Gebhart, J., and Stuck, B.,** Streulicht-Formfaktoren von Ellipsoiden, in *Aerosole in Naturwissenschaft, Medizin und Technik,* Jahreskongress der GAF, Böhlau, V. and Straubel, H., Eds., Bad Soden, Germany, 1976.
25. **Zerull, D.,** Formabhängigkeit des Lichtstreuverhaltens von Staubpartikeln, in *Kolloquium Aerosolmesstechnik,* Luck, H., Kraus, F. J., and Fissan, H. J., Eds., Rheinisch-Westfälische Technische Hochschule Aachen, Germany, 1975.
26. **VDI-Richtlinie 3491,** *Messen von Partikeln,* Beuth Verlag, Berlin, 1976.
27. **Green, H. L. and Lane, W. R.,** *Particulate Clouds,* 2nd ed., E. & F. N. Spon, London, 1964.
28. **Von Smoluchowski, M.,** Versuch einer mathematischen Theorie der Koagulationskinetik kolloider Lösungen, *Z. Phys. Chem.,* 92, 129, 1917.
29. **Zebel, G.,** Coagulation of aerosols, in *Aerosol Science,* Davies, C. N., Ed., Academic Press, London, 1966.
30. **Wagner, P. E. and Kerker, M.,** Brownian coagulation of aerosols in rarified gases, *J. Chem. Phys.,* 66, 638, 1977.
31. **Fuchs, N. A.,** Sampling of aerosols, *Atmos. Environ.,* 9, 697, 1975.
32. **Belyaev, S. P. and Levin, L. M.,** Techniques for collection of representative aerosol samples, *J. Aerosol Sci.,* 5, 325, 1974.
33. **Zenker, P.,** Untersuchungen zur Frage der nichtgeschwindigkeitsgleichen Teilstromentnahme bei der Staubgehaltsbestimmung in strömenden Gasen, *Staub Reinhalt. Luft,* 31, 252, 1971.
34. **Rüping, G.,** Die Bedeutung der geschwindigkeitsgleichen Absaugung bei der Staubstrommessung mittels Entnahmesonden, *Staub Reinhalt. Luft,* 28, 137, 1968.
35. **Narjes, L.,** Anwendung neuartiger Nulldrucksonden zur quasiisokinetischen Staubprobenahme in Dampfkraftanlagen, *Staub Reinhalt. Luft,* 25, 148, 1965.
36. **Zebel, G.,** Some problems in sampling of coarse aerosols, in *Recent Developments in Aerosol Science,* Shaw, D., Ed., John Wiley & Sons, New York, 1977, in press.
37. **Yates, F.,** *Sampling Methods for Censuses and Surveys,* Griffin, London, 1960.
38. **Cochran, W. G.,** *Sampling Techniques,* John Wiley & Sons, New York, 1965.

39.   *Air Sampling Instruments Manual,* 5th ed., American Conference of Governmental Industrial Hygienists, Cincinnati, 1977.

40.   Dennis, R., Handbook of Aerosols, National Technical Information Service, U. S. Department of Commerce, Springfield, Va., 1976.

41.   VDI-Richtlinie 2119, *Messungen von Partikelniederschlägen: Bergerhoff Verfahren,* Paper 2, Beuth Verlag, Berlin, 1972.

42.   Orenstein, A. J., *Proc. of the Pneumoconiosis Conference, Johannesburg, 1959,* Churchill, Livingstone, London, 1960.

43.   Boose, C., Ein Gerät zur fraktionierten Abscheidung von Staub für Korngrössenanalysen, *Staub,* 22, 109, 1962.

44.   Goetz, A., Stevenson, H. J. R., and Preining, O., The design and performance of the aerosol spectrometer, *J. Air Pollut. Control Assoc.,* 10, 378, 1960.

45.   Gerber, H. E., On the performance of the Goetz aerosol spectrometer, *Atmos. Environ.,* 5, 1009, 1971.

46.   Hochrainer, D. and Brown, P. M., Sizing of aerosol particles by centrifugation, *Environ. Sci. Technol.,* 3, 830, 1969.

47.   Hochrainer, D., A new centrifuge to measure the aerodynamic diameter of aerosol particles in the submicron range, *J. Colloid Interface Sci.,* 36, 191, 1971.

48.   Abed-Navandi, M., Berner, A., and Preining, O., The cylindrical aerosol centrifuge, in *Fine Particles,* Lin, B. Y. H., Ed., Academic Press, New York, 1976.

49.   Stöber, W. and Flachsbart, H., Size separating precipitation of aerosols in a spinning spiral duct, *Environ. Sci. Technol.,* 3, 1280, 1969.

50.   May, K. R., The cascade impactor, *J. Sci. Instrum.,* 22, 187, 1945.

51.   Mercer, T. T., Tillery, M. I., and Newton, G. J., A multi-stage, low flow rate cascade impactor, *J. Aerosol Sci.,* 1, 9, 1970.

52.   Andersen, A. A., A new sampler for the collection, sizing and enumerating of viable airborne bacteria, *J. Bacteriol.,* 76, 471, 1956.

53.   Berner, A., Praktische Erfahrungen mit einem 20-stufen Impaktor, *Staub Reinhalt. Luft,* 32, 315, 1972.

54.   Zebel, G. and Hochrainer, D., Zur Messung der Grössenverteilung des Feinstaubes mit einem verbesseten Spektralimpactor, *Staub Reinhalt. Luft,* 32, 91, 1972.

55.   Dorman, R. G., Filtration, in *Aerosol Science,* Davies, C. N., Ed., Pergamon Press, London, 1966.

56.   Davies, C. N., *Air Filtration,* Academic Press, London, 1973.

57.   Spurný, K. and Pich, J., Analytical methods for determination of aerosols by means of membrane ultrafilters, *Collect. Czech. Chem. Commun.,* 30, 2276, 1965.

58.   Frank, E. R., Spurný, K. R., Sheesley, D. C., and Lodge, J. P., Jr., The use of nuclepore filters in light and electron microscopy of aerosols, *J. Microsc.,* 9, 735, 1970.

59.   Pich, J., Theory of aerosol filtration by fibrous and membrane filters, in *Aerosol Science,* Davies, C. N., Ed., Academic Press, London, 1966.

60.   Zebel, G., A simple model for the calculation of particle trajectories approaching nuclepore filterpores with allowance for electrical forces, *J. Aerosol Sci.,* 5, 473, 1974.

61.   Aurand, K. and Bosch, J., Gerät zur kontinuierlichen Bestimmung der Konzentration staubförmiger Luftverunreinigungen, *Staub,* 27, 445, 1967.

62.   Sem, G. J. and Borgos, J. A., An experimental investigation of the exponential attenuation of beta radiation for dust measurement, *Staub Reinhalt. Luft,* 35, 5, 1975.

63.   Macias, E. S. and Husar, R. B., A review of atmospheric particular mass measurement via the beta attenuation technique, in *Fine Particles,* Liu, B. Y. H., Ed., Academic Press, New York, 1976.

64.   Morrow, P. E. and Mercer, T. T., A point to plane electrostatic precipitator for particle size sampling, *Am. Ind. Hyg. Assoc. J.,* 25, 8, 1964.

65.   Drinker, P., Alternating current precipitators for sanitary air analysis, *J. Ind. Hyg.,* 14, 364, 1932.

66.   Liu, B. Y. H., Whitby, K. T., and Yu, H. H. S., Electrostatic aerosol sampler for light and electron microscopy, *Rev. Sci. Instrum.,* 38, 100, 1967.

67.   Whitby, K. T., Electrical measurement of aerosols, in *Fine Particles,* Liu, B. Y. H., Ed., Academic Press, New York, 1976.

68.   Marlow, W. H., Reist, P. C., and Dwiggens, G. A., Aspects of the performance of the electrical aerosol analyzer under nonideal conditions, *J. Aerosol Sci.,* 7, 457, 1976.

69.   Watson, H. H., The thermal precipitator, *Trans. Inst. Min. Met.,* 46, 176, 1936.

70.   Walkenhorst, W., Ein neuer Thermalpräzipitator mit Heizband und seine Leistung, *Staub,* 22, 103, 1962.

71.   Kethley, T. W., Gordon, M. T., and Orr, C., Jr., A thermal precipitator for aerobacteriology, *Science,* 116, 368, 1952.

72.   Zahradnicek, A., Methoden zur Aerosolherstellung aus vorgegebenen Feststoffhaufwerken, *Staub Reinhalt. Luft,* 35, 226, 1975.

73.   Spurny, K. and Lodge, J. P., Jr., Herstellung hochdisperser Modellaerosole für Staubforschung und Filterprüfung, *Staub Reinhalt. Luft,* 33, 166, 1973.

74.   VDI-Richtlinie 3491, *Messen von Partikeln, Herstellungsverfahren für Prüfaerosole, Grandlagen und Übersicht,* Paper 2, Beuth Verlag, Berlin, 1976.

75.   Raabe, O. G., The generation of aerosols of fine particles, in *Fine Particles,* Liu, B. Y. H., Ed., Academic Press, New York, 1976.

Chapter 3
X-RAY FLUORESCENCE

D. E. Leyden

TABLE OF CONTENTS

I. INTRODUCTION

X-Ray fluorescence (XRF) is an emission spectroscopic method for the qualitative identification and quantitative determination of elements. The method is generally applicable to elements of atomic number between 11 and 92; with nonroutine equipment, this range may be extended. With few exceptions, XRF provides no information concerning the chemical species in which the sought element is found. For all practical purposes, it is a method to be used for the determination of the bulk elemental composition of a sample as opposed to those techniques which may provide information about the composition of the sample on a microscale or chemical speciation within the sample.

A brief introduction to the principles of XRF will be given. However, it is not the purpose of this chapter to dwell upon these principles, but to illustrate the potential of XRF for the characterization of airborne dust particles. The reader should consult one of the many monographs that provide various levels of introduction to the principles and practices of XRF.[1-4]

II. PRINCIPLES OF XRF

Atoms in the sample are converted from the ground state to an excited state by an energy source. The source used will depend upon the nature of the sample, the design of the spectrometer, and the goals of the analysis. The source may be an X-ray tube, a radioactive isotope which emits X-rays, an electron beam, or a beam of accelerated charged particles such as protons or alpha particles. The energy content of these sources must be sufficient to promote an electron from an inner orbital (usually the K or L shell) of the atom to a higher level. In fact, the atom is usually ionized; however, the ion is in an excited state because the orbital vacancy is in a lower shell than the valence electrons. The cascade transfer of outer shell electrons to fill this vacancy releases energy which is manifested as a series of emission spectroscopic lines in the X-ray region of the electromagnetic radiation spectrum. The wavelengths (energies) of these lines are characteristic of the element emitting the radiation, and their intensities are related to the amount (or concentration) of the element present. The detection of the radiation and the intensity-concentration relationships will be discussed below. Thick or thin solid or liquid samples may be used. However, because any matrix of relatively high atomic numbers will absorb both the source and emitted X-rays and those of relatively low atomic number will scatter the radiation, an ideal sample is one which is a very thin film. In the use of such thin film samples, total intensity is sacrificed to be free from various matrix effects upon the intensity.

XRF is a technique that offers many advantages for the quantitative determination of the average elemental composition of dust. When the samples are collected on filters, detection limits ranging from several hundred nanograms per square centimeter (ng/cm^2) for elements such as sodium to a few nanograms per square centimeter for the transition elements may be obtained. From the volume of air drawn through the filter, the amount of element per unit volume of air (viz. $\mu g/m^3$) may be calculated. The most obvious advantage of XRF is the fact that the dust may be analyzed directly on the filter. The speed and simplicity of such a scheme is attractive, and the lack of sample treatment is important in the reliability of the results. However, the method is not without fault. Even with the advantages of thin dust deposits on a filter, particle size effects can affect the results. Typically, X-ray photons of analytically useful energies penetrate 1 to 100 μm into a matrix. Therefore, particles of this diameter range will show an intensity which is dependent upon the diameter, and each emitted X-ray wavelength will be affected differently. This particle-size effect on the intensity usually limits the accuracy of elemental determinations in dust by XRF. There are mathematical corrections which may be made if the particle size is known.[5] Unfortunately, this is never the case as a particle-size distribution is obtained on the filter. In some cases, an upper limit of particle size is known. An important secondary effect is the serious difficulty in the preparation of standards which are meaningful representations of the samples. Because of the particle-size effect and standard preparation problems, the overall accuracy of X-ray fluorescence for the determination of average dust composition is probably ± 10% (relative).

III. SAMPLE COLLECTION METHODS

A detailed discussion of the philosophy and methods of airborne particle sampling was given in Chapter 2 earlier and further comments here will be limited to sample collection methods directly applicable to XRF analysis. Several practical considerations are important. With modern instrumentation, detection limits of a single element on a thin film range from one to several hundred nanograms per square centimeter. However, the definition of a thin film will vary with the element to be determined and the particle size effects mentioned earlier will cause signal losses, particularly if particles larger than a few micrometers in diameter are present. Thus, the combination of excitation efficiency, matrix effects, instrument design, and so forth will determine the amount of sample required. As a general statement, however, a thin film of dust on the order of 100 $\mu g/cm^2$ is satisfactory for analysis by current instruments. Frequently, a few milligrams per square centimeter of aerosol sample are collected.

Both impactor and filter samplers may be used. However, it must be remembered that XRF is a technique to be used for the determination of the average composition of a macroscopic sample. Therefore, the collected sample must be representative of the airborne dust. Cascade impactors have been found to be ill suited for sample collection when XRF is to be used for the analysis.[6] Dry surface impactors were found to bounce larger particles from the initial stages causing them to appear on subsequent stages mixed with smaller particles or to be lost on the walls of the impactor. Surfaces treated with greases reduced this problem, but were not suited for X-ray analysis because of the grease film. Also, the diameter of the sample deposit is usually small compared to the X-ray beam leading to a large substrate blank. Furthermore, with the possible exception of a specialized study, there is little advantage in using an impactor to collect samples for XRF analysis. The advantages of XRF provide for rapid, simple, and reasonably accurate analysis of the elemental composition of airborne dust, that is, a technique for monitoring environmental aerosols (Chapter 14). If a detailed transport study or other information is sought, then a technique such as electron microprobe should be considered. As a result of the above discussion and other factors, the major sample collection technique used for XRF analysis of airborne dust has been some type of filtration or virtual impactor using filters.

The selection of filter material is not a simple task. Some of the factors to be considered are

1. The filter must be of a pure composition which does not provide a high spectral blank for the elements to be determined in the dust. Furthermore, the level of those impurities which are present should be constant as the blank intensity must be substracted from the measured intensities of samples.

2. The filter material should not create an unduly high pressure drop at the sampling flow rate desired.

3. The filter material must be resistant to severe damage by X-radiation. Since most samples will be run under a vacuum in the X-ray spectrometer, the filter should not degass significantly in a vacuum.

4. The composition of the filter material should be such that it does not react with or catalyze reactions of components in the air, particulate or otherwise. For example, the determination of sulfur in aerosols has been suspect because of the possible conversion of SO_2 to sulfate on the surface of the filter.[7]

5. The dimensions of the filter should be stable to changes in temperature, relative humidity, and other environmental factors.

6. The filter should be of high porosity to prevent clogging upon loading.

7. Cost may be of concern if large numbers of samples are to be run.

Consideration of all these criteria coupled with practical considerations such as availability of the filters in the size and shape desired will determine the filter of choice. Filters made of fluorocarbons,[8] polycarbonate,[9] and filter paper[10,11] have been used. The fluorocarbon filters appear to offer several advantages including purity and stability. However, they are expensive. At the time of this writing, there appears to be no clear single choice of filter materials. Glass-fiber filters may be suspect as their surface may catalyze chemical reactions. These filters also show a large blank for several elements.

One important consideration in selection of a filter for aerosol sample collection is the choice of pore diameter of the filter. This parameter is intimately related to the design and purpose of the

sampling system. Therefore, selection of pore size requires a decision of the sampling system to be used. An example of the type of detailed study required is the work of Spurny et al.[12-14] These studies were performed using 10 μm thick sheets of polycarbonate (Nuclepore®) which are etched to form cylindrical pores. The filtration process was described mathematically; based upon this model and filtration properties,[12] aerosol sampling[13] and the effects of pore clogging[14] were studied. The theoretical models developed for these filters permit accurate calculations of the fractionation of particle size using tandem filters.[9] As will be discussed later, the fractionation of particle size reduces problems of data treatment when XRF is used for the analysis; the importance of particle size fractionation to increased understanding of the source and transport of the particulate has been well demonstrated.

Several types of sampling devices have been used. The simple high-volume sampler which was used early for the collection of airborne particles is ill-suited for the information sought by XRF analysis. There is no particle size discrimination in these devices except for the loss of small particles which pass through the filter. The concept of high volume sampling is not compatible with the membrane filters required for XRF analysis. There have evolved two main goals for sampling systems other than the obvious requirements of collecting reproducibly representative samples. These are (1) to provide information about the composition of the particles as a function of particle size and (2) to provide information about the variations in ambient aerosol composition with time. As a result, several different types of sampling systems have been developed; each has advantages and disadvantages. Three types of sampling systems will be briefly discussed here. These are a time-sequenced filter with no particle size discrimination, a *virtual* impactor filter with two-stage particle size discrimination, and an impactor filter which has both time and particle size resolution. In this brief space, all nuances on these designs cannot be discussed. However, these three were selected because they are representational models concerned with aerosol sampling and provide the reader with a degree of familiarization with the samples which have been used.

Nelson et al.[11] have described an automatic time-sequence filter sampler which has become commonly called the "streaker." This device holds

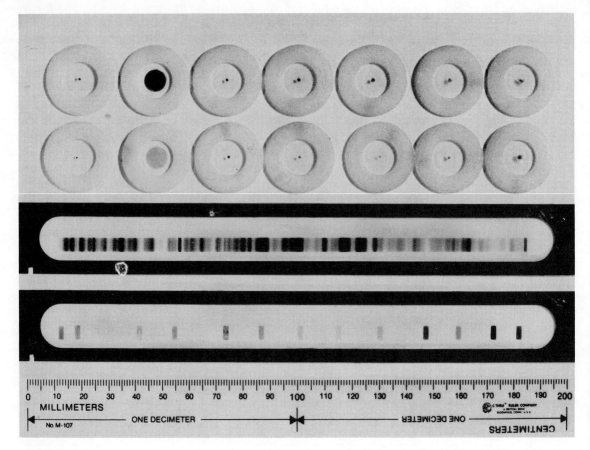

FIGURE 1.   Filters from "streaker" sampler. (From *X-Ray Fluorescence Analysis of Environmental Samples,* Dzubay, T. G., Ed., Ann Arbor Science, Ann Arbor, Mich., 1977, 22. With permission.)

a strip (ca 200 mm × 5 mm) of 0.4 μm pore size Nuclepore® filter. A movable Teflon® orifice (2 mm × 5 mm) located on top of and against the filter strip pulls air one liter per minute by suction through the bottom side of the filter. This arrangement excludes the filtration of particles larger than about 30 μm. The orifice may be moved continuously along the filter (1 mm/hr), or it may be moved in discrete increments so that a sample integrated over a time period may be taken. Figure 1 shows filters exposed in these two modes. These devices provide for flexible time resolution in sampling. However, they do not provide for particle size discrimination.

A sampler devised by Dzubay et al.[8] and Parker et al.[9] for particle size discrimination has been termed a "dichotomous sampler." This device has appeared in two forms: a simple tandem arrangement of 12 μm and 0.2 μm Nuclepore filters[9] and a rather complicated device using the concept of a virtual impactor.[8] In the second device, the

aerosol particles are impacted into a void where the path of smaller particles is aerodynamically disturbed.[15] The particles are then more gently (than impaction) drawn through a fine filter. The larger particles proceed to be impacted into a separate void where they are filtered on a coarse filter. A cross section of this device is shown in Figure 2. These two devices give results which are in good agreement in terms of particle-size fractionation.[9] A third device is the Lundgren-type multiday impactor.[16] This device has been used as the sampler for an extensive aerosol analysis program in the state of California.[17] It contains two tandem rotating drums using 1/8 mil Mylar® film as the impactor substrate. The mylar surface is coated with 50 μg/cm² paraffin wax to minimize particle bounce. A 0.8 μm Nuclepore filter is the final stage. As a result, three stages of particle size ranges are collected: 20 to 4 μm, 4 to 0.6 μm, and 0.6 to 0.1 μm. The device can automatically change the filter at preset time intervals. This

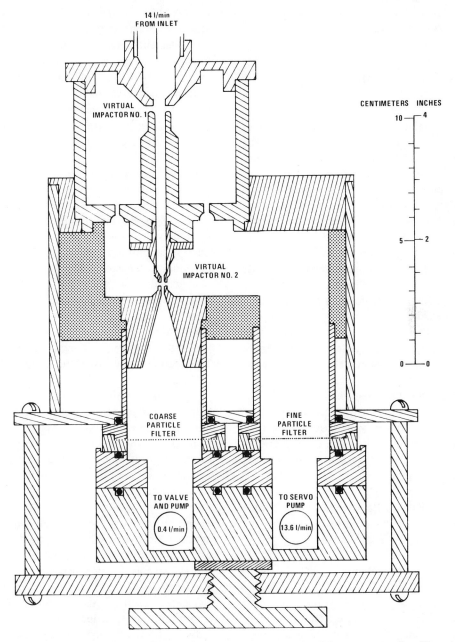

FIGURE 2. Cross-sectional view of the virtual impactor in the dichotomous sampler. (From *X-Ray Fluorescence Analysis of Environmental Samples*, Dzubay, T. G., Ed., Ann Arbor Science, Ann Arbor Science, Ann Arbor, Mich., 1977, 22. With permission.)

device is currently available commercially and offers the advantage of easily setting up a monitor station.

With all of these devices, one must be careful to account for various factors which may affect the sampling operation. Weather factors are important, especially wind velocity. As with any analytical

procedure, the sampling process is of prime importance. Although the ultimate sampling device has not been identified, there has been much research and development in the evaluation of aerosol sampling, and the reader should become very familiar with the results of this work before engaging in aerosol sampling for XRF analysis. Of particular importance is the work of Spurny et al.,[12-14] Ranz and Wong,[18] Marple and Liu,[19] and others.

IV. ANALYTICAL MEASUREMENTS

The previous discussions in this chapter have dealt with obtaining a representative sample of airborne dust particles for analysis by XRF. The sampling device to be used and goals to accomplish the sampling task require much thought and many decisions. Simply because XRF has been selected as the analytical method does not mean that there are no further decisions concerning the instrumentation, as many options still remain open. Only a brief description of the instrumental choices available are provided herein. The reader must consider each type of instrument in terms of cost, speed, number of samples to be run, number of desired elements in each sample, sensitivity, and level of automation. Although there are specialized instruments for particular applications, only those instruments commercially available for routine use will be discussed here.

The older, more conventional form of XRF spectrometers are known as wavelength-dispersive spectrometers. These instruments angularly disperse the radiation emitted from the sample by means of an analyzing crystal. Each excited element in the sample emits a series of lines of X-radiation; the wavelength of each line is characteristic of the element. Radiation from the sample is directed upon a crystal which diffracts the radiation according to Bragg's law:

$$n\lambda = 2d \sin \theta$$

where n is an integer (order); λ is the wavelength in Angstroms (Å); d is the interlattice spacing of the crystal (Å); and θ is the angle of incidence of the radiation on the crystal. In practice, the angle θ is selected by a goniometer which moves the crystal until Bragg's law is satisfied for the wavelength desired. Normally, a crystal is selected with a value of d such that $20° < \theta < 80°$. It is readily seen that this operation provides for single element determination at one setting; these parameters as well as crystal changes, sample changes, and other controls have been automated such that automated, sequential elemental analysis may be performed. Furthermore, multichannel wavelength-dispersive instruments with a dozen or more

crystals located radially about the sample are available. Such instruments permit simultaneous multi-element determinations to be performed on preselected elements. The elements to be determined may be changed, but not readily. These instruments operate with high radiation power (up to 3.5 kW) to gain signal intensity, yet protect the detector from saturation by selecting only the line which is to be measured.

The second major type of XRF spectrometer has been known as energy-dispersive X-ray spectrometer (or X-ray energy spectrometer). This design eliminates the need for an analyzing crystal by use of a solid state Si (Li) detector with greater resolution than the gas or scintillation detectors. The pulse output from these detectors is proportional to the energy of the X-ray photon which is detected. The resolution of these pulses is sufficient to discriminate the X-ray photons of elements adjacent in the periodic table. Using a multichannel analyzer, the spectrum of all elements in the sample is acquired simultaneously. This feature is a great advantage in multi-element determinations.

Currently, both types of spectrometers use either conventional (Coolidge-type) or transmission X-ray tubes as the source of radiation. Many of the energy-dispersive spectrometers make provisions for a pure element target which acts as a secondary source. The most efficient wavelength for excitation of an element is that which is just shorter than the absorption edge for the element. Therefore, when a few elements adjacent in the periodic table (or elements having different types of absorption edges near the same wavelength) are to be determined, it is much more efficient to use a characteristic emission line of an appropriate element to excite these elements. This lowers the background and reduces the count rate for radiation from elements of no interest.

The use of charged particles (viz. protons or alpha particles) has found increasing interest as a mode of inducing X-ray emission. These particles are usually in the 2 to 4 MeV energy range which requires an expensive source such as a van der Graaff accelerator. These installations are not installed for the purpose of X-ray spectrometry. However, it is quite easy to place X-ray equipment on an existing accelerator and rent time for this purpose. In almost every case, an energy-dispersive analyzer system is used in conjunction with the accelerator source. The program of aerosol analysis in California mentioned earlier is an example of the use of proton-induced X-ray emission (PIXE) on a routine basis for a high volume of samples.[17]

The advantage of particle excitation lies primarily in the fact that the excitation is efficient, and there is in principle no bremsstrahlung or continuous background of radiation with which to contend. This is in fact not the case in practice. An X-ray continuum is emitted at the low-energy portion of the spectrum. Furthermore, samples of insulating materials exhibit charging effects under charged particle bombardment, and the thermal energy produced as the particles are absorbed may create problems such as degassing or even charring of the sample. Fortunately, these problems are not severe with thin films. However, there are considerable difficulties in treating the experimental data, and a great deal of effort has gone into the preparation of computer programs to extract a signal which is reliably related to the amount of each element on the sample.[20]

A primary question facing the analyst, who is initiating a program to study aerosol samples and has determined that XRF is a technique of choice, is which type of instrument to acquire. This is not an easy question to answer. Frequently, sensitivity and detection limit are the first points raised. The reader must remember that sensitivity is the signal per unit amount of analyte, whereas detection limit is the lowest amount of analyte that may be detected as present using some previously established statistical criteria. The importance of considering both parameters is illustrated by remembering that an instrument using a high-wattage continuum source may provide a high sensitivity. However, because a high background is also generated, the detection limit is no better than that obtained with an instrument using a much lower power monochromatic source. Several comparison studies have been made on detection limits of the various instrument types. The results of one of these studies are given in Table 1.[21] A few selected elements are shown to illustrate that for single elements prepared on a thin film the detection limits are not drastically different between the various types of modern instrumentation. It should be cautioned that detection limits in real samples may be much less favorable.

One must consider the number of elements to be determined. If one is to routinely determine a dozen or more elements with large numbers of

samples, the energy-dispersive spectrometers should be seriously considered. With fewer elements, wavelength-dispersive instruments can compete favorably for speed, if automated. The best approach in making the final selection is to take samples to the vendors and give them the requirements. Then, if the analytical results are acceptable, consider factors such as analysis time, simplicity of operation, and cost. Keep in mind that the effects of the filter background, matrix, particle size, and other parameters influence the X-ray intensity regardless of whether wavelength-dispersive or energy-dispersive techniques are used. The extraction of the raw intensity data is usually more direct and simple with wavelength-dispersive methods; but it is more time consuming unless a multichannel instrument is used. On the other hand, the energy-dispersive spectrometers are normally equipped with a minicomputer that performs the necessary data manipulations. The

TABLE 1

Current Lower Limits of Detection (ng/cm^2) of Selected Single Elements as a Thin Film on Millipore® Filters[a]

Element	MC-WDXRF[b]	TE-EDXRF[c]	PIXE[d]
Al	2.2	–	–
Si	5.4	157	–
S	6.7	38	1100
K	3.9	13	27
V	9.5	22	8.4
Fe	17	10	3.7
Zn	6.9	6.2	2.2
As	17	3.0	1.6
Se	–	2.6	1.7
Sr	–	3.5	3.1
Cd	3.3	5.9	21
Ba	3.2	7.4	–
Sn	4.7	38	–
Pb	5.3	8.9	5.8

[a]Millipore filters were chosen for comparison, but in many cases do not represent the best substrate.
[b]Multichannel WDXRF at EPA Laboratory, Research Triangle Park, N.C.
[c]Tube-excited EDXRF at Lawrence Berkley Laboratory, Cal. secondary fluorescers were used to generate exciting radiation.
[d]Proton-induced X-ray emission system at Duke University, Durham, N.C.

Data adapted from Jaklevic, J. M. and Walter, R. L., *X-ray Fluorescence Methods for Analysis of Environmental Samples,* Ann Arbor Science, Ann Arbor, Mich., 1976, 63.

user must be alert as to whether the programs employed are adequate for his requirements.

To convert the raw X-ray intensity to meaningful analytical results requires several procedures. First, the blank/background factor of the filter must be considered. The filter material will contain some of the elements to be determined in the sample which emit radiation as a blank. In addition, radiation from the source will be scattered by the filter and detected as a background. If the filters are reproducibly manufactured, it would seem a simple task to subtract the intensity obtained from a clean filter from the raw data for a sample. This is in fact done; however, the presence of the particulate film will affect the intensity of the radiation from the filter material. This is one source of error.

Once the intensities have been corrected for the filter contributions, corrections for overlapping spectra lines must be made. For example, it is very common that a K_α line of an element of atomic number Z overlaps, at least partially, with the K_β line of the element of atomic number Z-1. Other examples are spectral interferences such as a K line of an element occurring at the same energy (wavelength) as an L line of a different element. Arsenic and lead are specific examples of the latter problem. These difficulties are overcome in at least two ways. First, if the intensity of a line with no interference for an element can be measured, the intensity of a different line for that element may be calculated from theoretical intensity ratios. This value may be subtracted from the intensity effected by overlap with the second line. In the case of energy-dispersive techniques, one may store spectra of all elements acquired from pure samples in the computer; the spectrum of the clean filter may also be stored. Using *stripping* routines, the blank/background is subtracted, and utilizing a least squares-fitting routine, the experimental spectrum is fitted to a linear combination of pure element spectra. There are a variety of mathematical ways to accomplish this result.

Once the intensities corrected for blank/background and overlap problems are obtained, it is necessary to account for interelement and particle size effects. Interelement effects arise from the fact that X-rays emitted by an element Z_1 in the sample may be absorbed by a second element, Z_2. The result is that the X-ray intensity from element Z_1 is lower than it would be if Z_2 were not present. Z_2 may be excited by the absorption of X-rays from Z_1; this may lead to an enhancement of the X-ray intensity from Z_2. These absorption/enhancement effects are frequently called interelement effects. There have been comparisons[22] of the many procedures devised to correct these interactions.

The most difficult problem in particulate analysis by XRF is the effect of variations in particle size. There are mathematical correction procedures which may be used to compensate for particle-size variations.[5] However, these can be very complicated if a broad particle size distribution of inhomogeneous particles is to be analyzed. When simple high-volume filter samplers are used, the particle size and distribution of the particles on the filter are not known. Fortunately, there is relief from the problem. There is considerable evidence that a bimodal distribution of particle size exists in atmospheric particles with a minimum between 1 and 2 μm and maxima near 0.3 and 10 to 20 μm.[23-25] This has led to the use of particle size discrimination during sampling using devices similar to those described earlier in this chapter. Using the low energy sulfur K_α line as a test, Hawthorne et al. have shown the result of attenuation of radiation within the particles[26] for particles in a monolayer below 1 μm diameter. Of course, the attenuation will depend upon the composition and shape of the particles as well as the energy of the X-ray to be measured.

The preparation of standards for XRF analysis of aerosols is of prime importance. The ideal standard should simulate the physical and chemical properties of the samples and include the range of concentrations of the elements to be found in the samples. This is not an easy task. Several approaches to the preparation of standards have been presented. Chessin and McLaren[27] used an enclosed bubble-chamber method to form aerosols from dilute solutions of salts of the elements to be determined. Although this method was simple in practice and gave excellent calibration curves, the deposit of the aerosol on the filter is not expected to resemble that of a natural sample collected from the atmosphere. Filter standards prepared from aerosols created by blown pulverized materials are commercially available.[28] These materials are prepared in bulk, and chemical analyses are performed on representative sets (see Chapter 13). A disadvantage with any of these standards is the great care with which they must be handled to avoid losses.

A trend which seems to be improving in reliability is the use of a limited number of single-element standards to determine an overall system efficiency for the spectrometer in use. This procedure calibrates the response of the entire X-ray spectrometer system for the intensity resulting from a given amount of each element. Once the relative intensities are known, a single-element standard may be used to calibrate day-to-day variations in instrumental conditions. Dzubay et al.[29] have recently reported the preparation of single-element standards by encapsulating organometallic compounds in thin polymer films. These materials show considerable promise for durable thin-film standards. However, as with any analytical measurement, the analyst is charged with the responsibility of finding standards which are of the highest possible reliability.

The above paragraphs tend to give a somewhat negative impression of the use of XRF for aerosol particulate analysis. However, the majority of the problems discussed are easily overcome by using modern sampling methods and computer programs for performing the necessary corrections. The widespread utilization of XRF for aerosol analysis is evidence to the novice in the technique that, with careful use, XRF is a viable method.

V. APPLICATIONS OF XRF

The above sections have illustrated the sampling methods and the methods that are conventionally used for the average elemental determinations of airborne particles using XRF. Problems associated with the technique were also explored. This section intends to provide the reader with an overview of the applications of XRF to aerosol analysis. To attempt an extensive review is not possible here. Key reports will be presented as an overview. Sources of current and more detailed information include journals such as *Environmental Science and Technology, Atmospheric Environment, X-ray Spectrometry,* and series such as *Advances in X-ray Analysis.* The first two journals are sources of studies of atmospheric particulates, whereas the last two publications tend to focus on new methods and mathematical procedures.

As mentioned earlier, selecting the type of instrument to be used is one of the first decisions to be made in utilizing XRF. This is frequently a controversial topic. Gilfrich et al.[30] have performed a comparison of several instrument types.

Although the author does not entirely agree with the conclusions presented in this report, there are valuable data given for comparison. The conclusion of the report is focussed on the limited resolution of energy-dispersive detectors and the fact that the detector must process all photons from the sample. For these reasons, wavelength-dispersive methods were recommended for analysis of atmospheric particulate samples; however, each intended application must be considered as an individual case. Although reports such as these provide valuable information, they must be considered critically.

There are two reports of interlaboratory comparisons of X-ray techniques, as well as other methods of particulate analysis, which should be read by those preparing to perform aerosol analysis.[31,32] Two sample types were tested: uniform solution deposits on filters and simulated and real aerosol samples prespared from powdered rock. Although the study was heavily biased by the number of results utilizing energy-dispersive XRF compared with other methods (including wavelength dispersive XRF), the information gained is useful. One can conclude from this report that XRF can be used for analysis of elements of atomic number 20 and above on filter substrates with about 10% accuracy and 1% precision. The study does, however, illustrate the difficulties in preparing standards for particle depositions on filters. Something as simple as mechanical losses of a particulate material from the filter can result in serious errors.

One of the most extensive undertakings has been that of the California Air Resources Board in conjunction with the University of California at Davis.[17] This installation uses proton-induced X-ray emission. However, more conventional modern instrumentation can compete. The important factor is the high degree of organization of the laboratory which produces about 10,000 analyses per year with an average of 20 elements determined per sample. The samples are taken from a network of sampling stations, and correlations of the results are made with meterological data. One advantage of the particle source technique is that methods such as α-scattering may be used for the analysis of light elements (H to F).

With conventional instrumentation, XRF is an extremely convenient method for following changes in atmospheric pollutants as a function of weather change. Chessin and McLaren[27] performed a study of the Ca, K, Cl, and S content of atmospheric particulates as a cold front passed through the sampling area. Samples were collected on glass-fiber filters although the authors stated this was a poor choice. However, at that time, several agencies were using glass filters. The sample flow rate was 2 m^3/min, and the analyses were performed using wavelength-dispersive XRF. Samples were collected at 15- or 30-min intervals prior to the front passing, and 1-hr intervals after its passage. The results showed that Ca, K, and S increased significantly as the front approached the sampling area, and the concentration of all four elements in the atmosphere decreased significantly after passage of the front. Although there was no interpretation of the meaning of these data, the results show that XRF is very useful and the importance of correlating the data with weather conditions.

Hammerle and Pierson[33] used an automatic energy-dispersive X-ray spectrometer that was specially constructed to investigate the origin of atmospheric aerosols. Nine elements (Ca, Ti, V, Mn, Fe, Ni, Zn, Br, and Pb) were used in the study. Some of the conclusions drawn from the investigation are

1. Gasoline engine exhaust is the main source of Pb and Br.
2. Soil is the major source of Ti, Mn, Ca, and Fe.
3. All elements show orders of magnitude fluctuations in time periods less than 2 hr.
4. There is a diurnal pattern with peak concentrations in the morning and evening.
5. Ni, Br, Zn, and Pb are mainly found in smaller particles, whereas Ca, Ti, Mn, and Fe are chiefly found in larger particles; V is found in particles of intermediate size.

These conclusions have several important implications. First, Pb and Br are found in the respirable particles and can be reduced in the atmosphere only by removing these elements as constituents of engine fuels. There is a tendency for the man-made pollutants to be present in the small respirable particles. This most likely occurs as a result of these particles being formed by gas-to-particle reactions. Studies of this type, using short sampling periods, may be used to identify local discrete sources of atmospheric particulates which

are indicated by sharp, intense concentration changes as a function of time.

These examples illustrate the real application of XRF as a sensitive, convenient tool for monitoring the average elemental composition of airborne dust particles. The information obtained utilizing XRF analyses is determined to a great degree by the sampling procedure. Both particle size fractionation and time resolution are important factors to consider if maximum information is to be obtained. The analyst must carefully evaluate his goals before selecting a sampling method. The XRF measurements may be performed using wavelength-dispersive (sequential or simultaneous multichannel), energy-dispersive, or proton-induced X-ray emission spectroscopy. The selection of instrumentation depends upon funds available, the number of samples to be measured per day, and the number of elements to be determined in each sample. Of extreme importance is a careful evaluation of the standards. This is a difficult task and should receive considerable attention.

XRF will become the method of choice for routine elemental analysis of atmospheric particles. It must be supplemented by other methods such as electron microprobe analysis if detailed particle formation and transport studies are to be conducted. There is little evidence of dramatic improvements in instrumentation, detection limits, etc. at this time. However, these are adequate for aerosol measurement. Compared with alternate methods, XRF has many advantages such as the simplicity of sample preparation, simultaneous multi-element, nondestructive analysis, and the capability of total automation. Routine analysis may be performed by technicians of minimal training.

In the near future, the major improvements in products of XRF spectrometer manufacturers will be computer software to reduce the data. This is an area of significant competition, which benefits the potential users. Vendors have made software available for matrix corrections, interelement corrections, semiquantitative analysis without standards, and other useful programs. XRF combined with X-ray powder diffraction of aerosol samples shows significant promise to aid in the identification of aerosol particulates. The use of long wavelength X-ray spectra (> 100 Å) may be developed to augment methods such as ESCA and Auger spectroscopy in the determination of oxidation states of the elements in particulate matter. These developments, if they are realized, will only supplement the major usefulness of X-ray fluorescence spectrometry as a routine analytical tool.

REFERENCES

1. Bertin, E. P., *Principles and Practice of X-ray Spectrometric Analysis,* 2nd ed., Plenum Press, New York, 1975.
2. Birks, L. S., *X-ray Spectrochemical Analysis,* 2nd ed., Interscience, New York, 1969.
3. Jenkins, R., *An Introduction to X-ray Spectrometry,* Heyden & Son, London, 1974.
4. Woldseth, R., *X-ray Energy Spectrometry,* Kevex Corporation, Burlingame, Cal., 1973.
5. Criss, J. W., Particle size and composition effects in X-ray fluorescence analysis of pollution samples, *Anal. Chem.,* 48, 179, 1976.
6. Dzubay, T. G., Hines, L. E., and Stevens, R. K., Particle bounce errors in cascade impactors, *Atmos. Environ.,* 10, 229, 1976.
7. Novakov, T., Harke, A. B., and Siekhaus, W., Sulfates in Pollution Particles, U.S. Atomic Energy Commission Contract, W-7405-ENG-48, Lawrence Berkley Laboratory, Cal., 1973.
8. Dzubay, T. G., Stevens, R. K., and Peterson, C. M., Application of the dichotomous sampler to the characterization of ambient aerosols, in *X-ray Fluorescence Methods for Analysis of Environmental Samples,* Dzubay, T. G., Ed., Ann Arbor Science, Ann Arbor, Mich., 1976, 95.
9. Parker, R. D., Buzzard, G. H., and Dzubay, T. G., A dichotomous sampler using Nuclepore filters in series, *Atmos. Environ.,* in press.
10. Beitz, L. and Rönicke, G., Spurennachweis schädlicher Elemente in Luft und Regenwasser mit der Röntgenfluoreszanalyse, *Proc. Third Int. Clean Air Congress,* VDI-Verlag, Düsseldorf, Germany, 1973.
11. Nelson, J. W., Jensen, B., Desaedeleer, G. G., Akselsson, K. R., and Winchester, J. W., Automatic time sequence filter sampling of aerosols for rapid multielement analysis by proton-induced X-ray emission, in *Advances in X-ray Analysis,* Vol. 19, Kendall/Hunt, Dubuque, Ia., 1976, 403.
12. Spurny, K. R., Lodge, J. K., Frank, E. R., and Sheesley, D. C., Aerosol filtration by means of Nuclepore filters: structural and filtration properties, *Environ. Sci. Technol.,* 3, 453, 1969.
13. Spurny, K. R., Lodge, J. P., Jr., Frank, E. R., and Sheesley, D. C., Aerosol filtration by means of Nuclepore filters: aerosol sampling and measurement, *Environ. Sci. Technol.,* 3, 464, 1969.
14. Spurny, K. R., Havlova, J., Lodge, J. P., Ackerman, E. R., Sheesley, D. C., and Wilder, B., Aerosol filtration by means of Nuclepore filters: filter pore clogging, *Environ. Sci. Technol.,* 8, 758, 1974.
15. Dzubay, T. G. and Stevens, R. K., Ambient air analysis with dichotomous sampler and X-ray fluorescence spectrometer, *Environ. Sci. Technol.,* 9, 663, 1975.
16. Lundgren, D. A., An aerosol sampler for determination of particle concentration as a function of size and time, *J. Air Pollut. Control Assoc.,* 17, 225, 1967.
17. Flocchini, R. G., Shadoan, D. J., Cahill, T. A., Eldred, R. A., Feeney, P. J., and Wolfe, G., Energy, aerosols and ion-excited X-ray emission, in *Advances in X-ray Analysis,* Vol. 18, Plenum Press, New York, 1975, 579.
18. Ranz, W. E. and Wong, J. B., Impaction of dust and smoke particles, *Ind. Eng. Chem.,* 44, 1371, 1952.
19. Marple, V. A. and Liu, B. Y. H., Characteristics of laminar jet impactors, *Environ. Sci. Technol.,* 8, 648, 1974.
20. Kaufmann, H. C. and Akselsson, R., Non-linear least squares analysis of proton-induced X-ray emission data, in *Advances in X-ray Analysis,* Vol. 18, Plenum Press, New York, 1975, 353.
21. Jaklevic, J. M. and Walter, R. L., Comparison of minimum detectable limits among X-ray spectrometers, in *X-ray Fluorescence Methods for Analysis of Environmental Samples,* Ann Arbor Science, Ann Arbor, Mich., 1976.
22. Rasberry, S. D. and Heinrich, K. F. J., Calibration of interelement effects in X-ray fluorescence analysis, *Anal. Chem.,* 46, 81, 1974.
23. Whitby, K. T., Husar, R. B., and Liu, B. Y. H., The aerosol size distribution of Los Angeles smog, *J. Colloid Interface Sci.,* 39, 177, 1972.
24. Willike, K., Whitby, K. T., Clark, W. E., and Marple, V. A., Size distribution of Denver aerosols: a comparison of two sites, *Atmos. Environ.,* 8, 609, 1974.
25. Lundgren, D. A., Mass Distribution of Large Atmospheric Particles, Ph.D. thesis, University of Minnesota, Minneapolis, 1973.

26.   **Hawthorne, A. R., Gardner, R. P., and Dzubay, T. G.,** Monte Carlo simulation of self-absorption effects in elemental XRF analysis of atmospheric particulates collected on filters, in *Advances in X-ray Analysis,* Vol. 19, Kendall/Hunt, Dubuque, Ia., 1976, 323.

27.   **Chessin, H. and McLaren, E. H.,** X-ray spectrometric determination of atmospheric aerosols, in *Advances in X-ray Analysis,* Vol. 16, Plenum Press, New York, 1973, 165.

28.   Columbia Scientific Instruments, Austin, Texas.

29.   **Dzubay, T. G., Lamothe, P. J., and Yasuda, H.,** Polymer films as calibration standards for X-ray fluorescence analysis, in *Advances in X-ray Analysis,* Vol. 20, McMurdie, H. F., Barrett, C. S., Newkirk, J. B., and Ruud, C. O., Eds., Plenum Press, New York, 1977, 411.

30.   **Gilfrich, J. V., Burkholter, P. G., and Birks, L. S.,** X-ray spectrometry for particulate air pollution – a quantitative comparison of techniques, *Anal. Chem.,* 45, 2002, 1973.

31.   **Camp, D. C., Cooper, J. A., and Rhodes, J. R.,** X-ray fluorescence analysis – results of a first round intercomparison study, *X-ray Spectrom.,* 3, 47, 1974.

32.   **Camp, D. C., VanLehn, A. L., Rhodes, J. R., and Pradzynski, A. H.,** Intercomparison of trace element determinations in simulated and real air particulate samples, *X-ray Spectrom.,* 4, 123, 1975.

33.   **Hammerle, R. H. and Pierson, B. M.,** Sources and elemental composition of aerosol in Pasadena, California, by energy dispersive X-ray fluorescence, *Environ. Sci. Technol.,* 9, 1058, 1975.

Chapter 4
EMISSION AND ATOMIC ABSORPTION SPECTROSCOPY

E. Pungor and L. Pólos

TABLE OF CONTENTS

I. METHODS OF ATOMIC SPECTROSCOPY: DEVELOPMENT AND APPLICATION

Emission and absorption of the photon can be described by the following equation:

$$M^* \rightleftharpoons M + h\nu$$

where M* and M equal the atom or molecule in excited and ground state, respectively, and $h\nu$ is the emitted or absorbed photon of frequency ν. The direction of the upper arrow indicates emission, while the direction of the lower arrow indicates absorption is taking place.

In the emission methods of atomic spectroscopy, the substance must be excited after evaporation. The excitation energy is applied in the form of heat, electric current, or light. The atoms remain in an excited state for a very short time, about 10^{-8} sec, only when a spontaneous emission occurs. The energy (frequency) of the emitted photon is characteristic of the quality of the material, while the intensity of the photon current is proportional to the concentration of atoms present in the excited space. The measuring system of the emission method can be demonstrated by the following simple scheme:

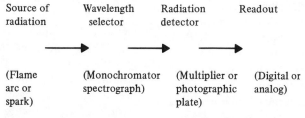

Source of radiation	Wavelength selector	Radiation detector	Readout
(Flame arc or spark)	(Monochromator spectrograph)	(Multiplier or photographic plate)	(Digital or analog)

The sample to be tested must be brought into the light source; in the case of emission spectrography, the powdered sample is evaporated, usually from a graphite electrode. However, in flame spectrometry, the sample solution is introduced into the flame by means of an atomizer.

In contrast to molecular spectroscopy, absorption methods of atomic spectroscopy use light sources which emit no continuous radiation. Usually, hollow cathode lamps are used since these emit very intensive lines of small half-width. As a result, measurements of high sensitivity can be achieved even if very simple monochromators are used. The measuring system of atomic absorption methods is shown in the following scheme:

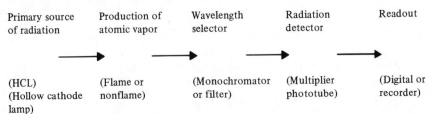

Primary source of radiation	Production of atomic vapor	Wavelength selector	Radiation detector	Readout
(HCL) (Hollow cathode lamp)	(Flame or nonflame)	(Monochromator or filter)	(Multiplier phototube)	(Digital or recorder)

An area of high temperature causes the substance to dissociate and form atoms. In the utilization of flame methods, the flame serves as a thermostat as well. In comparison to the flame gases, the atomic vapors are present in a very low concentration. The medium of high temperature may be strongly reducing, such as the acetylene nitrous oxide flame, which makes possible determination of elements forming thermally stable oxides. A graphite furnace, which involves a graphite tube glowing at 2500 to 3000 K, can be used as a high-temperature area as it allows substances to dissociate and form atoms. Certain elements such as mercury vapors can be measured also at room temperature, in which case there is no need for an atomizing source of high temperature.

In the measuring arrangement of atomic fluorescence spectrometry, the light source and the sensor are not aligned. However, the light source is generally perpendicular to the optical axis of the monochromator. The light source is either a hollow cathode lamp or a high-intensity lamp without electrodes. The excitation is performed with photons, and the spontaneous emission of the excited atom can be measured

$$h\nu + M \rightarrow M^* \rightarrow M + h\nu$$

Certain elements, such as Ag, Cd, Cu, Ni, Tl, and Zn, can be measured with higher sensitivity when applying atomic fluorescence spectrometry, than when using the atomic absorption method. Both methods are based on the absorption of ground state atoms. While the atomic absorption method measures the portion of absorbed radiation, the fluorescence technique determines the intensity of the emitted photons.

If the degree of atomization

$$\beta = \frac{M_m}{M_t}$$

(where M_m is the concentration of free atoms of an element and M_t is the sum of the concentration of atoms, ions, and molecules of the element in concern) does not change to a great extent with temperature, i.e., in the determination of compounds which easily dissociate, the temperature changes do not significantly influence the results of analysis. The half-width of the absorption lines changes linearly with the square root of temperature ($T^{1/2}$), as a consequence of the changes in the Voight profile of the line. From the analytical point of view, the development of atomic absorption methods in recent times has been mainly connected with the introduction of the sample and different atomizer systems. Several attempts have been made to develop "direct" atomic absorption methods using solid samples. Although the graphite furnace techniques are very sensitive and in principle are also suitable for direct determination of solid substances, they may be regarded primarily as solution methods.[1] The amount of the sample used is often only a few tenths of a milligram. Such small quantities represent the average composition of the material only in the case of very homogeneous samples. The matrix effect cannot be neglected either, so that standardization constitutes a rather difficult problem.

However, the sample amount can be increased by a very simple solid-sample technique that was developed by Kántor et al.[2] According to this method, the sample is placed in the boring of a graphite electrode, and the material is evaporated in a direct current arc. The aerosol is transported into the flame of an atomic absorption burner, and the absorption is measured. The principle of this technique is schematically shown in Figure 1. Different types of electrode heads are shown in Figure 2.

Development of measuring techniques in emission spectrometry has taken place in recent years

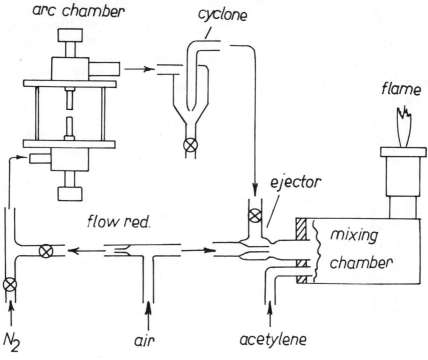

FIGURE 1a. Block diagram of solid-sample apparatus.

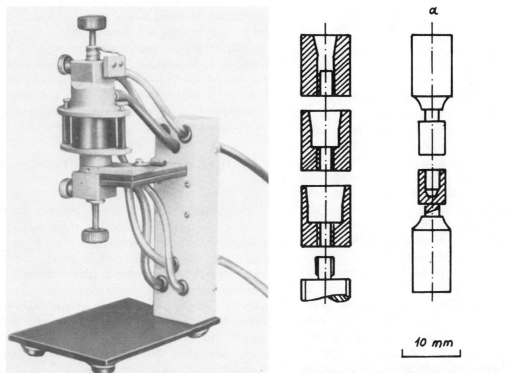

FIGURE 1b. Arc chamber. (From *Hung. Sci. Instrum.*, 38, 57, 1976. With permission.)

FIGURE 2. Changeable electrode holder heads and different electrode shapes used: (a) most commonly applicable for powdered as well as liquid samples; (b) covered electrode for vaporization studies.

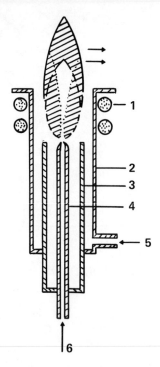

FIGURE 3.   Schematic drawing of plasma tube assembly. (1) Induction coil; (2) coolant tube (outer tube); (3) intermediate tube; (4) carrier gas tube; (5) inlet coolant gas; (6) inlet carrier gas. Total length of assembly is approximately 20 cm. (From Boumans, P. W. J. M. and De Boer, F. J., *Spectrochim. Acta,* 30B, 309, 1975. With permission.)

mainly in the field of the applied light sources. The inductively coupled plasma burner (ICP) is of the greatest importance. The principle of this equipment, according to Boumans and DeBoer,[3] is shown schematically in Figure 3. The burner consists of coaxial quartz tubes placed in a high-frequency coil. In the inner tube, argon flows and carries the sample as an aerosol into the plasma. The shaded area (marked with lines in the figure) illustrates the distribution of the plasma. Around the axis of the plasma, a lower temperature zone (a so-called "tunnel") can be found, into which the aerosol is introduced.

The method is suited for simultaneous multielement trace analyses. The electrical flames have greater stability than other classical light sources such as the direct current (DC) arc and the spark. They also have relatively high temperatures and operate in inert atmospheres. Chemical interference effects are minimal; the use of internal standards is not necessary. In general, the electrical flames require the sample be presented in an aerosol form, which may be seen as a limitation because of solution and dilution requirements. The best detection limits were obtained using an ultrasonic nebulizer. Use of this device increases the complexity of the technique. Although the introduction of powders directly into the plasma has been the subject of study in other fields, the application as a spectrochemical source has received relatively little attention. The accuracy of the ICP method is equal to that of the atomic absorption method, while its sensitivity exceeds even that of absorption flame photometry (in the case of certain elements), as shown in Table 1 by the comparative results of Fassel and Kniseley.[4]

The comparison of the methods of atomic spectroscopy shows that as far as simplicity of arrangement and price of equipment are concerned, atomic absorption methods are more favorable than the emission spectroscopic types. When considering accuracy, the application of absorption and emission flame spectroscopy as well as ICP emission spectrometry, i.e., the techniques working with solutions, seems to be the most advantageous. The main advantage of the emission methods lies in the fact that several elements can be measured simultaneously, although the optimum plasma zone is different for the various elements, as when using ICP.

Since emission spectrometry provides such simultaneous information on several elements, it is valuable in qualitative or exploratory analysis. However, if accurate and quantitative analysis of simple systems is desired then atomic absorption is the preferred technique.

II. QUALITATIVE ANALYSIS OF DUST SAMPLES

The qualitative analysis of airborne dust samples can advantageously be carried out by emission spectrometry. The sample is placed in a graphite carrier electrode with a boring, and a direct current arc excitation is applied. Since the volatility of materials may vary significantly, a selective evaporation must be considered in attempting to reach total evaporation of the sample. The photographic plate of the spectrograph integrates the signals obtained in the sequence of the evaporation periods. If from time to time the plate is forwarded during the measurement, a time study of the evaporation process can be carried out which may facilitate the evaluation as samples of complicated composition are tested.

TABLE 1

Comparison of Experimentally Determined Detection Limits

		Methods of detection		
		Flame (μg/ml)		
Element	ICP (μg/ml)	AAS	AFS	AE
Ag	0.004	0.005	0.0001	0.008
Al	0.002 (0.0004)[a]	0.03	0.005	0.005
As	0.04 (0.002)[a]	0.1	0.1	50
Ba	0.0001	0.05		0.002
Be	0.0005	0.002	0.01	0.1
Bi	0.05	0.05	0.05	2
Ca	0.00007	0.001	0.000001	0.0001
Cd	0.002 (0.00007)[a]	0.001	0.00001	0.8
Co	0.003 (0.0001)[a]	0.005	0.005	0.03
Cr	0.001 (0.00008)[a]	0.003	0.004	0.004
Cu	0.001 (0.00004)[a]	0.002	0.001	0.01
Fe	0.005 (0.0005)[a]	0.005	0.008	0.03
Hg	0.2	0.5	0.02	40
Mg	0.0007	0.0001	0.001	0.005
Mn	0.0007 (0.00001)[a]	0.002	0.002	0.005
Na	0.0002	0.002		0.0001
Ni	0.006 (0.003)[a]	0.005	0.003	0.02
Pb	0.008 (0.001)[a]	0.01	0.01	0.1
Sb	0.2	0.1	0.05	0.6
Se	0.03 (0.001)[a]	0.1	0.04	100
Si	0.01	0.1		5
Sn	0.3	0.02	0.05	0.3
Sr	0.00002	0.01	0.01	0.0002
Th	0.003			200
U	0.03			10
V	0.006 (0.00009)[a]	0.02	0.07	0.01
W	0.002	3		0.5
Zn	0.002 (0.0001)[a]	0.002	0.00002	50

[a] With ultrasonic nebulization. Reprinted with permission from Olson, K. W., Haas, W. J., and Fassel, V. A., *Anal. Chem.*, 49, 632, 1977. Copyright by the American Chemical Society.

Reprinted with permission from Fassel, V. A. and Kniseley, R. N., *Anal. Chem.*, 46, 1110A, 1974. Copyright by the American Chemical Society.

It is advisable to use a spectrograph of medium resolution by which the whole spectrum can be photographed on a single plate. Under the spectrum to be examined, an iron spectrum is also photographed since this makes simple evaluation possible.

Evaluation of the qualitative picture is performed by wavelength determination of the spectrum lines. Identification is usually made by means of spectrum atlases. Accordingly, the aim of an examination is to control the appearance of the sensitive line of a given element in the spectrum. The identification of all the lines present would entail much unnecessary work. In qualitative analysis, the most sensitive and characteristic lines should always be used. With the exception of noble gases — halogenes, oxygen, nitrogen, and hydrogen — each of the 92 natural elements can be excited in arc discharge. With the exception of alkaline and alkaline earth metals, ionization is not very significant at the temperature of arc plasma. Consequently, the spectrum obtained is simple and contains mainly atom lines. The optimum excitation temperature of elements of high excitation energy (such as S, Se, Te, P, C, As, Sb, and Zn) is higher than the temperature of arc plasma. However, since these elements are volatile due to the selective excitation created by arc discharge, they can be examined in a more sensitive way than could be expected on the basis of their excitation potential.

III. QUANTITATIVE ANALYSIS OF POWDERY SAMPLES

A. Preparation of Powdered Samples for Analysis

The preparation of the substance to be tested varies somewhat with the technique applied for sampling. In the most commonly used sampling methods, the air is pumped through:

    a. A membrane filter (Whatman® 42, 44; Millipore®, type HA, AA)

    b. A porous graphite layer

    c. A layer of sintered glass and porcelain

    d. An impinger (dry or wet)

    e. Between high-voltage (12 to 30 kV) electrodes (in the case of electrostatic dust precipitation)

    f. An impactor

Wet digestion techniques — The most frequently used sampling method consists of filtering through a cellulose filter. Metal contamination of the original filter may be regarded as homogeneous and when considering metals important from the point of view of air pollution control, can be neglected. In the so-called direct procedures, the powder sample on the filter paper is examined with the filter. In qualitative analysis, the filter paper containing the sample is rolled without any powder loss and placed in the cavity of the graphite electrode where the sample is examined by the total evaporation method. In quantitative solution analysis, the sample, which may contain organic and inorganic substances, is destroyed with the filter paper. For the destruction, different acid mixtures are used. The method suggested by Kotz et al.,[46] digestion with HNO_3-H_2F_2 in a Teflon® bomb, is very well suited to the fusion of substances containing silicate or those which are otherwise difficult to fuse. The amount of acid is contingent upon the amount of sample. For example, 2 ml of concentrated HNO_3 and 2.5 ml of concentrated H_2F_2 are added to a 200 mg sample in a Teflon bomb. The Teflon bomb is 80 mm high, 24 mm in diameter, and has a 6-mm wall thickness. This fits inside a stainless steel tube, the lid of which is fastened by means of a spring with a constant of 50 klb/mm. The bomb is closed and placed within a heating block, then heated to 190°C (the temperature of the solution within the bomb is about 170°C). In 2 or 3 hr, the organic matter decomposes and silicates dissolve.

If the sample contains a remarkable amount of calcium, CaF_2 may precipitate. The contents of the Teflon bomb are transferred to a polyethylene vessel containing approximately 20 ml of a saturated solution of boric acid, which serves to bind excess H_2F_2 in the form of a complex. The solution is then heated at 80°C for 10 min to dissolve any CaF_2 that possibly may have precipitated. The solution is then transferred to a volumetric flask filled to volume and used for atomic absorption analysis. The interference from boron tetrafluoride is noticeable in the determination of Ca, Mg, Al, and Ti. This interference must be accounted for by treating calibration standard solutions similarly to the samples. According to our experience, the HNO_3 + $HClO_4$ mixture is well suited for this purpose also. However, the use of this mixture may be dangerous if nitric acid evaporates before the complete destruction of the organic substances occurs as the perchloric acid oxidizes the organic material causing an explosion. Accordingly, the nitric acid must be in excess until any organic substance is present. In this way, no explosion will occur. In the case of flame atomic absorption methods, it is advantageous if the metal ions are present as chlorides since certain anions (e.g., sulfates and phosphates) may reduce the concentration of free atoms in the flame and the sensitivity of the determination. After the destruction, the material is evaporated to dryness and dissolved in hydrochloric acid. However, when using the graphite furnace technique, it is advantageous for the determination if the metals are present as sulfates, since chlorides, due to their volatility, yield signals that are difficult to reproduce. If this technique is applied, the destruction can advantageously be performed in a mixture of H_2SO_4 and H_2O_2. This acid mixture oxidizes the organic substances effectively and quickly so that preparation time can be shortened significantly.

A number of materials contain silicates as one of the major constituents. An examination of the reported methods for the atomic absorption analysis of siliceous materials reveals their categorization into two distinct groups: those in which the sample is fused with an alkali fusion salt or mixture and those in which the sample is brought into solution by way of acid attack. A further subdivision of these methods separates those in which silicon is retained in solution and determined by atomic absorption from those where it is removed, either completely by volatilization or precipitated as silica and perhaps determined by gravimetry.

Fusion methods — The main disadvantage of fusion methods is that they result in a prepared solution with high content of dissolved solids. On the other hand, fusion methods ensure that all silicon is present in solution. The most common fusion mediums are lithium metaborate and sodium carbonate.

Acid attack — Most solvent acid mixtures used in this method contain hydrofluoric acid. However, depending upon the method of utilization and the other components of the acid mixtures, if silicon is removed completely by vaporization as silicon tetrafluoride, AlF_3, TiF_4, SbF_3 become volatile. This loss can be avoided by the use of excess sulfuric acid or by keeping the silicon in solution for determination by atomic absorption.

When obtaining powdered samples, it is desirable to limit (as much as possible) the number of operations to which the sample must be subjected to avoid sample losses or contamination. For these reasons, an elaborate technique is followed when a porous graphite tube is used to filter air to obtain samples. Excessive treatment of the sample can be avoided since the graphite tube serves as filter and also as the tube which is ignited in the graphite furnace. In this way, the sample is readied for analysis. Woodriff and Lech[5] developed such a method in which an atomic absorption measurement is conducted after sampling. Similarly, Seeley and Skogerboe[6] used a porous graphite tube to obtain a sample and examined it with emission spectrography. The scheme of this tube, suitable for sampling, can be seen in Figure 4.

Filtering through sintered glass and porcelain filters is an effective method of sampling, but according to Luke et al.[7] and Kometani and co-workers,[8] these filter layers contain contaminants which affect the sample. To overcome these contaminating effects, large samples must be obtained. If the filter layer is heated with the sample (while reducing the organic material to ashes),

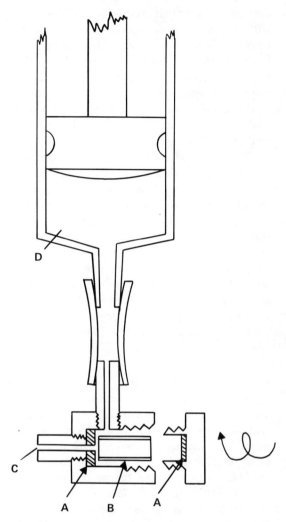

FIGURE 4. Air sampling device. (A) Buthyl rubber gaskets; (B) graphite tube; (C) air inlet; (D) 50 cm³ plastic syringe. (Reprinted with permission from Siemer, D., Lech, J. F., and Woodriff, R., *Spectrochim. Acta*, 28B, 470, 1973. Pergamon Press.)

then insoluble metal silicates are formed on the surface of the glass.[8] Even though paper and plastic membrane filters are less efficient, their application in sampling is more advantageous because minimal losses and contamination of sample occur. The destruction of the material together with the filter layer can be carried out as with the cellulose membrane. However, to determine the termination point of the destruction is somewhat more difficult in the case of cellulose filters than in the case of paper filters.

The destruction of the organic matter in the sample collected in the impinger or that obtained by electrostatic precipitation can be carried out in the same way as that used with cellulose filters. With the help of atomic spectroscopic methods, a number of constituents can be determined simultaneously if the selective observation of their emission or absorption intensity signals can be ensured. However, in the analysis of multicomponent systems, the selective observation is often difficult. In these cases, chemical separation methods are to be applied.

The sample collected with a cascade impactor is size-fractionated, thus the results of analysis also give the distribution of elements according to the particle size of the dust. Lee and co-workers,[47] carried out careful analyses on size-fractionated samples. If a cascade impactor is used for sampling, then a method of high absolute sensitivity (such as the graphite furnace technique) must be used for analysis, otherwise sampling time of considerable duration is necessary. In the Andersen cascade impactor, the sample is collected in petri dishes; thus, the sample preparation is the same as described for cellulose filters.

B. Enrichment by Extraction

Chemical separation and enrichment procedures are usually applied to increase the sensitivity of analytical methods. Although spectrochemical methods are generally known as "high sensitivity" methods (10^{-3} to $10^{-6}\%$), it must be stated that a sensitivity corresponding to the requirement of our age (10^{-6} to $10^{-10}\%$) can only rarely be attained by direct analysis of the samples. Accordingly, the application of chemical methods is important even to the analysis of unicomponent materials of high purity. Extraction methods used for the determination of metal ions are generally based on reactions connected with complex formation. Ionic compounds dissolve best in water. Metal ions must be transformed into electrically neutral ion associates or chelate complexes to render them soluble in apolar solvents having a lower dielectric constant than water. In extraction procedures, the most frequently used ion-association complexes are halides, thiocyanates, and nitrates. However, for the purposes of extraction methods, the chelate complexes are of greater importance than the ion-association complexes, since they enable highly selective separations to be performed. We wish to mention here only a few of the complexing ligands.

Dithiocarbamate derivatives are widely used as ligands, in spite of the fact that they easily decompose in acidic media. Due to this property, the extraction must be carried out quickly and at a suitable pH value. β-Diketone (e.g., acetylacetone) can equally be used as ligands or solvents. They are used most frequently by dissolution in chloroform or carbon tetrachloride. Dithizone is already a classical reagent that can be used for the extraction of a number of metals.

In selection of solvents, one has to consider that the organic solvent should be only slightly water-soluble. The solvent should have favorable burning properties when using flame spectrometric methods. For this reason, alcohols, ethers, and ketones are the most suitable solvents but methyl isobutyl ketone and amyl methyl ketone are also often applied. During spark-spectrometric and arc-spectrometric techniques, it is advantageous if the solvent is not inflammable. In this respect, carbon tetrachloride and chloroform are very suitable.

Extractions are generally preceded by a dissolution or destruction process. Care must be taken that the chemicals used do not interfere with the further course of the examination.

Application of the extraction technique in atomic absorption and flame photometric determinations, besides providing the advantages of enrichment and separation, ensures a further increase in sensitivity. This is due to the fact that with the application of an organic solvent, the efficiency of atomization also increases. Due to the finer vapors, the evaporation is more complete. The change in sensitivity of atomic absorption with the applied solvent is demonstrated on the example of lead in Figure 5. The application of organic solvents that have a lower heat of evaporation than water also ensures an increase in sensitivity in spark spectrometric methods.

C. Solution Methods

In emission and absorption flame photometric determinations, both the irregularity and efficiency of atomization play a very important role. The amount of the substance introduced into the flame depends on the method of atomization. With direct atomizers, the whole substance is consumed by the flame, while with indirect atomizers, only about 10% of the material is consumed. The bulk of it remains in the form of droplet aggregates in the spray chamber and is lost from the point of view of the determination. The sensitivity depends on the amount of sample

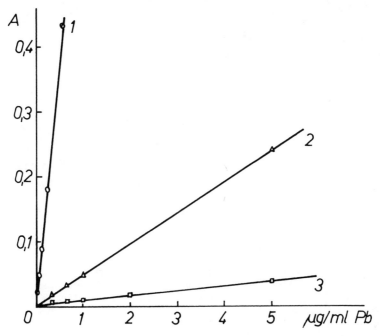

FIGURE 5. Calibration curves of Pb solutions. (1) Pb in normal amyl- acetyl-ketone; (2) Pb in methyl isobutyl ketone; (3) Pb in aqueous solution. (From Prugger, H. and Kling, O., *Z. Werkstofftech.*, 1, 142, 1970. With permission.)

introduced into the flame. However, the size of the drops is also a factor in the sensitivity of the determination. With direct atomizers, one cannot attain a tenfold increase in sensitivity (in comparison to indirect atomizers), in spite of the fact that the entire solution is introduced into the flame. The gases of the flame rise at a velocity of 10 m/sec. If the diameter of the drop is 1 μm, then after covering a distance of about 0.1 to 0.2 mm, the solvent evaporates from the drop. If drops larger than 30 μm get into the flame, they would pass through the flame without being volatilized in the imaged section. Thus, the average size of the drop is a deciding factor in sensitivity. In the course of atomization, energy must be applied to establish drop surfaces. The required amount of energy depends on the surface tension of the solvent. Organic solvents are then applied; the surface tension decreases while the efficiency of the atomization increases.[9] Along with this, the sensitivity of the flame photometric determination also increases.

1. The Properties of Flames

In flame spectrometry, mostly premixed flames are used, i.e., the gases are mixed in a chamber before they are brought into the flame. The combustible component is usually a hydrocarbon (acetylene, propane-butane) or hydrogen. The oxidizing component may be air or nitrous oxide (N_2O), etc. The flames ensure heat for evaporation and the dissociation processes. In flame spectrometry, the acetylene-air flame is widely applied since it can be used for the determination of about 35 elements with the help of atomic absorption methods.

Depending on the composition of the gas mixture, the temperature of the flame may vary between 2300 to 2500 K. The composition of the flame influences to a great extent the capability, sensitivity, and accuracy of flame spectrometric methods (see Figure 6). The determination of elements forming thermally stable oxides (such as Be, B, Al, Si, Ti, V, Zr, Ge, La, and W) can only be carried out if an appropriate reducing medium and a high temperature are produced in the flame. For this purpose the N_2O-C_2H_2 flame is well suited. The products of the reactions taking place in the flame depend on the initial composition. The burning area of the flame most suitable for analytical purposes is found in the red zone, having a height of about 15 to 20 mm. The reducing radicals C, C_2, CN, CH, and C_2H develop in the burning zone, forming a reducing sur-

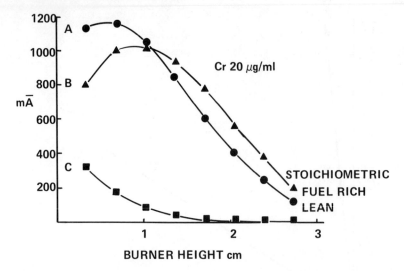

FIGURE 6. Flame profile of Cr in different flame conditions. Absorption intensities in terms of milliabsorbance are shown relative to burner height (cm). (From Hwang, J. Y. and Feldman, J., *Appl. Spectrosc.*, 24, 371, 1970. With permission.)

rounding at that point in the flame. By increasing the proportion of acetylene, the reducing effect increases. After a certain limit, however, one has to deal with soot formation. This sooty flame is not suited to analytical purposes since its reducing effect decreases. After recognizing the favorable properties of the nitrous oxide-acetylene flame, the properties of the similar nitrous oxide propane-butane flame were investigated.[48] It was found that the critical proportion between C:O at which the soot formation begins is significantly lower in the reducing flame when propane-butane is used.

The temperature of the N_2O-H_2 flame nearly reaches that of the nitrous oxide-acetylene flame, and it has a significantly lower reducing effect. Consequently, the above-mentioned elements can be determined at a much lower sensitivity when using nitrous oxide-hydrogen flames than when using nitrous oxide-acetylene flames. However, in the emission flame spectrometric determination of elements that are easy to reduce, the application of the N_2O-H_2 flame was found to be more favorable than that of the N_2O-C_2H_2 flame. Background radiation of the former is significantly lower since there are no hydrocarbon radicals present; thus, their spectra do not interfere. These reducing radicals decrease the partial pressure of oxygen in the flame because carbon monoxide has a great dissociation energy, and its formation consumes oxygen. Furthermore, the carbon-

containing radicals react directly with metal oxides to produce metals. The following general equations may be written:

$$C + MO \rightarrow CO + M$$

$$CN + MO \rightarrow CO + 1/2\, N_2 + M$$

$$CH + MO \rightarrow CO + 1/2\, H_2 + M$$

Due to the high flame temperature, one has to count on the ionization of easily ionizing elements. This can be reduced, however, by atomizing highly concentrated solutions of easily ionizing alkaline metals.

In the course of its expansion, nitrous oxide cools. Consequently, the signal and the flow rate of the gas fluctuate causing the detection limit to shift in the direction of higher concentrations. For safe operation, a sufficiently high rate of gas must be ensured. If the flame is sufficiently reducing, the flow rate of acetylene is relatively high, and no flashback can occur. However, if the flame is less reducing than at a low total flow rate, it will flashback; and at a high total flow rate, it may fly away.

2. Interfering Effects

In flame spectrometry, all changes in magnitude of the analytical signal (usually a decrease) are called interfering effects. Physical interfering effects are regarded as occurring in the course of atomization. Optical and spectral disturbances are effects originating from light scattering and the disturbing effects of ionization. Chemical interfering effects are those in which the element to be determined forms a compound, thus causing the concentration of free atoms in the flame to decrease. Compounds formed in such a manner include oxides, hydroxides, mixed oxides, phosphates, aluminates, titanates, and silicates. Upon increasing the temperature, processes described by the following general equation can take place:

$$\text{M-O-X} \overset{\text{heat}}{\rightleftharpoons} \text{MO} + \text{X} \overset{\text{heat}}{\rightleftharpoons} \text{M} + \text{O} + \text{X} \overset{\text{C}}{\rightleftharpoons} \text{M} + \text{CO} + \text{X}$$

As seen in the equation, increase of both temperature and concentration of the reducing components in the flame favor an increase in the concentration of free atoms. Chemical disturbing effects can generally be eliminated or controlled in two ways, i.e., either by the application of a reducing flame of higher temperature or by using so-called releasing agents. If R is a metal that forms a compound of similar or greater thermal stability with the oxy-acid, then according to the law of mass action:

$$\text{M-O-X} + \text{R} \rightleftharpoons \text{R-O-X} + \text{M}$$

The reaction is shifted in the direction of the upper arrow if R is in excess and M is liberated. The releasing agent will react either with the element to be examined or with the interfering ion. For example, if in acetylene-air flame, the determination of calcium is disturbed by the presence of phosphate ions, this interference can be eliminated by adding lanthanum salt to the solution which will bind the phosphate ions and render calcium free. The other possibility is to use C_2H_2-N_2O flame for the determination.

Among the solution methods, the graphite furnace technique is the most suitable if the absolute amount of the component to be determined is very small. In general, atomization with graphite furnace is carried out in three steps (see Figure 7):

1. Removal of the solvent, "drying"
2. Removal of the organic material, "ashing," as well as removal of the bound water and volatile inorganic salts

3. Evaporation and dissociation, "atomization"

This process, occurring in three different temperature ranges, can be carried out by the gradual heating of the graphite tube furnace. In the course of the heating program, the temperature as well as the time period of the separate steps can be altered. The program is determined experimentally for every kind of material and matrix.

The regulation of temperature stages can be performed by adjusting the cell voltage (current) or the product of current and voltage, that is, the power. In principle, the difference between graphite cell and filament atomizers lies in the fact that with the latter no further energy supply is available after atomization. Evaporation and atomization take place almost simultaneously. Therefore, the concentration decreases more rapidly than with the graphite furnace technique. Consequently, a greater matrix effect should be expected in the case of filament technique. In the application of flameless atomizer cells, the background compensation is of greater importance owing to molecular absorption and diffuse radiation.

Another advantage of the graphite furnace technique over the flame technique is that the gases of the flame do not dilute the atom vapor, and the atoms spend a longer time in the analysis instrument. Furthermore, the oxydizing effect of air is smaller, and the analysis can be performed in a vacuum that makes possible the determination of elements whose most sensitive lines fall into the far ultraviolet (UV) region (e.g., the noble metals). A disadvantage of the method is that elements forming stable carbides can be determined with a significantly lower sensitivity. The technique requires relatively expensive equipment.

Another solution method of emission spectrography is the rotating disc technique in which the lower superficies of a rotating graphite disc (10 to 15 mm in diameter) are immersed in the solution to be examined. To produce a light source, a high voltage spark discharge must be established between the upper side of the disc and the graphite electrode. The solution to be investigated is introduced into the imaged spark gap in the form of a continuously renewed thin film.

High sensitivity can be attained with the ICP burner mentioned earlier. An advantage of both emission spectrographic methods is that the analy-

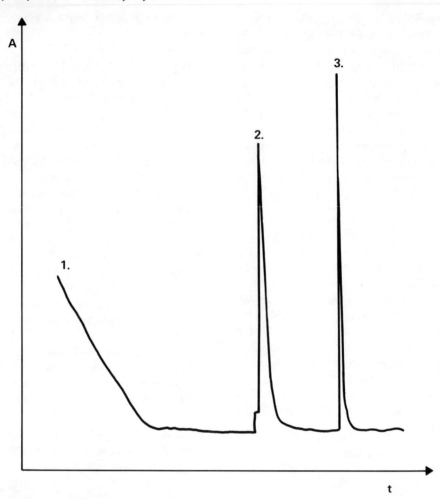

FIGURE 7.   Atomization with graphite furnace is carried out in three steps: (1) drying, (2) ashing, (3) atomization.

sis of several elements can be carried out simultaneously. Their disadvantage, in respect to the atomic absorption method, lies in lower accuracy.

D. Direct Analysis of Solid Samples

From among the solid sample methods, the emission spectrographic method, the arc-flame cell atomic absorption method, and the graphite furnace technique will be mentioned here. In the direct quantitative analysis of solid samples, standardization (i.e., preparation of reference materials) is problematic. In the case of airborne particles, an informatory analysis must be carried out first to determine the type of matrix. The matrix effect, which plays an important part in every analysis, must be taken into consideration by modeling it on the basis of the results obtained in the preliminary analysis, e.g., by mixing oxides

in appropriate ratio. Another approach is to prepare the calibration curves with solutions and to carry out the analysis with solid samples, although this is not quite correct in principle because the sign may significantly differ in these two cases. This technique should always be controlled using a solid sample of known composition. Even under ideal conditions, erroneous results are often obtained. In the emission spectrographic measurement and in the arc-flame cell atomic absorption method of Kántor et al.,[2] the solid sample must be placed in a crucible-shaped graphite electrode. The sample amount changes between 2 and 10 mg. The shape of the electrode is shown in Figure 2. The boring of the electrode is 4 to 5 mm in diameter and 6 to 8 mm in depth. DC arc excitation is applied. In emission spectrographic methods, the arc gap will be the light source; this will be imaged at the entrance slit of the spectrograph. In the arc-flame cell method, the arc serves solely to evaporate the substance, and the aerosol produced is carried into the flame of the atomic absorption equipment[2] (see Figure 1).

With the graphite furnace method, the amount of the sample investigated is smaller. It amounts only to about 0.1 to 1.0 mg. The shape and size of graphite tubes in different equipment may vary widely as seen from the few examples shown in Table 2. The absolute sensitivity varies inversely with the cross-sectional area of atomization chamber. The absolute sensitivity is independent of the length of the tube, provided the tube is long enough so that all of the substance evaporates before the first portion leaves the light path. The relative sensitivity is directly proportional to length of the tube; if the length of the tube is doubled, it will hold twice as much sample.[45]

Papers published in recent years dealing with

techniques of determination of air pollutants using atomic absorption methods are summarized in Table 3. Frank and Dreher[10] prepared a detailed compilation of contributions in the field of atomic absorption and emission flame spectrometry.

TABLE 2

Comparison of Graphite Tube Sizes

Type	Interior atomizer dimensions (mm)		Atomizer volume (cm³)
	Length	Diameter	
L'vov	40	2.5	0.2
Massmann	55	8	2.76
Woodriff	300	7	11.5
Varian®	9	3	0.06
Perkin-Elmer® HGA 74	28	6	0.79

TABLE 3

Atomic Absorption Methods Used in Determination of Air Pollutants

Elements	Treatment	Atomization	Techniques	Ref.
As	Reduction with $NaBH_4$ to AsH_3	Heated SiO_2 tube	NF AAS	11
AgJ	After extraction with acetone	Graphite furnace	NF AAS	12
Be	After filtering through porous graphite	Graphite furnace	NF AAS	13, 14
Cd	Cellulose filters or glass-fiber filters	Flame, graphite furnace	Flame AAS NF AAS	15, 16
Cd	Direct determination	R.F. carbon bed atomizer	NF AAS	17
Cd + Hg	The technique involved passing a slow flowing air stream over carbon rods heated to 1350°C by induction with a radio-frequency field	Carbon furnace	AFS	18
Cd + Hg	Direct determination of Hg + Cd			19
Hg	$Hg^{2+} + Sn^{2+} \rightarrow Hg + Sn^{4+}$	Cold vapor technqiue	NF AAS	20
Hg	After collection on carbon loaded paper	Cold vapor technique	NF AAS	21
Hg	Adsorbed on active carbon	Ta boat	Flame AAS	22

TABLE 3 (continued)

Atomic Absorption Methods Used in Determination of Air Pollutants

Elements	Treatment	Atomization	Techniques	Ref.
Hg	Porous graphite filter tube	Graphite furnace	NF AAS	23
Hg	Amalgamated with Ag wool	Cold vapor technique	NF AAS	24
Pb	Aerosols were collected by passing air through a cellulose filter	Graphite furnace	NF AAS	25
Pb	Glass wool filter was used to separate the atmospheric dust	Flame	AAS	26
Pb	Particulate matter was collected in a Millipore® filter (pore size 0.22 μm)	Graphite furnace	NF AAS	27
Pb	Membrane filter for collecting particulate lead and iodine monochloride solution for collecting organic lead species	Flame	AAS	28
Pb	Cellulose filter without further treatment	Flame, graphite furnace	AAS	29
Pb	Porous graphite filter medium	Graphite furnace	NF AAS	5, 30
Pb	Samples were collected on Millipore® organic membranes	Flame	AAS	31
Pb	Air contaminants were collected on MF Millipore filters	Graphite furnace	NF AAS	32
$Pb(CH_3)_3$ and $Pb(C_2H_5)_4$	Air was conducted through a solution of iodine monochloride in HCl	Graphite furnace	NF AAS	33
Pb	Samples were collected on membranes	Flame	AAS	34
Pb	Sampling was accomplished by using porous polymer filters	Graphite furance	NF AAS	35
Se, Be, Cd, Pb, Ag, Hg	Filtration on porous graphite	Carbon rod	NF AAS	36
As, Ba, Cd, Ca, Cr, Cu, Fe, Mn, Ni, Pb, Tl, V, Zn	Atmospheric particulate matter was collected for metal analysis on glass filters and membrane filters	Flame	AAS	37
Cr, Ni, Mn, Cu	Metallic elements in atmospheric samples were collected on glass filters	Flame	AAS	38

TABLE 3 (continued)

Atomic Absorption Methods Used in Determination of Air Pollutants

Elements	Treatment	Atomization	Techniques	Ref.
Al, Si, Ca, Fe, K, Na, Mg, Pb, Cu, Ti, Zn, Sn, Ni, V, Mn, Cr, Rb, Li, Bi, Co, Cs, Be	Atmospheric particulate samples were collected on polystyrene filters	Flame	AAS	39
Al, Cd, Ca, Cr, Cu, Fe, Pb, Mg, Ni, Si, Sn, Zn	The airborne dust was collected on 25 mm dia filter paper/Schleicher Schule	AC spark	ES	40
Al, Be, Cr, Co, Hg, Mg, Mn, Mo, Ni, Pb, Ti, V, W, Zn	Spectroscopic graphite electrodes were used as filters	Flame 28A DC arc	AAS ES	41
Be, Bi, Cd, Cr, Cu, Mn, Ni, Pb, Sn, Ti, Vi, Zn	Air was drawn through a membrane filter (Gelman DM-800)	15A DC arc	ES	42
Zn, Cd	Samples were deposited on small tantallum substrates placed into the furnace	Radio frequency furnace as spectrometric source	ES AAS	43
V	Glass filters	Flame, graphite furnace	AAS	44

Note: AAS, atomic absorption spectroscopy; NF AAS, nonflame atomic absorption spectroscopy; AFS, atomic fluorescence spectroscopy; ES, emission spectroscopy.

REFERENCES

1. **Massmann, H.,** Entwicklungsstand der Atomabsorptionsspektrometrie, *Angew. Chem.,* 86, 542, 1974.
2. **Kántor, T., Fodor, P., Youssef, Y. S., and Pungor, E.,** Arc nebulization of samples for flame emission and atomic absorption spectrophotometry, *Hung. Sci. Instrum.,* 36, 19, 1976.
3. **Boumans, P. W. J. M. and De Boer, F. J.,** Studies of an inductively-coupled high-frequency argon plasma for optical emission spectrometry. II. Compromise conditions for simultaneous multielement analysis, *Spectrochim. Acta,* 30B, 309, 1975.
4. **Fassel, V. A. and Kniseley, R. N.,** Inductively coupled plasma — Optical emission spectroscopy, *Anal. Chem.,* 46, 1110A and 1155A, 1974.
5. **Woodriff, R. and Lech, J. F.,** Determination of trace lead in the atmosphere by furnace atomic absorption, *Anal. Chem.,* 44, 1323, 1972.
6. **Seeley, J. L. and Skogerboe, R. K.,** Combined sampling-analysis method for the determination of trace elements in atmospheric particulates, *Anal. Chem.,* 46, 415, 1974.
7. **Luke, C. L., Kometani, T. Y., Kessler, J. E., Lormis, T. C., Bove, J. L., and Nathanson, B.,** X-ray spectrometric analysis of air pollution dust, *Environ. Sci. Technol.,* 6, 1105, 1972.

8. **Kometani, T. Y., Bove, J. L., Nathanson, B., Siebenberg, S., and Magyar, M.,** Dry ashing of airborne particulate matter on paper and glass fiber filters for trace metal analysis by atomic absorption spectrometry, *Environ. Sci. Technology,* 6, 617, 1972.

9. **Szivós, K., Pólos, L., and Pungor, E.,** The effect of nebulizer parameters on the enhancement of flame spectrometric sensitivity by organic solvents, *Spectrochim. Acta,* 31B, 289, 1976.

10. **Frank, C. W. and Dreher, G. B.,** Air Pollution, in *Flame Emission and Atomic Absorption Spectrometry,* Dean, J. A. and Rains, T. C., Eds., Marcel Dekker, New York, 1975, chap. 24.

11. **Vijan, P. N. and Wood, G. R.,** An automated submicrogram determination of arsenic in atmospheric particulate matter by flameless atomic absorption spectrophotometry, *At. Absorpt. Newsl.,* 13, 33, 1974.

12. **Lacaux, J. P., Pham Van Dinh, and Beguin, J.,** The determination of silver iodide in air by flameless atomic absorption spectrophotometry, *At. Absorpt. Newsl.,* 13, 49, 1974.

13. **Siemer, D., Lech, J. F., and Woodriff, R.,** Direct filtration through porous graphite for A.A. analysis of beryllium particulates in air, *Spectrochim. Acta,* 28B, 469, 1973.

14. **Griggs, K.,** Toxic metal fumes from mantle-type camp lanterns, *Science,* 181, 842, 1973.

15. **IUPAC,** Applied Chemistry Division Toxicology and Industrial Hygiene Section, Determination of airborne particulate cadmium by atomic absorption spectrophotometry, *Pure Appl. Chem.,* 40, 37-1, 1974.

16. **Brodie, K. G. and Matousek, J. P.,** Determination of cadmium in air by non-flame atomic absorption spectrometry, *Anal. Chim. Acta,* 69, 200, 1974.

17. **Robinson, J. W., Wolcott, D. K., Slevin, P. J., and Hindman, G. D.,** The determination of cadmium by atomic absorption in air, water, sea water and urine with a R.F. carbon bed atomizer, *Anal. Chim. Acta,* 66, 13, 1973.

18. **Robinson, J. W. and Araktingi, Y. E.,** Study of the application of atomic fluorescence spectrometry to the direct determination of mercury and cadmium in the atmosphere, *Anal. Chim. Acta,* 63, 29, 1973.

19. **Christian, C. M., and Robinson, J. W.,** The direct determination of cadmium and mercury in the atmosphere, *Anal. Chim. Acta,* 56, 466, 1971.

20. **Hwang, J. H., Ullucci, P. A., and Malenfant, A. L.,** Determination of mercury by a flameless atomic absorption technique, *Can. Spectrosc.,* 16, 2, 1971.

21. **Janssen, J. H., Van Den Enk, J. E., Bult, R., and De Groot, D. C.,** Determination of total mercury in workroom air by atomic absorption or X-ray fluorescence spectrometry after collection on carbon-loaded paper, *Anal. Chim. Acta,* 84, 319, 1976.

22. **Moffitt, A. E. and Kupel, R. E.,** A rapid method employing impregnated charcoal and atomic absorption spectrophotometry for the determination of mercury in atmospheric, biological and aquatic samples, *At. Absorpt. Newsl.,* 9, 113, 1970.

23. **Lech, J. F., Siemer, D. D., and Woodriff, R.,** The determination of mercury in air samples by flameless atomic absorption, *Spectrochim. Acta,* 28B, 435, 1973.

24. **Long, S. J., Scott, D. R., and Thompson, R. J.,** Atomic absorption determination of elemental mercury collected from ambient air on silver wool, *Anal. Chem.,* 45, 2227, 1973.

25. **Janssens, M. and Dams, R.,** Determination of lead in atmospheric particulates by flameless atomic absorption spectrometry with a graphite tube, *Anal. Chim. Acta,* 65, 41, 1973.

26. **Hantzsch, S., Kaffanke, K., and Nietruch, F.,** Lead determination in street dust by means of atomic absorption spectrophotometry, *Staub Reinhalt. Luft,* 38, 34, 1973.

27. **Matousek, J. P. and Brodie, K. G.,** Direct determination of lead airborne particulates by nonflame atomic absorption, *Anal. Chem.,* 45, 1606, 1973.

28. **Purdue, L. J., Enrione, R. E., Thompson, R. J., and Bonfield, B. A.,** Determination of organic and total lead in the atmosphere by atomic absorption spectrometry, *Anal. Chem.,* 45, 527, 1973.

29. **IUPAC,** Applied Chemistry Division Toxicology and Industrial Hygiene Section, Determination of airborne particulate lead by atomic absorption spectrophotometry, *Pure Appl. Chem.,* 40, 35-1, 1974.

30. **Lech, J. F., Siemer, D., and Woodriff, R.,** Determination of lead in atmospheric particulates by furnace atomic absorption, *Environ. Sci. Technol.,* 8, 840, 1974.

31. **Hwang, J. Y.,** Lead analysis in air particulate samples by atomic absorption spectrometry, *Can. Spectrosc.,* 16, 1, 1971.

32. **Omang, S. H.,** The determination of lead in air by flameless atomic absorption spectrophotometry, *Anal. Chim. Acta,* 55, 439, 1971.

33. **Hancock, S. and Slater, A.,** A specific method for the determination of trace concentrations of tetramethyl-and tetraethyllead vapours in air, *Analyst,* 100, 422, 1975.

34. **Pólos, L., Fodor, P., Szivós, K., Kántor, T., and Pungor, E.,** Determination of lead in floating dust by the atomic absorption technique, *Hung. Sci. Instrum.,* 38, 45, 1976.

35. **Begnoche, B. C. and Risby, T. H.,** Determination of metals in atmospheric particulates using low-volume sampling and flameless atomic absorption spectrometry, *Anal. Chem.,* 47, 1041, 1975.

36. **Siemer, D. D. and Woodriff, R.,** Direct AA determination of metallic pollutants in air with a carbon rod atomizer, *Spectrochim. Acta,* 29B, 269, 1974.

37. **Thompson, R. J., Morgan, G. B., and Purdue, L. J.,** Analysis of selected elements in atmospheric particulate matter by atomic absorption, *At. Absorpt. Newsl.,* 9, 53, 1970.

38. **Hwang, J. Y. and Feldman, J.,** Determination of atmospheric trace elements by atomic absorption spectroscopy, *Appl. Spectrosc.,* 24, 371, 1970.

39. **Ranweiler, L. E. and Moyers, J. L.,** Atomic absorption procedure for analysis of metals in atmospheric particulate matter, *Environ. Sci. Technol.,* 8, 152, 1974.

40. **Lander, D. W., Steiner, R. L., Anderson, D. H., and Dehm, R. L.,** Spectrographic determination of elements in airborne dirt, *Appl. Spectrosc.,* 25, 270, 1971.

41. **Seeley, J. L. and Skogerboe, R. K.,** Combined sampling-analysis method for the determination of trace elements in atmospheric particulates, *Anal. Chem.,* 46, 415, 1974.

42. **Sugimae, A.,** Emission spectrographic determination of trace elements in airborne particulate matter, *Anal. Chem.,* 46, 1123, 1974.

43. **Talmi, Y.,** Determination of zinc and cadmium in environmentally based samples by the radiofrequency spectrometric source, *Anal. Chem.,* 46, 1005, 1974.

44. IUPAC, Applied Chemistry Division Toxicology and Industrial Hygiene Section, Determination of airborne particulate vanadium by atomic absorption spectrophotometry, *Pure Appl. Chem.,* 40, 38-1, 1974.

45. **Woodriff, R.,** Atomization chambers for atomic absorption spectrochemical analysis: a review, *Appl. Spectrosc.,* 28, 413, 1974.

46. **Kotz, L., Kaiser, G., Tschöpel, P., and Tölg, G.,** Aufschluss biologischer Matrices für die Bestimmung sehr niedriger Spurenelementgehalte bei begrenzter Einwaage mit Salpetersäure unter Druck in einem Teflongefäss, *Z. Anal. Chem.,* 260, 207, 1972.

47. **Lee, R. E., Jr., Patterson, R. K., and Wagman, J.,** Particle size distribution of metal components in urban air, *Environ. Sci. Technol.,* 2, 288, 1968.

48. **Butler, L. R. P. and Fulton, A.,** Nitrous oxide supported flames for atomic absorption spectroscopy, *Appl. Opt.,* 7, 2131, 1968.

49. **Prugger, H. and Kling, O.,** *Z. Werkstofftech.,* 1, 142, 1970.

Chapter 5

RADIOACTIVITY MEASUREMENTS

K. Buchtela

TABLE OF CONTENTS

I. RADIOACTIVE PARTICLES IN THE ATMOSPHERE

During tests of nuclear weapons in the years after the Second World War, a considerable amount of radioactive contamination of the air was detected throughout the world. Surveillance of radioactive particles in the air provided one of the earliest indications of a release of radioactive fission products in the environment. World-wide assessment of airborne radioactivity attracted public attention to the serious consequences of uncontrolled handling of fission materials.

Methods have been developed to detect radioactive material in the air and to measure the naturally occurring daughters of radon and the artificially produced radionuclides. Radioactive contamination of the biosphere not only takes place as a consequence of the tests of nuclear weapons, but during the peaceful use of radionuclides, radioactive material is also released into the environment.[1-17]

A. Sources of Radioactive Contamination of Airborne Dust

Although there is a high degree of sophistication in the design of nuclear operations, the release of radioactive particulate matter in the air cannot be completely eliminated in all working areas. The nuclear operations from which radioactive airborne dust may result are: mining, milling, and chemical processing of natural radioactive ores, fabrication of nuclear fuel, reactor operation, reprocessing of nuclear fuel, production and processing of radionuclides, and treatment of radioactive wastes. In all of these operations, measures for air monitoring and for control of radioactive contamination of airborne particles must be taken.[18-20]

If, in the future, a larger scale program of nuclear energy operation is developed, all of the above-mentioned processes would multiply in magnitude and numbers, and radioactive contamination of airborne dust will become a serious problem. The present control practices are adequate for the scale of nuclear operations of the past years and perhaps for the present scale of nuclear operation. However, for a better survey and control of the environment, the development of sensitive and rapid methods to monitor and control radioactive particulate contamination is a challenging problem.[21-25] Usually in laboratories the chemical processing of highly radioactive materials is done in closed areas such as "hot cells" and "glove boxes". It is always possible that, due to leakages and failures in the ventilation system, radioactive materials are released into the environment. During reactor operation, a release of fission products into the air may take place as a result of a failure in fuel element cladding. The airborne particles, which consist of radioactive material or adsorb radionuclides, cover a wide size range, from a fraction of a micrometer to a few micrometers in diameter. Therefore, a large assortment of instruments has been developed for measuring radioactive air contaminants, including sampling devices, filter materials, and radioactivity detector systems. The most important types will be described in this chapter.[25]

The surveillance is frequently confined to gross beta measurements, and it is of primary interest to report unexpected changes in activity levels. In order to assess total human radiation exposure, identification and quantitative measurement of specific radionuclides are required. Therefore, qualitative and quantitative radiometric analyses of the dust samples must be carried out. Examples for such procedures are described in this chapter.

B. The Radioactivity in the Atmosphere and Inhalation Exposure

In the atmosphere, natural and artificial radionuclides can be detected. Naturally occurring radioactive materials are usually the daughter products of radon. This radioactive noble gas is exhaled from the soil into the atmosphere. The activity of air in levels near the ground is given in Table 1.[24]

Residues of nuclear explosions were the main source of artificially produced radionuclides in the atmosphere. Presently, however, this radioactivity has decayed to small quantities, and a great portion of airborne radioactive dust has already sedimented to the ground. Figure 1 illustrates the sources of airborne radioactivity and the potential hazards.[25] Airborne radioactivity can be harmful either due to external irradiation or after inhalation and incorporation. External irradiation takes place either directly from the atmosphere or from deposits of dust on the ground. The external radiation dose can be easily measured using conventional dosimeters. The assessment of inhaled and incorporated radioactive particles is more complex.[26,27]

TABLE 1

Average Concentration of Natural Radionuclides and Fallout Nuclides in the Atmosphere Near Ground Level[24]

Nuclide	Concentration (μCi/cm^3)
Radon-222 and short-lived daughter nuclides	$5 \cdot 10^{-10} - 5 \cdot 10^{-11}$
Radon-220 and short-lived daughter nuclides	$4 \cdot 10^{-12}$
Lead-210, bismuth-210, polonium-210	10^{-15}
Hydrogen-3	$5 \cdot 10^{-15}$
Carbon-14	$1 \cdot 10^{-12}$
Uranium	$10^{-16} - 10^{-17}$
Radium	$10^{-16} - 10^{-17}$
Potassium-40	10^{-17}
Fallout nuclides, Spring 1958	$2 \cdot 10^{-12}$
Fallout nuclides, Spring 1960	$1 \cdot 10^{-13}$
Strontium-90, Spring 1958	$1 \cdot 10^{-14}$

From Kiefer, H. and Maushart, R., *Überwachung der Radioaktivität in Abwasser und Ablaft,* 2nd ed., B. G. Teubner, Stuttgart, 1967, 81. With permission.

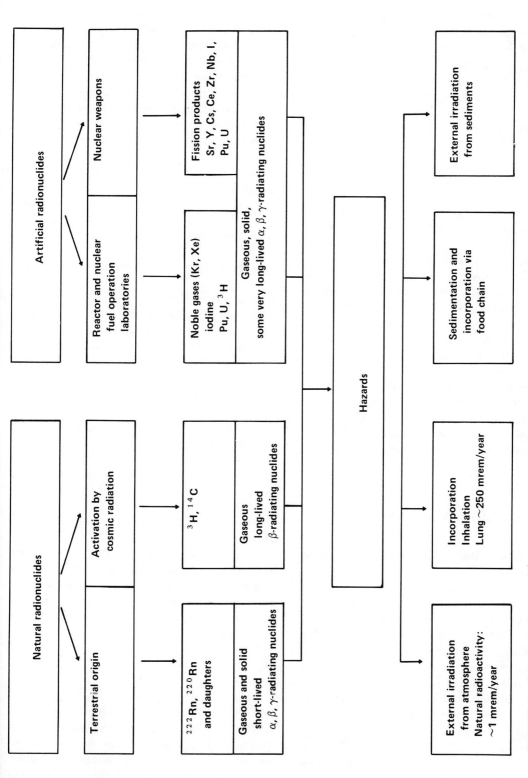

FIGURE 1. Sources of airborne radioactivity and their potential hazards. (From Tschirf, E., *Analytische Chemie und Luftschadstoffe*, Committee for Environmental Protection, Technical University, Vienna, 1973, 147.

Incorporation takes place via either inhalation or intake of food contaminated with radioactive dust. The inhalation dose depends on the concentration of the particulate radioactive matter in the air, the fraction of inhaled activity deposited in the lung, the effective energy of the radiation emitted, and the length of exposure. To calculate the dose rate (R_t) at any time (t) after an individual has been continuously exposed to a concentration (D) in the air, the following formula can be used:[27]

$$R_t = \frac{1.6 \cdot 10^{-8} \cdot D \cdot f_a \cdot V \cdot E \, (1 - e^{-\lambda \cdot t})}{M} \text{ rem/sec}$$

where D is the concentration of radioactivity in the air in disintegrations per second per milliliter; f_a is the fraction of inhaled activity deposited in the lung; V is the inhalation rate in milliliters per second (ml/sec); M is the mass of the lung in grams (g); E is the effective energy of radiation in million electronvolts (MeV); t is the time of exposure in seconds (sec); λ = 0.693 / $T_{1/2}$ where $T_{1/2}$ is the effective halflife in seconds; and $1.6 \cdot 10^{-8}$ is the conversion factor for the dose in rems.

Incorporated radioactive particles that are consumed with food will be deposited in different organs of the human body according to the chemical properties of the radionuclides, e.g., strontium −90 in the bone, plutonium in the liver, etc. Usually the amount of airborne radioactive dust taken in with food is low compared with inhaled material.[28,29]

II. AIR MONITORING PRACTICE

A. Principles of Dust Collection and Measurement of Radioactivity

The most commonly used methods for measuring radioactivity of airborne particles depend on sampling, either using the sedimentation of particles on foils or using suction of air through a suitable selected filter medium. The activity that is collected on the foils or on the filter is measured by a suitable radioactivity detector. Filters are either in the form of a stationary disc (fixed-filter air monitor) or in the form of a moving tape (moving-filter air monitor).[30] After discussion of the instruments, filters, and sampling techniques, a short description is given of the detectors that are usually applied for measuring radioactive dust samples.

B. Instrumentation

For the sake of discussion, the air sampling instruments can be grouped into direct reading and accumulating instruments. Direct reading instruments have been developed only for radioactive gases. Accumulating samplers have been developed that indicate the occurrence of radioactive particles almost instantaneously or after a delay. Instantaneously recording instruments are best suited to provide a time record of dust radioactivity at a particular location or to give warning in the event of an excessive dust radioactivity. Instruments that measure radioactivity after a delay are suited to detect low levels of radioactive dust concentration or to provide samples for radiometric and radiochemical analyses. Accumulating instruments use either fixed-filter discs or moving-filter tapes.[25]

1. Accumulating Instruments with Fixed Filter Discs
a. Measurement after Delay

Accumulating instruments are commonly used for radioactivity measurements of dust. During operation, they draw air at a metered flow rate through a collecting device that separates the dust from the air in such a manner that the sample can be measured at a later time. The particulates from the air are separated by filters or electrostatic precipitators. A variation of an accumulation sampler is direct collection of a volume of air in a container that is retained for analysis. A collector, an air mover, and an air flow meter comprise the basic components of a sampler. The air mover must be matched to the collector both with respect to air flow rate and resistance to flow. (Figure 2).

A few fundamentals apply to all accumulating air samplers. Three factors govern the capacity requirements: the minimum concentration that is to be measured, the period of sample collection, and the radiometric sensitivity.[31] These factors are related by

$$r = \frac{s}{c \cdot t}$$

where r is the sampling rate in air volume per unit of time; s is the analytical sensitivity in units of radioactivity; c is the minimum concentration to be measured in radioactivity units per unit volume of air; and t is the sampling time. Air flow and sampling time must be adjusted to suit the

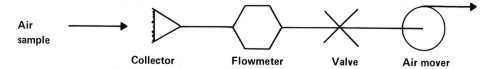

FIGURE 2. Basis components of an air sampler for collecting airborne particles for radioactivity measurements.

TABLE 2

Halflives of Short-lived Daughter Nuclides of Radon-222 and Radon-220

Nuclide	Halflife	Type of decay
Radon-222	3.82 days	α
Polonium-218	3.05 min	α, β
Lead-214	26.8 min	β
Bismuth-214	19.7 min	α, β
Polonium-214	$1.64 \cdot 10^{-4}$ sec	α
Thallium-210	1.3 min	β
Radon-220	55 sec	α
Polonium-216	0.15 sec	α
Lead-212	10.64 hr	β
Bismuth-212	60.6 min	α, β
Polonium-212	$3.0 \cdot 10^{-7}$ sec	α
Thallium-208	3.10 min	β

conditions set by radiometric sensitivity and minimum radioactivity.

With mechanical filters 70 to 90% of the particles, depending on their size, can be collected. Electrostatic collecting filters collect approximately 30%, depending on many factors, such as humidity of the air. They are not frequently used nowadays.

Radioactivity is measured using Geiger-Müller tubes, windowless proportional counting tubes, sodium iodide crystals, or solid state detectors. For special purposes, autoradiographic techniques are used.[27] Recent investigations determined the usefulness of liquid scintillation counting techniques in the measurement of filter samples.[32,33] Using conventional radioactivity detectors, 10^{-12} μCi / cm^3 can be detected if an air sample of 10 to 20 m^3 is filtered. During the detection of artificially produced radionuclides, one must face the problem that the concentration of natural radioactive material in air is so high that the measurement of low-level artificial radioactivity is affected. Fortunately, the daughters of radon and

thoron have relatively short halflives (see Table 2).[34]

After the decay of the short-lived natural radionuclides, long-lived artificially produced radionuclides can be detected. In many routine measurements, long-lived radionuclides are measured after a decay period of several days. Frequently, allowance cannot be made for such a delay during rapid surveillance of working areas. In this case, other discriminations of natural and artificial radioactivity must be made. Gamma-spectrometric measurements must be applied.

b. Measurement without Delay

These methods use filters whereby the activity of the filter is measured during the collection period. Figure 3 illustrates an example of this type of equipment. A relatively large counting tube with thin walls is used, which can be penetrated by β-particles. The counting tube is surrounded by a filter of cylindrical shape and air passes the filter in such a way that the dust is collected at that side of the filter that faces the counting tube. Counting tube and filter are shielded with lead to reduce the natural background radioactivity. After passing the filter, the air flow is measured by the flow rate meter. After air flow begins, an increase of the pulse rate of the radioactivity detector is recorded. According to the short halflife of the naturally occurring radioactivity, a stationary state is reached corresponding to the collected and decaying natural radionuclides (after approximately 30 min). If a long-lived artificial radionuclide is collected at the filter, a marked increase of activity is recorded. The increase depends on the concentration of the radionuclide in the airborne dust. An alarm level can be adjusted with the system. With this equipment, a long-lived β-radioactivity of 10^{-9} μCi/cm^3 can be detected in presence of 10^{-10} μCi/cm^3 of natural radioactivity after 30 min.[25] The disadvantage of this apparatus is that the air flow decreases and the self-absorption of

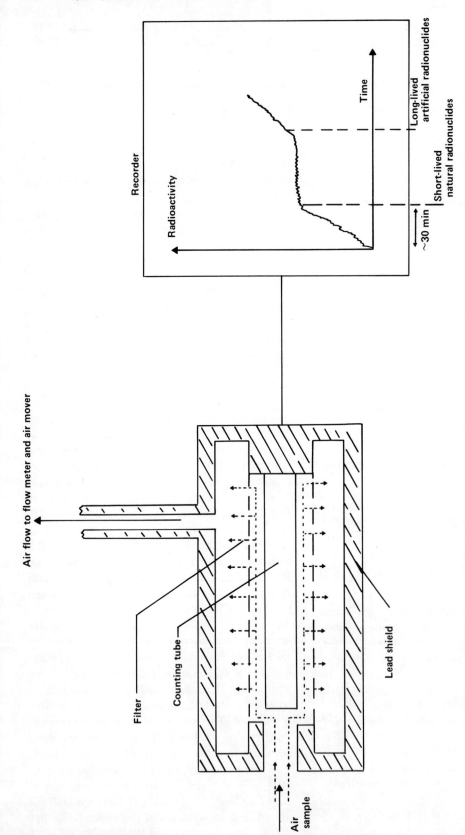

FIGURE 3.   Apparatus for collecting and measuring radioactivity of airborne particles (schematic diagram). Measurement without delay.

the sample for low energy radiation increases after some loading of the filter. Therefore, the filter must be changed periodically, which results in interruptions of the monitoring.

2. Accumulating Instruments with Moving Filter Tape

This type of instrument is equipped with a filter tape that initially passes the dust collection area and the first radioactivity detector. There the radioactivity is measured immediately after dust collection. After a suitable delay, a second detector measures the long-lived radionuclides. Nowadays, such systems are usually equipped with three detectors. The delay of the third detector is about three days. A schematic picture of this type of instrument is given in Figure 4.[25] Some systems prefer a stepwise movement with overlapping of the filter positions. A stepwise movement ensures a better connection between the filter holding system and the filter itself. The activity levels which can be detected by such an instrument are given in Table 3.[36]

3. Dust Collecting Media
a. Filter Media

In nearly all dust sampling instruments which are used in preparing samples for radioactivity measurement, the airborne particles are collected on filter discs and filter tapes. Many types of filter media are available for this purpose; for specialized work, it is frequently possible to select a filter with nearly ideal characteristics. The characteristics of principle concern for monitoring radioactive dust are: efficiency of particle collection, resistance to air flow, chemical composition, and cost. A comprehensive list of filter media that are useful for sampling radioactive dust, their characteristics, and sources of supply is contained in the *Handbook of Air Sampling Instruments.*[37]

For routine air sampling, Whatman® 41 filter paper is a good compromise with respect to performance characteristics. It is a cellulose fiber paper with very low ash content and low resistance to air flow. Its collection efficiency is satisfactory (>90%) for particles greater than a few tenths of a micrometer at normal sampling velocities (0.5 to 1 m/sec air flow). Moreover, it is inexpensive and presents no problem in later radiochemical procedures. When efficient collection of small particles is required, glass-fiber, cellulose-acetate, or cellulose-nitrate filters are often used. Cellulose-ester filters are somewhat fragile and must be supported in the filter holder against the stress of air flow. Resistance to air flow of glass-fiber papers

TABLE 3

Detection Limits for Radioactivity of Airborne Particles Collected with an Instrument Having Moving-filter Tape

Time span of measurement (min)	Detection limits (Ci/m^3)	
	Immediately after dust collection	After a 5-day delay
α-emitting nuclides		
10	$4.3 \cdot 10^{-12}$	$1 \cdot 10^{-12}$
30	$8 \cdot 10^{-13}$	$1.8 \cdot 10^{-13}$
60	$2.9 \cdot 10^{-13}$	$6.4 \cdot 10^{-14}$
180	$4.5 \cdot 10^{-14}$	$1.2 \cdot 10^{-14}$
360	$1.9 \cdot 10^{-14}$	$4.2 \cdot 10^{-15}$
β-emitting nuclides		
10	$2.8 \cdot 10^{-12}$	$3.6 \cdot 10^{-13}$
30	$5.5 \cdot 10^{-13}$	$7.3 \cdot 10^{-14}$
60	$1.9 \cdot 10^{-13}$	$2.3 \cdot 10^{-14}$
180	$3.6 \cdot 10^{-14}$	$5 \cdot 10^{-15}$
360	$1.3 \cdot 10^{-14}$	$1.7 \cdot 10^{-15}$

From Gebauer, H., *Raum-und Abluftüberwachung an kerntechnischen Einrichtungen,* Berthold Friesecke Co. for Measurement Techniques, Karlsruhe, 1973, 8. With permission.

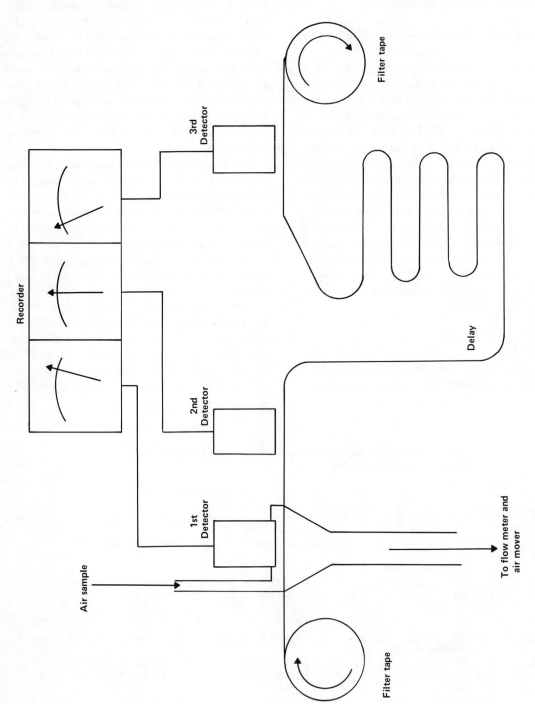

FIGURE 4. Accumulating sampling instrument with moving filter tape. Measurement with and without delay after collecting the airborne particles.

is substantially greater than that of Whatman filters, and the resistance of cellulose-ester filters is the highest of all filter media.[37] Cellulose filters are practically ashless and can be made transparent for microscopic examination of the dust or for use in liquid scintillation measurement of the radio-activity.[32,33] Other available filters are made of combinations of glass, cellulose, or asbestos fibers, providing a wide selection of collection efficiencies, resistance to air flow, corrosion, heat resistance, and cost.

b. Dust Collection Foils

When using dust collection foils, only the precipitating airborne particles can be collected and investigated. Dust collecting foils are not suitable for measuring the total airborne radio-activity since only that fraction of particles can be determined that sediment, i.e., in the unit of time at a given surface area.[24] The only advantage of this procedure is that a great number of such foils can be applied simultaneously without any difficulties.

Dust collection foils consist of paper, textile, or plastic materials covered with an adhesive substance. Their size can range up to 1 m². They were applied during the early testing of nuclear weapons to determine radioactive fallout. Of course, results depend on meteorological conditions, and therefore, their use is rather limited. Table 4 gives some results obtained by dust collecting foils.[24]

Methods have also been described in the literature that use a silicon grease covered pot where air is sucked over the surface of this pot. The efficiency is 70 to 80% at an air flow rate of 80 m³/hr. The advantage of this method may be that radon daughter nuclides are collected with a rather low efficiency. Obviously, only rather heavy particles are collected.[24]

4. Radioactivity Detectors for the Measurement of Airborne Particles
a. Proportional Counting Tubes

Operation and construction of proportional counting tubes are described extensively in the handbooks of nuclear measurement techniques.[38-40] For measurement of radioactivity of filters, proportional counting tubes offer the advantage of a good counting yield for β-particles. Windowless tubes are very useful for detection of α-particles and low energy β-particles. Some problems may arise according to the self-absorption of

TABLE 4

Fission Products Sedimented on Dust Collection Foils

Nuclide	Sedimention per day on an area of 1 km² (mCi)
Strontium-90	0.08
Strontium-89	0.72
Cesium-137	0.31
Zirconium-95	37.5
Cerium-155	9.36
Yttrium-91	10.0

Note: From Kiefer, H. and Maushart, R., *Überwachung der Radio-aktivität in Abwasser und Abluft,* 2nd ed., B. G. Teubner, Stuttgart, 1967, 88. With permission.

the sample. This can be overcome by calibration procedures.[42] Difficulties can be caused by particulate radioactive material that is not very strongly adsorbed on the filter, if these materials produce radioactive contamination of the tube.

b. Geiger-Müller Counting Tubes

Function and operation of Geiger-Müller counting tubes are described in many books dealing with nuclear measurement techniques.[38-40] Geiger-Müller counting tubes are most frequently used in instruments for monitoring radioactivity of airborne dust because this type of detector is powerful and operates with great reliability even during long time intervals.

Disadvantages include the adsorption of the window and the lack of discrimination between particles of different energies. All particles entering the tube and causing an ionizing event result in the production of an ion avalanche and an electronic pulse, independent of the type and energy of the radiation to be detected. The yield for β-particle detection is high, because all β-particles entering the tube are counted. Because of the low specific ionization of γ-radiation, this type of radiation is counted with a yield of less than 1%. Alpha particles can only be detected if they penetrate the window of the tube.

c. Scintillation Counters

When radiation interacts with certain substances (so-called fluors), a small flash of visible or ultraviolet light (a scintillation) is produced. This process constitutes the basis for the operation of all scintillation detectors.[41,42]

The fluorescent substances used are especially suited to detection of specific types of radiation. Alpha fluors consist mostly of zinc sulfide. This material was one of the first fluorescent substances ever used in the detection of radiation. Beta fluors are large crystals of anthracene or similar organic compounds. Plastic fluors are also used; however, the most commonly used β-detectors are liquid scintillation counters, because these devices allow incorporation of the radioactive sample directly in the scintillator, thereby increasing the efficiency of the detector. Theoretically, a 4π-geometry is hereby obtained. Gamma fluors are large, single crystals of sodium iodide containing trace amounts of thallium iodide as an activator. A photocathode with photomultiplier tube is used to detect the small light flashes. By means of a photomultiplier tube, an electrical pulse is produced by each scintillation. If it is assumed that the efficiency of light collection by the photocathode is independent of the location of the light flash in the scintillator, and that an electron is emitted from the photocathode for each photon striking it, then the pulse height is proportional to the scintillation intensity. Electronic pulse height analyzers sort the pulses according to their height. Since pulse height is proportional to the incident radiation energy, the pulse height analyzer is in fact analyzing the radiation energy. By this method, a spectrum of the radiation energy can be obtained. This allows a quantitative and qualitative analysis of a mixture of radioactive substances. Usually these radiation analyzing devices are not used for monitoring the radioactivity of airborne dust but for detailed investigation in a laboratory.

d. Solid-state Detectors

The basic mode of operation of solid-state detectors is similar to the operation of an ion chamber. In an ion chamber, an incident particle passing through the gas produces ion pairs; in a solid-state detector, an incident particle produces electrons and holes in the crystal structure. When a potential difference is applied between opposite surfaces of the crystal, electrons migrate to the positive surface. Solid-state detectors have become increasingly more important as radiation detectors. Among their advantages are their high efficiency for conversion of incident energy into ions, which

gives an excellent energy resolution of the spectra. Unfortunately, most of the solid-state detectors must be kept at a constant temperature, that of liquid nitrogen. Therefore their use is restricted to laboratory investigations.[43-45]

e. Background Reduction by Anticoincidence

During measurement of low levels of radioactivity, a low and stable counter background is an important requirement. Measures must be taken to reduce the background. If shielding of the detector does not result in a sufficiently low background counting rate, the detector has to be surrounded by guard detectors, which operate in anticoincidence with the main counter. The main counter records only events that are not in coincidence with pulses from the guard detector. Pulses that are registered simultaneously, e.g., in coincidence with both the guard and the main detector, are considered to be caused by cosmic radiation.[44-46]

The most common system is probably that using Geiger-Müller detectors. They are inexpensive, reliable, and modest in their needs for ancillary electronic equipment, although those incorporating an organic quenching agent have only a limited life time (order of 10^9 counts). Usually, the main detector is surrounded with a complete ring of guard counters or with a suitably shaped guarding tube to eliminate cosmic ray muons, which traverse the space between the active volumes of adjacent counters. To reduce possible contributions to the background from radioactive materials in the anticoincidence counters, a close-fitting shield may be placed between the material and the main counter. Doubly or triply distilled mercury is often used.

As a result of the disadvantages of the Geiger-Müller tubes, other types of counter for the anticoincidence guard may be more suitable. Organic scintillators are admirably suited to this task because of their availability in large sizes and in virtually any shape.

f. Autoradiographic Techniques

To identify radioactive particles in the presence of a large number of inactive particles normally found in an air sample, an autoradiographic stripping film technique has been developed that permits an exact localization of sources of α-radioactivity from α-track observations.[48] Araldite on thin polyethylene foils is used to secure the dust layer on the filter. The polyethylene layer, which does not adhere strongly to the araldite mixture, can be peeled off, leaving a surface of partially embedded dust particles. A stripping film, floated on distilled water, is made to adhere to the prepared slice, which is then dried. The exposure time depends on the sensitivity required, i.e., on the minimum particle size that is to be detected. These limits are also determined by the visibility of the track star formed by the α-radioactive particle. The ideal exposure time should result in obtaining a well-formed star of 10 to 30 tracks. A particle of plutonium oxide having a diameter of 0.5 μm at the surface of the araldite gives a star of 10 tracks in an exposure time of only 5 hr.[48] The efficiency of the detection of particles depends on their depth in the araldite. In the case of plutonium-239, no particle lower than approximately 33 μm in the plastic layer can be detected.

This method gives unambiguous information on the location, size, and number of α-active dust particles. A count of the number of tracks per star permits an estimation of the activity distribution of the dust particles. Although this technique is valuable for examining individual air samples, it is too time-consuming for wide-scale routine use. A more rapid method of autoradiography, which measures only the particle activity, has also been developed. In this technique, the α-particle energy is first converted to visible light by means of a zinc sulfide screen. The light flashes are recorded by a photographic plate. The zinc sulfide is spread onto adhesive tape which is attached to the sample, and a photographic plate is placed on the sample under pressure and exposed. After developing the photographic plate, black spots are observed. The diameters of these spots are related to the activity of the dust particles. Spots with a diameter of 0.2 mm are obtained with an α-activity of 2 dpm or a β-activity of 700 dpm within 16 hr.[48]

For mixed α- and β-emission, a light-sensitive plate mounted as described above with an X-ray film in contact with the clean side of the paper will differentiate between the two types of radiation. Autoradiographic methods permit exact location of radioactive dust particles in the filter. The sensitivity of autoradiographic techniques is inferior compared with low-level measurement techniques utilizing guard counting tubes. With these latter techniques, less than 1 dpm can be detected.

Only if the radioactive material is restricted to a very small and well-defined area, can low amounts of radionuclides be detected within a rather short exposure time of the autoradiographic picture. To achieve this, Weizs used the ring oven technique for concentrating radioactive material into a narrow ring zone on filter paper.[49] With this technique, a drop of the solution to be analyzed is placed in the center of a filter disc. The filter is then mounted on a heated ring oven (on top of a heated cylinder with a central hole having a diameter which corresponds to the diameter of the desired ring zone). Suitable washing solutions are added to the center of the filter paper by means of a small pipette. The material to be analyzed is washed to the heated area of the filter disc where the solvents evaporate, rendering the dissolved material concentrated in a narrow ring zone. This technique was applied to the analysis of dust samples by Loley and Malissa.[50] They also separated mixtures of radionuclides.

For quantitative autoradiographic determinations the chronoautoradiographic technique turned out to be very useful. With a ring oven, a sample can be concentrated into an area of approximately 20 mm^2. An autoradiographic picture is taken by placing the filter and the film together. However, the exposure time is governed by a device that is constructed in such a way that the filter is gradually peeled off the film during exposure. The filter disc carrying the ring of the concentrated material is provided with a central hole and cut along a radial line; it is then placed on a slit plate and a similarly cut piece of film is placed on top. Both paper and film are held together by a plastic cylinder fitted with fine needles on the lower side. The cylinder is fixed to an axle with a toggle, so that stepwise rotation shifts the filter with the radioactive ring zone beneath the slit plate while the film remains above the plate. If the rotation is done stepwise with increasing velocity, a circular stepped grey wedge is obtained. The time of each exposure is known, hence one of the steps with a suitable degree of blackening can be used for comparison with a series of standards and to estimate the activity of the unknown sample.

Ottendorfer carried out detailed investigations in the field of chronoautoradiography. His work determined that this method has the advantage of using simple apparatus and uncomplicated operation for the determination of low amounts of radioactive material down to the range where results obtained by simple Geiger-Müller counting devices become questionable and complicated coincidence-counting systems are required.[51-55]

C. Analysis of Radionuclides on Air Filters

1. Radiometric Analysis

The method of accumulating the radioactive dust on single filter discs offers the possibility of collecting a large sample and measuring not only the gross α- and β-radioactivity but carrying out a detailed qualitative and quantitative radiometric analysis. For this purpose, the spectrometric methods described above, using either scintillation counters or solid-state detectors, are used. Background reduction is also necessary to increase radiometric sensitivity. Anticoincidence devices must be used.

Figure 5 illustrates the γ-spectra of a dust sample. For these measurements, a highly sensitive γ-ray spectrometer was used, consisting of a Ge(Li) detector and a NaI(Tl) detector to detect coincident and noncoincident γ-rays. The sensitivity of that device is referred to be about 1 dpm for coincident γ-ray emitting radionuclides. Figure 5 illustrates the detection of 22 radionuclides.[44]

Figure 6 is a graphic representation of the analysis of a sample of dust containing α-emitting radionuclides. Natural radionuclides and plutonium-239 were detected immediately after dust collection. An accumulating air sampler was used. Solid-state silicium detectors were used for α-spectrometry.[36]

2. Sequential Analysis of Radionuclides on Air Filters

When the activity levels in a given sample are low, a sequential radiochemical analysis in the largest available sample is performed rather than a parallel assay. Procedures for such analyses are described, i.e., in the manuals of the Health and Safety Laboratories of the U.S. Energy Research and Development Administration.[31] As an example, a scheme is described that is designed to determine pCi picocurie amounts of the radioisotopes of the following elements: antimony, barium, cadmium, cerium, cesium, iron, manganese, plutonium, strontium, yttrium, and zinc.

Filters of cellulose materials are used for dust collection, and the filters are wet ashed with nitric acid and hydrochloric acid in the presence of inactive carrier materials of the radionuclides to be determined. The silicate residue is rendered soluble by treatment with hydrofluoric acid. The wet-ashing procedure prevents losses of volatile components. Such losses may occur under the condition of dry ashing and high-temperature fusion. The following is a list of the separation procedures involved:

1. Plutonium: precipitation of plutonium as a fluoride.

2. Antimony: coprecipitation with AgCl.

3. Barium and strontium: precipitation of the nitrates of barium and strontium; precipitation of barium as chromate; precipitation of strontium as carbonate.

4. Cesium: precipitation as bismuth iodide complex.

5. Cadmium: precipitation as nitrosylchloride.

6. Manganese: precipitation as oxide.

7. Cerium, yttrium, zirconium and iron: precipitation as hydroxides; separation on ion exchange column.

A scheme is also described for sequential analysis of total uranium, radium-226, thorium-230, and lead-210 in dust samples.[37] For this procedure, tracers and carriers are added, and the sample is fused in sodium carbonate or dissolved in a mixture of perchloric and nitric acid. Uranium is determined fluorimetrically from an aliquot of the original solution. Radium and lead are separated from thorium by precipitation in strong nitric acid. Thorium is purified by ion exchange procedures. Ethylenediaminetetraacetic acid (EDTA)-chelation separates the lead from radium. The radionuclides are collected and measured as sulfates. Comparisons are also made of the results of radiometric parallel assay and sequential radiochemical analysis. The values shown in Table 5 have been determined by calculating the three-sigma (3σ) levels for the instrument backgrounds, i.e., Ge(Li) detector and low background beta counter) when the samples were counted for 1000 min on the Ge(Li) and 400 min on the low background beta counter.[56]

Loley and Malissa analyzed fission products in dust samples using the ring oven technique.[50] A solution of carrier material (containing cesium, strontium, cerium, zirconium, and ruthenium) was added to the loaded dust filter, and then the filter was moistened with a solution of sodium borate and sodium carbonate. After drying and ashing the

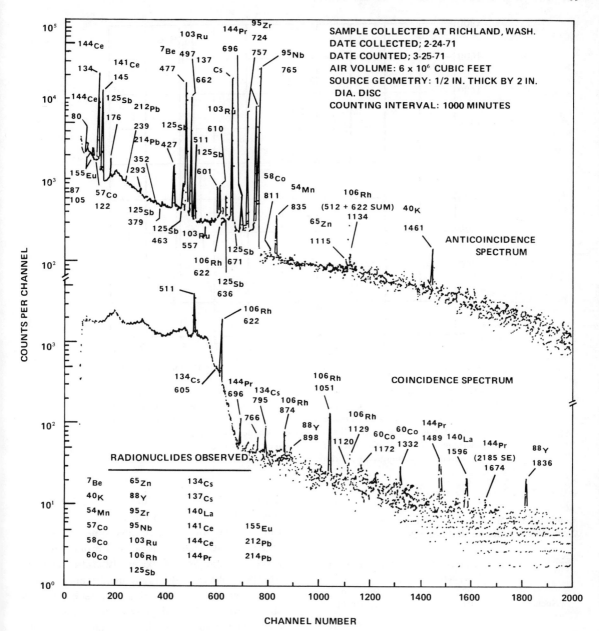

FIGURE 5. Gamma spectra of dust samples. 22 radionuclides can be observed using a Ge(Li) detector. (From Cooper, J. A. and Perkins, R. W., *Nucl. Instrum. Methods*, 99, 125, 1972. With permission.)

filter in an electrically heated platinum spiral, sodium borate beads containing the inorganic dust materials are formed. The beads are dissolved in diluted hydrochloric acid, and the dissolved nuclides are concentrated on a filter paper disc using the ring oven technique. Specific washing and precipitation procedures are applied to separate and concentrate the radionuclides in separate sections on the filter paper. Cesium is precipitated by hexachloroplatinic acid, strontium

by sulfuric acid, and for cerium, oxalic acid is used. Zirconium is precipitated by p-dimethyl-aminophenyl-azophenyl-arsinic acid and ruthenium by ammoniumsulfide solution.

Identification and determination are done by autoradiography. Experiments were carried out using dust samples that had been collected during the testing period of nuclear weapons in 1961/62.[50] In 1.7 mg of dust, cesium-137, strontium-90, and ruthenium-106 have been

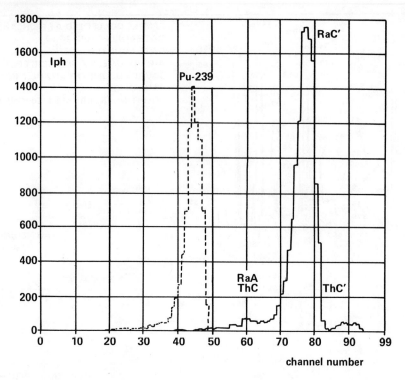

FIGURE 6.   Alpha spectrum of a dust sample containing natural radionuclides and plutonium-239. The dust was collected with a stepwise moving filter tape instrument. The measurement was made with the second detector of a three detector instrument immediately after collecting. (From Gebauer, H., *Raum- und Abluftuberwachung an kerntechnischen Einrichtungen,* Berthold Friesecke Co. for Measurement Techniques, Karlsruhe, 1973, 7. With permission.)

TABLE 5

Comparison of Minimum Detection Levels Using Spectral Analysis and Radiochemical Separation

Nuclide	Nondestructive γ-spectral analysis (fCi/m³)	Measurement after radiochemical separation (fCi/m³)
Ruthenium-103	0.3	0.05
Ruthenium-106	1.2	0.02
Cesium-137	0.2	0.02
Strontium-90	—	0.01

From Krieger, H. L., Martin, E. R., and Frishkorn, G. W., *Health Phys.,* 30, 456, 1970. With permission.

detected. About 1 pCi of strontium-90 was assessed. The above-mentioned separation method can be combined with instrumental nuclear measurement techniques also. Chronoautoradiographic techniques are extremely useful for quantitative analysis of the radionuclides concentrated on a filter disc using the ring oven technique. For all of these autoradiographic techniques, only common laboratory equipment and no electronic instrumentation is necessary. These methods should be of considerable interest under circumstances where sophisticated laboratory equipment is not available, and radioactivity determinations must be carried out by analyzing mixtures of radionuclides.

REFERENCES

1. Jacobs, D. G., Environmental behaviour of radionuclides released in the nuclear industry, *At. Energy Rev.*, 11, 691, 1973.

2. Beattie, J. R., *An Assessment of Environmental Hazard from Fission Product Release*, Report AHSB R 64, Authority Health and Safety Branch, Risley, Lancashire, England, 1963.

3. Welty, C. G. and Biles, M. B., The U.S. Atomic Energy Commission Program for Monitoring the Behaviour of Radionuclides Released to the Environment, Paper IAEA/SM-160/36, Proc. Symp. Environmental Behaviour of Radionuclides Released in the Nuclear Industry, Vienna, 1973.

4. Blanchard, R. L. and Kahn, B., Pathways for the Transfer of Radionuclides from Nuclear Power Reactors through the Environment to Man, presented at Int. Symp. Radioecology Applied to the Protection of Man and his Environment, Rome, Sept. 7—10, 1971.

5. Beattie, J. R. and Bryant, P. M., *Assessment of Environmental Hazards from Reactor Fission Product Releases*, Report AHSB R 135, Authority Health and Safety Branch, Risley, Lancashire, England, 1970.

6. Vogt, K. J., Dispersion of Airborne Radioactivity Released from Nuclear Installations and Population Exposure in the Local and Regional Environment, Paper IAEA/SM-181/39 Proc. Symp. Physical Behaviour of Radioactive Contaminants in the Atmosphere, Vienna, 1973.

7. Peirson, D. H., Worldwide deposition of long-lived fission products from nuclear explosions, *Nature*, 234, 79, 1971.

8. Peirson, D. H. and Cambray, R. S., Interhemispheric transfer of debris from nuclear explosions using a simple atmospheric model, *Nature*, 216, 755, 1967.

9. Holt, F. B. and Brunskill, R. T., Aerosol Studies in Plutonium and Uranium Plants at the Windscale and Springfields Works of the United Kingdom Atomic Energy Authority, Paper IAEA/SM-95/30 Proc. Symp. Assessment of Airborne Radioactivity, Vienna, 1967.

10. Junge, C. E., *Air Chemistry and Radioactivity*, Academic Press, New York, 1963

11. Lockhardt, L. B., Behaviour of Airborne Fission Products, *Am. Sci.*, 52, 301, 1964.

12. Anspaugh, L. R., Phelps, P. L., Kennedy, N. G., and Booth, H. G., Wind-driven Redistribution of Surface-deposited Radioactivity, Paper IAEA/SM-172/71, Proc. Symp. Environmental Behaviour of Radionuclides Released in the Nuclear Industry, Vienna, 1973.

13. Bruce, R. S., Agricultural Aspects of Radioactive Contamination of the Environment by the Nuclear Power Industry, Paper IAEA/SM-172/47, Proc. Symp. Environmental Behaviour of Radionuclides Released in the Nuclear Industry, Vienna, 1973.

14. Sherwood, J. D., Moore, D. T., and Kuroda, P. K., Radioactive cerium isotopes: the fallout from recent French and Chinese nuclear weapons tests, *Health Phys.*, 24, 491, 1973.

15. Izrael, Y. A., Ter-Saakov, A. A., and Kazakov, Y. E., Radioactive contamination of the atmosphere and ground by single and multiple underground nuclear cratering explosions, Report UCRL-Trans-10517, University of California Radiation Laboratory, 1970.

16. Herrman, G., Abgabe radioaktiver Stoffe bei Normalbetrieb aus Leichtwasserreaktoren in der Bundesrepublik Deutschland, Report IRS-W-1, Institut für Reaktorsicherheit, Cologne, Germany, 1972.

17. Wright, J. H., Environmental radiation from pressurized water reactors, *Trans. Am. Nucl. Soc.*, 13, 221, 1970.

18. Fritze, G. and Herrmann, R., The Interpretation of Aerosol Release Measurements at Nuclear Power Plants, Paper IAEA/SM-180/30, Proc. Symp. Physical Behaviour of Radioactive Contaminants in the Atmosphere, Vienna, 1973.

19. USAEC, The Potential Radiological Implications of Nuclear Facilities in a Large Region in the U.S. in the Year 2000, Report WASH-1290 U.S. Atomic Energy Commission, Washington, D.C., 1973.

20. Eisenbud, M., Review of USA Power Reactor Operating Experience, Paper IAEA/SM-146/35, Proc. Symp. Environmental Aspects of Nuclear Power Stations, Vienna, 1971.

21. Kempken, M., Verzeichnis der Kernkraftwerke der Welt, *Atomwirtschaft*, 18, 487, 1973.

22. Kautsky, H., Radioaktivität im Meer zur Zeit unbedenklich, *Umschau*, 73, 527, 1973.

23. Gebauer, H., Überwachungsanlagen für radioaktive Aerosole, *Atomwirtschaft*, 19, 81, 1974.

24. Kiefer, H. and Maushart, R., *Überwachung der Radioaktivität in Abwasser und Abluft*, 2nd ed., B. G. Teubner, Stuttgart, 1967.

25. Tschirf, E., Messmethoden der Radioaktivität in der Atmosphäre, in *Analytische Chemie und Luftschadstoffe*, Committee for Environmental Protection, Technical University, Vienna, 1973.

26. *Inhalation Risks from Radioactive Contaminants*, Technical Report Series No. 142, IAEA, International Atomic Energy Agency, Vienna, 1973.

27. Vohra, K. G., Nair, P. V. N., and Singh, A. N., New Concepts in Air Monitoring and Control of Inhalation Dose in Nuclear Operations, Paper IAEA/SM-95/21, Proc. Symp. Assessment of Airborne Radioactivity, Vienna, 1967.

28. Russell, R. S., Deposition of strontium-90 and its content in vegetation and in human diet in the United Kingdom, *Nature*, 182, 834, 1958.

29. Benett, B. G., Estimation of Sr-90-Levels in the Diet, Report HASL-246, Health and Safety Laboratory, U.S. Atomic Energy Commission, New York, 1972.

30. Sanders, M., Innovations in Air-Monitoring Techniques for Large-Scale Programs, Paper IAEA/SM-95/45, Proc. Symp. Assessment of Airborne Radioactivity, Vienna, 1967.

31.   **Harley, J. H.,** Health and Safety Laboratory Procedures Manual, U.S. Atomic Energy Research and Development Laboratories, 1972.

32.   **Buchtela, K., Tschurlovits, M., and Unfried, E.,** A sensitive method for early determination of the artificial activity of aerosolfilters, *Nucl. Instrum. Methods,* 120, 203, 1974.

33.   **Buchtela, K. and Tschurlovits, M.,** Bestimmung der Gesamt-($\alpha + \beta$)-Aktivität von grossen Aerosolfiltern im flüssigen Szintillator, *Health Phys.,* 27, 491, 1974.

34.   **Lederer, C. M., Hollander, J. M., and Perlman, I.,** *Table of Isotopes,* 6th ed., John Wiley & Sons, New York, 1967.

35.   **Tschirf, E. and Schuh, H.,** Messanalyse zum direkten Nachweis von radioaktiven Aerosolen, *Acta Phys. Austriaca,* 18, 347, 1964.

36.   **Gebauer, H.,** Messung radioaktiver Aerosole, in *Raum- und Abluftüberwachung an kerntechnischen Einrichtungen,* Berthold Friesecke Co. for Measurement Techniques, Karlsruhe, 1973.

37.   Handbook of Air-Sampling Instruments, Am. Conf. of Governmental Industrial Hygienists, 1973.

38.   **Siegbahn, K.,** *Alpha-, Beta- and Gamma-Ray Spectroscopy,* 2nd ed., North-Holland, Amsterdam, 1966.

39.   **Overman, R. T. and Clark, H. M.,** *Radioisotope Techniques,* McGraw-Hill, New York, 1960.

40.   **Fünfer, E. and Neuert, H.,** *Zählrohre und Szintillationszähler,* 2nd ed., G. Braun, Karlsruhe, 1959.

41.   **Birks, J. B.,** *Scintillation Counters,* McGraw-Hill, New York, 1953.

42.   **Bell, C. G. and Hayes, F. N.,** *Liquid Scintillation Counting,* Pergamon Press, London, 1958.

43.   **Miller, G. L., Gibson, W. M., and Donovan, P. F.,** Semiconductor particle detectors, *Annu. Rev. Nucl. Sci.,* 12, 189, 1962.

44.   **Cooper, J. A. and Perkins, R. W.,** A versatile Ge(Li)-NaI(Tl) coincidende-anticoincidende gamma-ray spectrometer for environmental and biological problems, *Nucl. Instrum. Methods,* 99, 125, 1972.

45.   **Lewis, S. R. and Shafrir, N. H.,** Low-level Ge(Li) gamma-ray spectrometry in marine radioactive studies, *Nucl. Instrum. Methods,* 93, 317, 1971.

46.   **Camp, D. C., Gatrousis, C., and Maynard, L. A.,** Low-background Ge(Li) detector systems for radioenvironmental studies, *Nucl. Instrum. Methods,* 117, 189, 1974.

47.   **Billard, F., Madelaine, G., and Parnianpour, H.,** Penetration des Aerosols dans les Filtres – Determination de l'Autoabsorption, Paper IAEA/SM-95/11, Proc. Symp. Instruments and Techniques for the Assessment of Airborne Radioactivity in Nuclear Operations, Vienna, July 3–7, 1967.

48.   **Lister, B. A. J.,** Development of Air Sampling Technology by the Atomic Energy Research Establishment, Harwell, Paper IAEA/SM-95/21 Proc. Symp. Instruments and Techniques for the Assessment of Airborne Radioactivity in Nuclear Operations, Vienna, July 3–7, 1967.

49.   **Weisz, H.,** *Microanalysis by the Ring Oven Technique,* 2nd ed., Pergamon Press, Oxford, 1961.

50.   **Loley, F. and Malissa, H.,** Beiträge zur mikrochemischen Analyse langlebiger Uranspaltprodukte in Staubproben, *Anal. Chim. Acta,* 34, 278, 1966.

51.   **Ottendorfer, L. J.,** Chronoautoradiography by the ring oven technique, *Proc. Feigl Anniversary Symposium, Birmingham,* Elsevier, Amsterdam, 1963, 100.

52.   **Weisz, H. and Ottendorfer, L. J.,** Semiquantitative Autoradiographie mit Hilfe des Ringofens. I. Allgemeines. Apparatur und Handhabung, *Mikrochim. Acta,* 191, 1961.

53.   **Ottendorfer, L. J. and Weisz, H.,** Semiquantitative Autoradiographie mit Hilfe des Ringofens. II. Genauigkeit und Empfindlichkeit, *Mikrochim. Acta,* 725, 1962.

54.   **Weisz, H. and Ottendorfer, L. J.,** Semiquantitative Autoradiographie mit Hilfe des Ringofens. III. Expositionsgerät mit kontinuierlicher Drehbewegung, *Mikrochim. Acta,* 818, 1962.

55.   **Ottendorfer, L. J.,** Über die Auswertung von Autoradiogrammen auf Grund einer linearen Beziehung zwischen Aktivatät und Schwärzung als Methode zur mikroanalytischen Bestimmung langlebiger β–Aktivitäten, *Z. Anal. Chem.,* 217, 81, 1966.

56.   **Krieger, H. L., Martin, E. R., and Frishkorn, G. W.,** Sequential Analysis for Ruthenium, Strontium and Cesium in Environmental Air, *Health Phys.,* 30, 456, 1976.

Chapter 6
SPARK SOURCE MASS SPECTROMETRY

I. Cornides

TABLE OF CONTENTS

I. INTRODUCTION

When characterizing any physical object, it is of primary importance to have at least a rough idea about the atomic species of which the substance is composed. With pollution problems, establishing the presence or absence of many elements or isotopes may be required (some at very low concentration levels), and usually, more or less accurate quantitative data are also desired. The complete list of polluting elements of interest is frequently not known at the beginning of the investigation, and as a general rule, few atomic species can be excluded a priori. It may be said, with good reason, that the object of the investigation is defined in detail by the investigation itself.

To meet demands of this type, mass spectrometry is a highly effective approach. It is basically a physical method, identifying atomic species by measuring their mass, their most fundamental property. Since high mass resolving power (i.e., separation) is attainable as well and the mass range can be extended to considerably high values, mass spectrometry offers a wide range of applications. In elemental analysis, for instance, it covers practically the whole periodic system.

The quantitative approach of mass spectrometry is again prominent, as a direct counting of individual particles (atoms or molecules in the form of ions) is used to obtain concentration data. It provides high sensitivity and a very low detection limit; both are also very much uniform. The elements of the periodic system, for example, can be detected with an average detection limit of 1 to 10 of 10^9 atoms (1 to 10 ppba) by spark source mass spectrometry. Most of the individual detection limits are within a range of only one order of magnitude. Similarly, the range of the molecular detection limits of organic mass spectrometry is not very wide.

The accuracy of the mass spectrometric quantitative data is also considerably high in most cases. The mass spectrometric chemical analysis of gas mixtures provides data with errors between about ±0.2 to 2%. The relative isotope ratios can be determined with an accuracy of better than ±0.01%. Unfortunately, the most common method presently used for mass spectrometric solid analysis (application of the spark ion source) is much less accurate, with reproducibility averaging ±20 to 30% in routine work and not better than ±2 to 3% when special precautions are taken.

As is well known, mass spectrometric separation and mass measurement require the atomic and/or molecular constituents of the sample to be in ionic form; the ions must be accelerated and collimated in a suitably shaped ion beam. Therefore, the first highly essential unit of any mass spectrometer is the ion source, in which the ionization of the sample takes place and the ion beam is formed.

The classical ion source of solid mass spectrometry is the spark ion source (SS) initiated by Dempster in 1935.[1] It is still universally used for the analysis of metals, alloys, semiconductors, inorganic insulators, rocks, minerals, inorganic residues of waters, ashes of organic materials, etc. This is a well defined field of applied mass spectrometry, called Spark Source Mass Spectrometry (SSMS). The practical importance of SSMS is comparable to that of other mass spectrometric analytical fields, although the spark source itself compares quite unfavorably with the electron impact source that is used, e.g., in mass spectrometric gas and organic analysis.

For optimum performance, any mass spectrometer ion source should supply an ion beam to the analyzer that:

1. Properly represents the composition of the sample, i.e., contains ions of all constituents at sufficiently high and, as far as possible, similar relative concentration levels (condition for qualitative analysis)

2. Exhibits high stability as well as sufficiently high intensity, i.e., the total ion current intensity is constant in time, involving the constancy of the partial ion currents of each constituent also (condition for quantitative analysis)

3. Is properly collimated and exhibits small kinetic energy spread of the ions (condition for good resolution)

Requirements (2) and (3) are insufficiently satisfied by the spark ion source. The considerable energy inhomogeneity of its ion beam necessitates the use of a rather costly double focusing analyzer. The excessive ion current instability, on the other hand, requires an integrating-type ion detector. Ion-sensitive photoplates, commonly used for this purpose, make some inconveniences inevitable, as compared to the detection of ions by electrical means. They introduce an additional source of irreproducibility and, at the same time, do not sufficiently eliminate the problems of accuracy

caused by the inherent instability of the spark source.

SSMS, therefore, provides the least accurate results of all mass spectrometric analytical methods, even though its equipment is relatively more expensive. However, since most of the advantages of the mass spectrometric method can be retained, SSMS is still highly competitive in solids analysis. It is quite often indispensable when survey analyses are required and most advantageous when several trace elements are to be simultaneously determined quantitatively.

During the last decade, many efforts have been made to obtain increased accuracy in the mass spectrometric analysis of solids. In addition to improvements of the spark source method, new types of ion sources for ionization of solid samples have also been developed. While laser ion sources are still being tested in several laboratories, instruments equipped with sources utilizing secondary ionization are already commercially available; secondary ionization mass spectrometry (SIMS) has become a useful complementary method to SSMS.

SSMS can be advantageously used in pollution investigations to characterize dust particles, even those of completely unknown origin, by performing survey analyses at a very high and uniform sensitivity level. As a result, almost complete information on the elemental composition of the samples is made available. In view of this capability and of the paramount importance of organic mass spectrometry (utilizing mostly electron impact ion source and a gas chromatograph as a sample inlet system) in the field of organic pollution detection and research, it is highly regrettable that SSMS has not yet been established as a standard method of pollution control.[2] In fact, this method is not familiar to the great majority of analytical chemists except relatively few staff members of some special laboratories (e.g., metallurgical, semiconductor, or geochemical). Moreover, the SSMS method is presently in a state of new development. For these reasons, it is necessary to present some essential information on the method itself within this chapter, including a short description of the typical setup of the spark source mass spectrometer. Facts especially important to pollution analytics are emphasized. More general and detailed reviews of the field can be found in the books of Ahearn,[3] Chupakhin et al.,[5] and Dietze.[4]

II. THE SPARK SOURCE MASS SPECTROMETER

Mass spectrometers used for spark source work are very similar to each other, since both the analyzer and the ion detector are strictly predetermined by the properties of the ion source. The spark ion source is followed by a double focusing analyzer of the Mattauch-Herzog type consisting of an electrostatic and a magnetic analyzer. At the exit of the latter, the ions of the resolved beam strike an ion-sensitive photoplate, used as the ion detector. The diversity among the various types of instruments produced by different manufacturers is restricted to differences in dimensions, constructional details, and availability of various accessories. Figure 1 presents a general scheme of the SS instruments.

The construction of the spark ion source is relatively simple. The ions are generated by a high-frequency, high-voltage vacuum spark in the gap between two electrodes, at least one of which is made of the sample. Rod-shaped electrodes are most frequently used, manufactured from the solid material to be analyzed. If analyses of powdered materials are to be performed, the electrodes are prepared by using a pressing technique. Electrically nonconducting materials are first pulverized and blended with a high-purity conducting powder. A sketch of the spark source is presented in Figure 2. The ions are drawn out and accelerated by the electric field between the planes of the accelerating slit and the main slit, penetrating to the spark gap through the accelerating slit. The ion energy is defined by the accelerating potential difference U_a maintaining the field. The spark housing protects other parts of the source (e.g., the high-voltage insulator, etc.) from contamination by the sputtered electrode material.

A different electrode system was developed by Chupakhin and co-workers[6] for the analysis of nonconducting materials. The powder of the pulverized sample is pressed into an aluminum plate (AlP in Figure 3), and sparking is maintained between this plate and a needle-shaped counter electrode (CE) of tantalum through the 0.1 to 0.3-mm thin layer of the sample. By moving the plate, the sample layer can be scanned by the spark. A decrease of the sample inhomogeneity problem, better direction homogeneity of the ions, and smaller sample quantity required (not more than 20 mg as compared to about 200 mg used to prepare rod-shaped electrodes) are benefits of this system.

Sparking in a vacuum is brought about by applying a high alternating-voltage, up to 100 kV

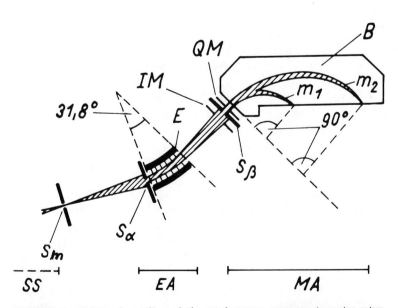

FIGURE 1.   Schematic outline of the spark source mass spectrometer setup. S_m = main slit; S_α, $S\beta$ = α-, β-slit; E = electric field; B = magnetic field; IM = ion current monitor; QM = ionic charge monitor; m_1, m_2 = low and high mass ions; SS = spark source; EA = electrostatic analyzer; MA = magnetic analyzer.

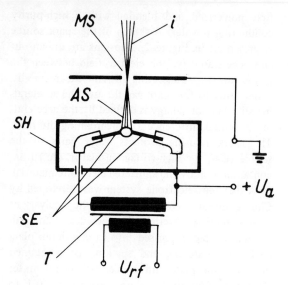

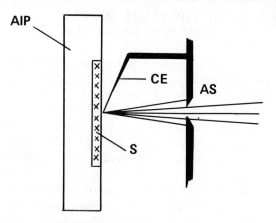

FIGURE 3. Sketch of the Chupakhin source. AlP = aluminum plate; S = sample layer; CE = counter electrode; AS = accelerating slit.

FIGURE 2. Sketch of the spark ion source. T = Tesla transformer; U_{rf} radiofrequency high voltage; $+U_a$ = ion accelerating voltage; SE = sample electrodes; SH = sample housing; AS = accelerating slit; MS = main slit; i = ion beam.

(peak-to-peak), across the electrodes, the frequency being typically near to 1 MHz (hence the names "radiofrequency (rf) spark", "rf ion source"). At a certain, voltage dependent upon the gap width of the electrodes, considerable current starts flowing through the gap, and the voltage breaks down. Consequently, a spark discharge occurs. This breakdown phenomenon is an abrupt process, usually taking place in less than 10 nsec, and its mechanism is not fully understood. It is generally accepted that the spark is initiated by field emission of electrons from micron size whiskers (microprotrusions) of the cathode surface. The emitted electrons, accelerated to high energies, may evaporate a small quantity of the anode material, but the whiskers may also be vaporized by the resistive heating caused by the electron flow. Some of the vaporized atoms are then ionized by electron impact, and the ions formed and accelerated start sputtering new atoms from the cathode surface as they strike.

Accordingly, there are several differing mechanisms simultaneously available to atomize the bulk material of both electrodes and introduce it into the electrode gap. Here this vapor is partly ionized, and a fraction of the ions is drawn out through the accelerating slit to form the output ion beam of the source.

A more detailed description is practically im-

possible, due to the fact that the details of the breakdown process depend on too many experimental parameters, several of which cannot be controlled sufficiently.

For example, it is known that, in addition to singly charged ions, many other kinds of ionized particles are produced by the vacuum spark, as multiply charged ions, ionized polymers, and various heteroatomic clusters, called complex ions. There is, however, no satisfactory information available about the parameters influencing the formation, abundance, etc. of these ions that may considerably decrease the reliability of SSMS analyses.

After breakdown, the discharge rapidly ceases, and the voltage starts to build up again until a new breakdown occurs if conditions for arc formation do not exist. This is shown by the oscillograms in Figure 4 for two different gap widths. To avoid excessive heating of the whole body of the electrodes, which would lead to selective vaporization of the components of differing volatility, the rf high voltage is applied in pulses. Both pulse length and repetition frequency of the pulses can be selected according to the heat and electrical conductivity of the electrodes to be sparked.

When selecting appropriate sparking parameters, the vaporization of the electrode material is confined at any moment to a very small volume of the electrodes, and its contents are then totally vaporized. Since the electron energy available in the gap is sufficient to ionize the atoms of any element, the ion yield for the various elements is quite similar; accordingly, the elemental com-

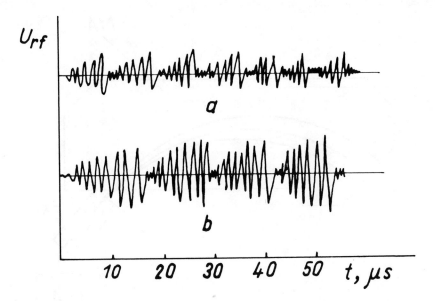

FIGURE 4. Oscillogram of the electrode gap voltage (a) for narrow and (b) for wide gap; U_{rf} = radiofrequency high voltage; t = time.

position of the ion beam is similar to that of the electrodes, i.e., to that of the sample. This is an important advantage of the spark source. On the other hand, the high electric field in the gap is also responsible for the wide energy spread of the ions (up to 1000 V) constituting a considerable disadvantage. The other serious drawback mentioned earlier, the instability of the spark source, is a result of both the inherently erratic character of the vacuum discharge and its necessarily pulse-like operation.

It may be mentioned that arc discharge sources, e.g., the triggered low voltage source of Franzen and Schuy,[7] though displaying much less energy inhomogeneity, have not come into general use.

The principle setup of the Mattauch-Herzog analyzer is well known. There are a few considerations concerning practical operation.

The elemental composition of samples of collected airborne particles usually covers the whole periodic system. Ions of different elements having the same mass number may therefore often be present in the ion beam and produce doublet or even multiplet lines in the mass spectrum. To obtain separation, a sufficiently high-resolving power is needed, M/ΔM = 5000, or even more in some cases. Resolution is increased by decreasing the width of α- and β-slit, i.e., by reducing the direction and velocity spread of the ions. At the same time, the main slit that is the object of the

analyzer as imaging system is narrowed to obtain as sharp and narrow spectrum lines (i.e., images of the main slit) as possible.

With narrow slits, the ion intensity becomes lower, and the sensitivity of the analysis decreases. Reasonably wide slits should be used, therefore, if high resolving power is not required. Typical slit width values for the main slit are 10 to 20 μm, and 50 to 60 μm, for the α-β-slit, 0.5 to 0.7 mm and 1.2 to 1.5 mm. The narrow slits are for high resolution, the wide ones for low.

The classical means to detect the ions, mass separated by the analyzer, is the ion-sensitive photoplate. The protective gelatin layer of this is much thinner than that of the usual plates. Presently, the Ilford® Q2 plates are used almost exclusively throughout the world for SSMS work. Plate development and fixing procedures have been investigated and described by McCrea,[8] Franzen et al.,[9] Kennicott,[10] and Cavard.[11] In a SSMS laboratory it is very important to use a highly standardized plate-processing method.

Recently, the use of electrical detection was introduced in the field of SSMS. Electrical measurement of ion currents, obviously the most natural and direct method of ion detection, is vastly superior to other methods so that the use of photoplates in mass spectrometry has been restricted (except in SSMS) to some very specific problems. In spark source (SS) work, photo-

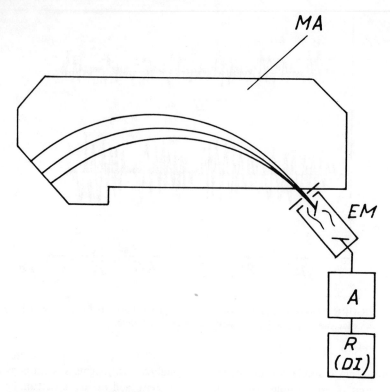

FIGURE 5.   Schematic outline of the electrical ion detection system. MA =
magnetic analyzer; EM = electron multiplier; A = amplifier; R(DI) = analog
recorder or digital integrator.

detection was used exclusively until recently when
electrical detection started to gain in popularity. A
comprehensive survey on the methods and
problems of electrical ion current measurements in
SSMS work is presented by Conzemius and
Svec.[12]

Ion currents can be measured with high sensitivity by the application of an electron multiplier
at the exit of the magnetic analyzer, as shown
schematically in Figure 5. By magnetic field
scanning, currents of ions of different masses can
be measured in succession.

In SSMS, both methods have particular merits
and drawbacks. To aid in deciding which is to be
used in a given case, the advantages and disadvantages are summarized below.

Photoplate — Due to the simultaneous detection and high-quality integration of all ions
(measured under exactly the same conditions),
data for a large number of components can be
recorded, even in the case of extremely small
sample quantities. The high resolution obtainable
(3000 to 10,000) makes analysis possible, even if
several major components exist, giving rise to a

variety of complex ions. On the other hand,
sensitivity is relatively low; linearity is restricted to
about 1½ orders of magnitude; reproducibility
(accuracy) is routinely only about 30%; and data
evaluation is delayed by the plate processing.

Electron multiplier — Because of nonsimul-
taneous   detection   and   integration   of
ions of differing masses (never measured under
exactly the same conditions), frequently only a
limited number of components of small quantity
samples can be determined. The relatively low
resolving power (500 to 1500) makes complete
analysis impossible if the sample and its mass
spectrum are highly complex. However, the ion
current sensitivity is high; linearity is excellent at 3
orders of magnitude, and the multiplier sensitivity
can be easily changed; reproducibility is routinely
5 to 10%; and data evaluation may be nearly
instantaneous.

As may be seen, the two methods are to a great
extent complementary. Presently, because of the
more frequent use of the photoplate and the
frequent occurrence of complex samples requiring
high resolution, photographic detection is pre-

ferred. Survey analyses of samples of air pollution particles with several major components (such as Na, Mg, Al, Si, etc.) can be performed by utilizing the advantages of the photoplate. However, if only a selected number of polluting elements is found and detection is not disturbed by other ions (atomic and complex ions), electrical detection is highly effective in controlling the concentration changes of these elements throughout a series of samples. In pollution analytics, it is therefore advantageous if both methods of detection are available.

Finally, the importance of a good vacuum system must be emphasized. A vacuum of 10^{-7} to 10^{-8} torr is required in the analyzer, with 10^{-6} torr or better in the ion source. If samples containing organic matter are analyzed, as often occurs in pollution analytics, a high pumping speed of the ion source pump is necessary, and additional use of cryopumping in this case is fully justified.[13]

III. SAMPLE PREPARATION AND TAKING THE MASS SPECTRA

Samples of airborne particles, which are practically powder samples, should be analyzed in pressed electrode form. Most suitably, a 1:1 mixture with graphite powder is used to prepare electrodes; 200 mg of sample is sufficient for three pieces, one serving as reserve. Before mixing with graphite, the samples should be dried at 105°C for about 1 hr to remove occasional humidity. Since organic content may give rise to excessive deterioration of the vacuum in the ion source during sparking and to production of oxide and hydroxide complex ions, it must be eliminated. This is most conveniently accomplished by wet digestion, with high-purity perchloric acid, after which the sample is dried under an infrared lamp. Baking the sample at 450°C to ash the organics would cause great losses in some inorganic compounds, as in the case of high quantities of volatile ammonium salts contained in urban aerosols.

The sample-graphite mixture must be homogenized carefully. This is most conveniently done in a mixer mill for about 30 min (if one is not available, it can be done manually in an agate mortar). For briquetting into electrodes, placing the mixture in the borings of a polyethylene plug which is then compressed in a pressing machine with 5 to 10 tons of pressure is recommended. The

plug may be 10 mm in diameter and have a length of 15 mm with three diametral borings 3 mm in diameter. The electrodes, ready for use, can be stored in a desiccator.

If the amount of the sample is less than 200 mg, only two electrodes are prepared, and the dimensions may be diminished. The electrodes may even be partially composed of pure graphite, the sample contained only in the tips that form the spark gap. For small sample amounts, however, the electrode system described by Chupakhin seems to be especially suitable.[6] A 10- to 20-mg sample is spread over the bottom of a small aluminum pot and then pressed into the aluminum surface. Since samples of airborne particles may be highly heterogenous, the application of this method may effectively decrease inhomogeneity errors.

To take and record the mass spectra, some typical SSMS procedures are generally used, irrespective of the character of the sample. This is to be considered when the parameters of the procedure are selected. Generally, the solid air pollution samples can be sparked in a similar manner to geochemical and biological samples.

If photoplate detection is used, a minimum number of ions hitting the plate per unit area is necessary to produce measurable blackening. Once a maximum number of ions arrive at the plate, saturation is attained, and further increase of the number of ions is not clearly indicated by a change in the blackening. The ratio of these extremes is 30 to 100; the measuring range of the photoplate is only 1½ to 2 orders of magnitude. Since the components of the samples may differ in concentration within a range of 6 orders of magnitude or more, several mass spectra of the same sample must be taken at appropriately different exposures, if this range is to be covered. Due to erratic behavior of the SS ion current, the total ionic charge reaching the photoplate is taken as a measure of the exposure instead of its time duration. In practice, a fraction of the total (unresolved) ion current is picked up by a monitor electrode in the front of the magnetic analyzer and measured by an integrating current meter (coulomb meter). Exposures are given in units of charge called nanocoulombs (nC). For example, graded exposures of 0.001, 0.003, 0.001, to 0.100, 300 nC are needed to cover the concentration range of about 0.03 ppm to 1%. At the same time, most spectral lines display measurable blackening in three or four different mass spectra (lines of elements with very low or very high concentrations are the exceptions) that is useful for increasing the precision of the analysis.

There are two basic methods utilizing electrical detection. Simple scanning, the whole mass range of interest, is the method most similar to those used in other fields of mass spectrometry. Even recently, however, due to the instability of the SS (including frequent complete breakdown of the ion current), the scanned spectra were found mostly unreliable even for qualitative analysis. The automatic gap control systems now used in many laboratories have considerably improved the SS performance to the extent that rough changes of ion current have been eliminated.[14,14a] The effect of the remaining instability is sufficiently decreased if the ratio of the ion currents of individual spectral lines ("peaks") to the total ion current (monitor current) is measured by the use of a ratio amplifier. In Figure 6, a portion of a scanned mass spectrum is shown, with the trace of the monitor current (above the peaks) recorded by a second channel.

The accuracy of this peak scanning method is approximately the same as that of photoplate detection, if errors caused by the lower resolution do not occur. Its greatest advantage is the high speed of measurement. Accuracy also can be improved considerably by using the peak switching method. Due to various inherent and uncontrollable changes inside and outside the spark, even the ratio of the collector current to the monitor current (i.e., the ratio of the current of a single peak — spectral line — selected for analysis of a given element to the total ion current) is not sufficiently constant. The resulting spread may be reduced by simultaneously integrating both the peak and the monitor current over time intervals much longer than that for scanning a single peak. In this case, the integration time is again defined by a preselectable total monitor charge. All peaks selected for analysis, including that of the reference element (internal standard), are to be switched sequentially to the multiplier slit for integration. This peak switching may be carried out by appropriately changing either the electrostatic or the magnetic field, and the whole procedure can be automatized, including the digital print-out of the normalized peak height data.[15] An on-line computer may be added to the system for data analysis and to control mass spectrometer activity.[16]

The peak switching method attains a reproducibility and accuracy better than ±10%, and under favorable conditions, ±2 to 3%.[17] Moreover, the detection limit is also lowered to 10^{-3} ppma or even to 10^{-4} ppma.[18]

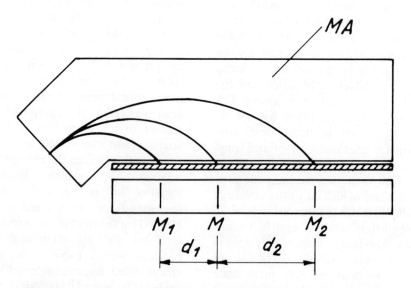

FIGURE 6. Schematic diagram of mass measurement in a photographically recorded mass spectrum. MA = magnetic analyzer; M_1, M_2 = known mass values; M = unknown mass value; d_1, d_2 = distances measured perpendicularly to the spectral lines.

To obtain optimum performance, the sparking parameters should be selected and/or adjusted carefully, regardless of the detection method used. The electrodes must not be overheated, even the darkest red glow is to be avoided. In samples of air pollution particles mixed with graphite, 20- μsec pulse length and 1000 Hz repetition frequency are, in general, suitable sparking parameters. Increasing either of these two parameters results in higher temperature; this, in turn, is dependent upon gap width and consistency (heat and electrical conductivity) of the electrodes. The spark (i.e., the gap) should be positioned on the optical axis of the ion optics. Adequate gap position is indicated by a rather sharp total ion current maximum in response to changing the accelerating voltage, which is to be set to this maximum. The total ion current can be measured by another monitor electrode. As an additional effect during these adjustments of the sparking conditions, the adsorbed gases are removed from the electrode surface, which are also cleansed of other occasional impurities. This "presparking" of the electrodes should last 5 to 10 min, corresponding to about 50 nC.

IV. COMPONENT IDENTIFICATION (QUALITATIVE ANALYSIS)

The valuable performance of SSMS in the field of survey analyses is due to its capability of identifying many individual components in a comparatively simple and reliable way.

Obviously, the analyzer of the mass spectrometer performs mass analysis of the ion beam, and this is converted to chemical (elemental) analysis of the sample. Relating the mass spectrum to the sample composition is, therefore, a two-step process: first, one must determine from the spectrum what kinds of ions exist in the beam (at what relative abundances). Then one must conclude what elements are present in the sample, and calculate their concentrations.

For identification of the ions, the mass number of a given ionic species (i.e., of a given peak — spectrum line) is easily determined. There are peaks of identifiable ions (peaks of known mass) in any part of the spectrum, facilitating determination of all neighboring mass numbers by simple counting of the peaks, including apparent vacancies, if there are any. One must take into account that the distance per mass unit is approximately constant only over a short range, about 5 mass units. If necessary, the unknown mass of a peak is calculated according to the square root law of the mass scale:

$$M = \left[\frac{d_1}{d_1 + d_2} (\sqrt{M_2} - \sqrt{M_1}) + \sqrt{M_1} \right]^2 \tag{1}$$

This is done after d_1 and d_2 distances of the unknown peak from the peaks of known masses M_1 and M_2 are measured (see Figure 6). Graphite mixed with powder samples supplies excellent reference lines (peaks) by its C_n^+ polymer ions at the mass numbers n $\times$ 12 (12, 24, 36, ...) for almost the whole mass range of the periodic system.

The identification of the ions of a given mass number M is often more difficult because different ions may have the same mass number. First, the table of the isotopic masses must be considered. Fortunately, there are no isobars in the range below M = 40. In the higher mass range, any isotope of the polyisotopic elements can be identified (or excluded) by finding (or not finding) the other isotopes of the given element in the mass spectrum at the expected abundance levels. Again, the nuclei of the monoisotopic elements have no atomic isobars; unfortunately, there may be various doubly (multiply) charged and complex ions of the same mass number. In order to make a reliable determination, it is necessary to exclude all possibility of existence of nonatomic and multiply charged ions at this mass number, by considering all major and sometimes minor components of the sample.

Identification of the $^{60}Ni^+$ line in the spectrum of an air pollution sample is presented as an example (Table 1). The most abundant ^{58}Ni isotope cannot be used to determine the nickel concentration, since it cannot be resolved from the ^{58}Fe isotope of quite high iron content (1.4%). The abundance of the ^{61}Ni, ^{62}Ni, and ^{64}Ni isotopes is too small for trace element analysis. Therefore, the ^{60}Ni isotope is used as if nickel were monoisotopic. There are, however, four lines found at mass number 60 (Figure 7). That of the highest mass and intensity is easily recognized as the C_5 carbon polymer. One of the remaining three weak lines may be the ^{60}Ni line. To determine which line this is, all must be identified. There are no more atomic ions of mass number 60. Since the tin concentration is too low, doubly charged $^{120}Sn^{2+}$ ions must be excluded, and the

TABLE 1

Results of SSMS Survey Analysis of a Sample of Airborne Particles

Element	Isotope used for analysis (mass number)	Concentration		Element	Isotope used for analysis (mass number)	Concentration	
		% (mass)	ppm (mass)			% (mass)	ppm (mass)
Na	23, (+2)	0.95		Y	89		9.3
		1.75 AAS		Zr	90		31.1
Mg	25, 26	4.04		Nb	93		3.0
		4.06 AAS					
Al	27	2.04		Mo	98		1.6
		1.87 AAS		Sn	118		1.7
Si	29, 30	3.71		Sb	121		1.5
P	31		610	Cs	133		0.3
				Ba	137		156
S	34	0.10					
Cl	35, 37	0.14		La	139		10.0
K	41	0.28		Ce	140		14.0
Ca	42, 43, 44	19.4		Pr	141		2.9
		18.4 AAS		Nd	143		8.6
Sc	45		3.7	Sm	147		3.2
Ti	49	0.12					
V	51		16.5	Eu	151		0.4
Cr	52		8.9	Gd	155		2.1
Mn	55		130	Tb	159		0.9
			500 AAS	Dy	163		1.2
Fe	57	1.38		Ho	165		0.45
		1.43 AAS		Er	167		0.78
Co	59		3.7	Tm	169		1.9
Ni	60		7.7	Yb	173		5.4
Cu	63, 65		39.3	Lu	175		0.4
			140 AAS	W	184		10.0
Zn	67		118	Pb	207, 208		472
Br	81		1.6				460 AAS
				Bi	209		0.38
Rb	85		6.6	Th	232		3.5
Sr	88		89.1	U	238		1.8

Note: The AAS results were kindly presented by Dr. L. Pólos, Institute for General and Analytical Chemistry, Technical University of Budapest; the SSMS analysis was performed by T. Gál, research worker in the author's laboratory.

appearance of $^{180}W^{3+}$ or $^{180}Ta^{3+}$ ions is even more impossible. On the other hand, the formation of CaO^+ and SiO_2^+ complex ions is quite probable, caused by the rather high concentration of both Si and Ca in the sample. Inspecting the exact mass values, the sequence of the three types of ions is obvious (see Figure 7). The calculated values of the resolving power necessary to separate any two neighboring lines are in agreement with the separations obtained by visual inspection alone. A resolving power of about 7000 was actually used.

Similar investigation is advisable in any case where there is uncertainty of identification, for instance, if the isotope ratios found are not those expected. It is essential in the case of single peaks, but also important if the peak is a member of a multiplet. In the latter case, it is preferable if all peaks are not identified, not only those of atomic ions representing components of the sample. Their relative positions within the multiplet must be determined to correctly reflect the mass differences. The method of mass measurement indicated in Figure 6 can be used to determine exact mass values or mass value differences also, if a high-precision comparator is available for measuring distances between spectral lines on the photoplate. This may help the identification of problematical ions.

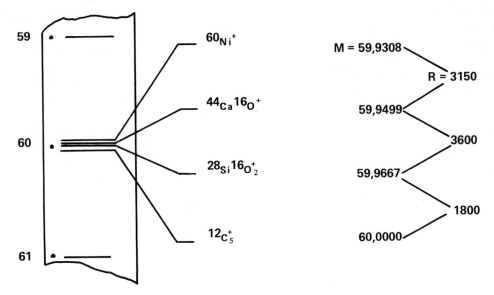

FIGURE 7. Multiplet at mass number 60. M = exact mass value; R = resolving power necessary.

In principle, if a complete analysis is required, all peaks of the mass spectrum must be identified. However, this is often quite difficult, especially if several major components exist, creating a variety of multiply charged and complex ions. Fortunately to some extent, the abundance of complex ions is much lower than that of the parent atoms (the singly charged atomic ions). Consequently, only major components or a major component plus a minor component are considered as parents. However, their levels may be as high as or higher than that of some trace element ions.

Since solid air pollution samples consist mostly of several major components, the problem of distinguishing between atomic and nonatomic ions must be resolved as with environmental analytics. Unfortunately, very little is known about the reactions occurring in the plasma of the spark ion source. Greater knowledge in this field will probably lead to safer methods of identification and possibly to suppression of complex ions, which possess little useful analytical information.

The final step in qualitative analysis is the most simple: all elements, the isotopic ions of which can be identified in the mass spectrum, are present in the sample. Presumably, any contamination during sample preparation and memory effects inside the ion source are negligible. For this reason, the general rules of microchemical analysis must be observed, and the surfaces of the ion source region surrounding the main slit must be thoroughly cleaned regularly. This must be done more often if the samples analyzed contain relatively volatile elements as major components.

The qualitative SSMS analysis can be easily developed into a semiquantitative one. Since the spark source and the SSMS measuring process in general are relatively nonselective, many of the elemental sensitivities are within a half order of magnitude, and most of them do not exceed a range of 1 order of magnitude. The ratios of the electrically detected peak heights (h) may be, therefore, approximately equal to the ratios of the elemental concentrations (c), if corrected for the different isotope abundances (I):

$$\frac{c_j}{c_s} \sim \frac{h_j}{h_s} \cdot \frac{I_s}{I_j} \qquad (2)$$

Denoting by indices s and j elements of known and unknown concentrations, respectively, the concentration c_j of the element to be analyzed can be calculated by the use of this equation and the concentration value c_s of the internal standard or reference element. The results obtained in this way may be correct on the average within a factor of three or less.

For semiquantitative analyses of this type, the peak scanning method is satisfactory, and if the peak heights are recorded logarithmically at two different sensitivity levels, a wide range of concentrations can be covered. To eliminate errors caused by unresolved superpositions of two or more peaks of a multiplet, high resolution photoplate detection is needed. After taking a series of mass spectra at graded exposures, again for semi-quantitative analysis, the weakest (nearly disappearing) spectral line of both the element to be determined and the internal standard element must be identified.

Assuming these weakest line blackenings are approximately equal, the ratio of exposures (E_o) at which the mass spectra of these lines were taken is inversely proportional to the ratio of the respective elemental concentrations. If corrected for the differing isotope abundances:

$$\frac{c_j}{c_s} \sim \frac{E_{so}}{E_{jo}} \cdot \frac{I_s}{I_j} \tag{3}$$

By the use of this equation (in which the indices j and s are used as indicated above), the unknown concentration c_j can be calculated at a similar reliability level.

V. QUANTITATIVE EVALUATION OF MASS SPECTRA

The spark source mass spectrometer, as any other analytical instrument, generates output signals quantitatively related to the concentration values of the various components of the sample introduced into the ion source (input of the instrument) for analysis. This quantitative relationship, often called calibration function, is used to calculate the elemental concentrations.

There are several output signals available for each element: line blackenings on the photoplate or peak heights on the recorder chart produced by the ions of each isotope of the given element. The monoisotopic elements are, of course, exceptions; the output signals of multiply charged ions are very rarely used. To clearly indicate which isotope is being measured its mass number is used. The isotope is chosen for analysis, the line peak of which is not disturbed by isobar ions. If several such isotopes are available, they can be used under parallel circumstances for control.

The calibration function is simple if electrical detection is used:

$$h_{jM} = s_j \cdot c_j \cdot I_{jM} \tag{4}$$

The h_{jM} peak height of the M mass isotope of the jth element is proportional to the c_j concentration of this element, decreased by the I_{jM} isotope

abundance. (I_{jM} is always less than unity, except in the monoisotopic elements.) The s_j elemental sensitivity factor, incorporating the ion source yield and analyzer transmission for the given element and the sensitivity of the measuring system (multiplier-amplifier-recorder), can be defined only as an average because of unstable ion source performance. The isotope ion currents are, therefore, integrated over a suitable time interval that is set by a preselected value of the total monitor ion charge. If this is done successively for all elements and for the internal standard as well by peak switching, each unknown c_j concentration can be calculated:

$$c_j = c_s \frac{q_{jM}}{q_{sM'}} \frac{I_{sM'}}{I_{jM}} \frac{1}{s_{js}} \tag{5}$$

M' indicates the mass number of the selected isotope of the internal standard element with known c_s concentration, and the q-values are the integrated peak heights, i.e., isotope ion currents presented by an integrating digital current meter instead of a recorder.

$$s_{js} = \frac{s_j}{s_s} \tag{6}$$

is the relative sensitivity factor of the jth element, relative to the internal standard, and can be experimentally determined, if a standard sample in which the concentrations of all elements are known, is analyzed in the same manner. Equation 5 is then used to calculate the s_{js} values instead of the c_j values. If semiquantitative analyses are performed by the peak scanning method or by the method utilizing the weakest lines (as described above), the s_{js} values of all elements are taken as unity.

When measuring ion currents, the charge of the ions and their number are determined. The concentration data obtained by the use of the above Equation 5 are, therefore, relative abundances of the atoms (ppma) of the given elements in the sample. To convert these data to mass concentration values, the ratio of the relative atomic masses (atomic weights) of the element to internal standard element (A_j and A_s) is used as an additional factor in the above Equation 5:

$$c_j = c_s \frac{q_{jM}}{q_{sM'}} \frac{I_{sM'}}{I_{jM}} \frac{A_j}{A_s} \frac{1}{s_{js}} \tag{7}$$

The quantitative evaluation of the photographically recorded mass spectra is much more complicated. Because of the nonlinearity of the photoplate detector, an indirect method of calibration has been generally adopted. Blackening (B) is dependent on the number of ions hitting the plate per unit area (i.e., ionic charge density). This is proportional to the concentration of the selected isotope of the element to be determined ($c_j I_{iM}$) and also to the exposure (E) as the ionic charge hitting the plate in the area of any individual spectral line is proportional to the total ionic charge. With the peak switching method, E was omitted from Equation 5 c_j, being present identically in both the numerator and the denominator. In the present case, however, it is retained as a controllable and measurable additional variable of the nonlinear calibration function:

$$B = f(s_j \cdot E \cdot c_j \cdot I_{jM}) \tag{8}$$

The measure B of the blackening in mass spectrometry is known as fractional blackening and is directly proportional to the number of blackened AgBr grains hit by the ions. The microdensitometers directly present the optical transmittance T, related to the fractional blackening by the equation:

$$B = 1 - T \tag{9}$$

Instead of using the direct $c_j - B$ (E = constant) relationship, the indirect method is more convenient and, therefore, most generally adopted i.e., the use of the relationships $E - B$ (c_j = constant) and $c_j - E$ (B = constant), successively. The latter relationship is the very simple linear function:

$$s_j \cdot E \cdot c_j \cdot I_{jM} = \text{constant (B = constant)} \tag{10}$$

The calibration for concentration is quite easily carried out with the help of an internal standard element of known c_s concentration.

The procedure in practice is as follows: for both the unknown and the internal standard element, the E_o exposures necessary to produce the same preselected B_o blackening must be determined, E_{jo} and E_{so}, respectively. The equation:

$$s_j \cdot E_{jo} \cdot c_j \cdot I_{jM} = s_s \cdot E_{so} \cdot c_s \cdot I_{sM'} \tag{11}$$

is used to calculate the unknown c_j concentration:

$$c_j = c_s \cdot \frac{E_{so}}{E_{jo}} \cdot \frac{I_{sM'}}{I_{jM}} \cdot \frac{A_j}{A_s} \cdot \frac{1}{s_{js}} \tag{12}$$

In Equation 12, c_j is again presented as mass concentration. To find the E_{jo} and E_{so} values, the B vs. log E curves are plotted for both the unknown and the internal standard element, using data from the graded mass spectra. These plots can be constructed most easily and reliably if the f calibration function is linearized. As an effective way of linearization the Seidel transformation is quite suitable:

$$W = \log \left(\frac{B}{1 - B} \right) = \log \left(\frac{1}{T} - 1 \right) \tag{13}$$

Instead of B itself, the quantity represented by the equation is plotted against log E, as seen in Figure 8. Usually, the medium blackening $B_o = 0.5$ is taken as the preselected identical value for B for all elements, corresponding to W = O in the Seidel scale.

Some blackening on the exposed and developed photoplates results from factors other than ions hitting the area of the spectral lines. There are several effects responsible for the "background" blackening, including positive ions caused by charge exchange, secondary positive ions and tertiary electrons from the plate and the surrounding metal parts, respectively, and defocusing effects caused by the space charge of the ion beam and surface charge on the plate. Since most of these effects are associated with high ion beam currents, the background is most intense near the lines of the high concentration elements (matrix element, major components), usually called fog or halo.

The background adversely affects the accuracy of the SSMS analysis. In intensive fog, the weak lines may completely disappear. Therefore, the detection limit is much higher for elements of isotopic masses near those of the matrix element, or even of several major elements. Moreover, even a moderate background changes the blackening of the spectral lines used for analysis: the measured blackening of the line is higher than the "true" blackening. There are two ways to deal with these harmful effects. First, the background should be reduced as much as possible, e.g., by improving the instrumental vacuum thereby decreasing the rate of the charge exchange reactions between ions and

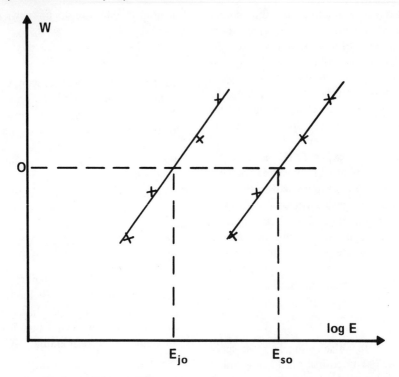

FIGURE 8.   W-log E curves plotted to determine the E_{jo} and E_{so} exposures necessary to produce blackening B = 0.5 (W = 0) by the ions of the element of unknown concentration and the internal standard, respectively.

neutral molecules; by limiting the ion beam density and the time of exposure; finding the best compromise for the given analytical problem; and by applying some modified developing techniques, e.g., the internal development method, etc.

The second approach is to correct calculations for background that could not be eliminated. The most simple method of background correction is to set the zero of the microdensitometer (i.e., the 0-point of the Seidel scale) at the background level near the line transmittance to be measured. A more satisfactory method calls for measurement of the apparent line blackening (B_l), the background blackening (B_{bg}) near the line, and the saturation blackening (B_s). These are to be used in the formula of Shuy and Franzen:[19]

$$B_{l,\ corr} = \frac{(B_l - B_{bg})\ B_s}{B_s - B_{bg}} \qquad (14)$$

More details about background correction in SSMS can be found in the reviews of Mai,[20] Ahearn and Paulsen,[21] and Kennicott.[22]

Two more methods of quantitative evaluation

of photographed mass spectra must be mentioned. Direct calibration without a set of samples containing the same element at various known concentration levels is possible, if the internal standard element, a known quantity of which is added to the sample to be analyzed, has four or five isotopes with abundances in the range of approximately 2 orders of magnitude (tin is ideal). Obviously, isotopes represent various known concentrations of an element and can be used to plot the log c vs. B or W calibration curves. These curves are useful in determining approximate values of unknown concentrations of other elements contained in the sample. Accuracy of the results may be increased by using appropriate relative sensitivity factors. This method is useful only if elements of similar concentrations (within 1 order of magnitude) are to be determined.

Determination of the relative sensitivity factors can be dispensed with if the external standard method is employed, i.e., the E_o exposures of the elements of unknown concentration in the sample are compared directly to those of the same element of known concentration in the standard. The mass spectra of sample and standard should be taken concurrently as far as possible, under the same conditions. Consequently, since the same isotope lines of the same element are compared, the formula for the calculation of the unknown c_j concentration is very simple:

$$c_j = c_j^* \cdot \frac{E_{jo}^*}{E_{jo}} \qquad (15)$$

The asterisk refers to data of the external standard.

Despite its straightforward means of measurement and calculation, the external standard method is much less popular than the application of internal standard. Generally, it is used only if single or a very small number of samples of the same type are to be analyzed, in which case it is unreasonable to determine relative sensitivity factors. Although the extent of matrix effects in SSMS is less significant,[23] it is advisable to use standards, whose chemical composition is similar to the samples being analyzed.

There is one problem that may, in some cases, seriously affect the accuracy of the analysis by the external standard method: the conditions of taking the mass spectra may not be exactly reproduced for both the sample and the standard. On the other hand, when using an internal standard, the spectral lines of the elements being analyzed are taken simultaneously with the reference (internal standard) lines, i.e., under exactly the same conditions. The benefit of minimal computational work is not a significant advantage in this era of computers.

Computerization in SSMS has advanced significantly in the last 5 years. Obviously, electrical detection is highly advantageous in this respect: the electrical signals produced by the mass spectrometer are directly processed by the computer after conversion to digital signals. If the output of the instrument is recorded in analog form (e.g., by an analog tape recorder), an analog to digital (A/D) converter is connected between the recorder and computer when computation is to be started. However, data collection may occur immediately in digital form when using a digital tape recorder or magnetic disc if the A/D converter is connected directly to the output of the mass spectrometer. On-line systems may also be built and operated, thus consisting of a spark source mass spectrometer, a computer, and a fast A/D converter as interface.

If photographic detection is used, data collection with the photoplate is simultaneous. However, it is, delayed by plate processing from the point of view of the computer. In addition, the data on the photoplate are in optical form, and an optical to electrical converter is needed; this is the microdensitometer. The output of this converter is digitized for the computer.

Data reduction is usually required as the first step in processing data obtained by analytical instruments. If electrical peak switching is used, data reduction is incorporated into the program governing the measuring process. Scanned or photographed mass spectra, however, contain redundant and even spurious data that cannot be eliminated a priori. In performing survey analyses, trace impurity investigations, etc., most of the peaks (lines) detected by the mass spectrometer are suspected of possessing information on some component than cannot be neglected. Selection of the isotope peaks used for the quantitative determination of the elements required can be decided only after thorough examination of many possibilities (isotope peaks of various elements, peaks of complex or multiply charged ions, or those of ions of the residual gas of the mass spectrometer). The problem of data reduction — data selection — is more difficult and complex, with less a priori information available about the sample to be analyzed. To deal with this problem successfully, a computer would need a rather complicated program. Therefore, the manual computations of the experienced analyst are preferred, entrusting the more clearly defined, detailed tasks to the computer, such as computation of exact ionic masses, comparison of the results obtained by the use of different isotope peaks of given elements, etc.

Determination of quantitative analytical results from electrically detected data is an extremely simple task for the computer, as the electrical data are directly proportional to the ion current intensities. Again, results are not as straightforward if photoplate detection is used. Electrical reading of the microdensitometer provides transmittance data. The relationship of this data with ion current intensities (and with elemental concentrations) is, as explained above, nonlinear and rather complicated. This fact presents mathematical difficulties, thus creating a typical computer task. There are many ways to deal with this task, the difference being mainly in the method by which the algorithm of the nonlinear calibration function of the photographic detection is expressed. The most elaborate method has been proposed by Franzen and Schuy,[24] based on a flexible analytical expression of their calibration function. Because of the complexity it is, however, not yet generally accepted. Considering the many intricate sources of error connected with the present-day operation of the spark source, it is more reasonable to use a less high-level approximation of the calibration function. Satisfactory results may be achieved using a linear approximation with the Seidel transformation as explained above.

Finally, three examples of pollution analysis by SSMS are presented. In Table 1, results of SSMS analysis of airborne particles collected in urban areas are summarized. The fraction of < 50 μm was analyzed. A 1:1 sample of graphite mixture was used for preparation of electrodes. Prior to mixing, 400 ppm indium was added to the graphite as internal standard. Mass spectra with graded exposures of 0.001 to 300 nC were taken on an Ilford Q2 photoplate, with sparking parameters set at 20 μsec pulse lengths and 1000 Hz repetition frequency. The gap width was automatically controlled. With a main slit of 20-μm width, a resolution better than 8000 was obtained.

For quantitative evaluation, indirect calibration was used: first, the E_o (W = 0) exposures were determined for all constituent elements of the sample and the internal standard; the concentrations were then calculated. The relative sensitivity factors used were determined earlier with the help of international geochemical standards. Analysis covered nearly the whole periodic system, lightest and gaseous elements excepted. The detection limit was set at about 0.1 ppm (mass). Results obtained by atomic absorption spectrometry were added to the list, most are in strong agreement with the SSMS data. Sodium is an exception, probably due to the fact that its doubly charged line must be used, the analytical value of which is apparently less than that of singly charged lines. Being monoisotopic, sodium has no low-abundance isotope with line blackenings below the saturation level.

As a second example of SSMS application in pollution analytics, a water reservoir investigation by the U.S. Environmental Protection Agency is detailed below.[25] This reservoir has been the site

of a number of recurring fish kills, and survey analysis was conducted to establish the background level to which the water impurities should be compared with future fish kills. Sample graphite mixture was again used, pressed into electrodes. Yttrium was added as an internal standard, and electrical detection with the peak scanning method and computer evaluation was used. The results obtained for 16 elements are summarized in Table 2, to which atomic absorption spectrometry data were added. The lowest concentration value determined by SSMS was 1 ppb. This detection limit could not be attained by the AAS method.

Brown and Vossen were the first to publish results of SSMS analysis of airborne particles.[26] A sample of 2 mg was collected from 10.8 m³ air on a nitrocellulose filter pad that was later burnt and the sample mixed with 5 μg silver as internal standard and high-purity graphite. Due to the small sample quantity, sample-tipped electrodes were prepared. Spectra were electrically detected; the peak scanning method was used. The results are summarized in Table 3.

Inspection of the results discussed above gives

TABLE 2

SSMS Analysis of Reservoir Water Sample

Element	SSMS result (mass) ppm	ppb	AAS result (mass) ppm	ppb
Pb		2		50
Ba		160		
Ce		2		
Te		1		
Sn		10		
Cu		5		12
Ni		8		50
Co		5		
Cr		20		50
V		3		
Fe	2		3.5	
Zn		60		40
Ti		20		
Mn		60		
P		30		
Sr	2			

From Taylor, C. E. and Taylor, W. J., EPA-660/2-74-001, Environmental Protection Technology Series, 1974.

TABLE 3

Results of SSMS Analysis of an Air Pollution Sample

Element	Concentration ($\mu g/m^3$)
B	0.004
F	0.11
Na	>5.5
Mg	11
Al	1.1
Si	63
P	1.1
S	2.3
Cl	0.28
Ca	2.8
K	40
Ti	0.23
Cr	0.30
V	1.9
Mn	0.07
Fe	2.4
Co	0.007
Ni	0.32
Cu	0.26
Zn	1.1
As	0.005
Br	0.12
Sr	0.05
Zr	0.004
Mo	0.01
Sn	0.07
Ba	0.02
Pb	4.3

Reprinted with permission from Brown, R. and Vossen, P. G. T., *Anal. Chem.*, 42, 1820, 1970. Copyright by the American Chemical Society.

reason to consider some important practical matters: (1) the attainable detection limit, i.e., the least detectable concentration value and (2) the minimum sample quantity. SSMS was introduced as a typical trace-analytical method to determine small concentration values in samples; the quantity of these samples were rarely limited. To produce a detectable line on the photoplate, at least 4000 ions (equal to 6.4×10^{-7} nC charge) are needed, and the highest utilizable exposure is about 300 to 500 nC; thus, the detection limit is approximately 1 ppba under optimum conditions of low back-ground and well-focused, sharp lines of monoisotopic elements. It is higher for polyisotopic elements and for those of low sensitivity factor, the mass ppm limit is higher for elements of higher atomic weight. In routine analytical work, the detection limit is between 10 and 100 mass ppb.

With electrical detection, detection limits lower by approximately 1 order of magnitude can be attained. The data in Table 3 do not demonstrate this improvement, because of the small amount of sample which was highly diluted by graphite to obtain sufficient quantity of mixture to prepare the electrode tips.

An increase in the detection limit may be expected with small sample amounts (below the 10 to 20 mg mentioned earlier), depending upon the way in which the sample is incorporated into the electrode and introduced into the spark gap. Since the electrode material consumption is roughly 1 μg/1 nC exposition, 0.3 to 0.5 mg may be sufficient without loss in detection capability, corresponding to a detectability of 1 to 10 pg per element. Some valuable methods of preparing electrodes with microsamples and limiting loss or excessive dilution of the sample are described by Ahearn.[3]

Precision and accuracy of the SSMS results are diminished with low concentrations or small sample amounts. This is tolerable however. More serious are errors exceeding 20 to 30% in "normal" range concentrations and sample amounts. One possible source of systematic error must be carefully noted: use of incorrect relative sensitivity factors. For instance, the potassium concentration given in Table 3 is certainly too high (by approximately 200%) due to the fact that in this preliminary investigation a relative sensitivity factor of one was used, although it should be higher for potassium.

To determine an appropriate set of relative sensitivity factors, reliable standards are required. There are many geochemical standards available, and establishing similar standards for pollution analysis is desirable.

VI. SUMMARY AND OUTLOOK

From the perspective of pollution analytics, this short outline of spark source mass spectrometry is certainly imperfect. It is hoped, however, that it may help the reader find answers to his questions, hopes, or doubts concerning the applicability of SSMS to his field.

According to Taylor,[2] the spark source mass spectrometer will assume an increased respected role as a complete survey method for trace elements in materials that are in or affect the existing environment. Conclusions drawn from this chapter definitely supported this opinion. The following is a summary of characteristic features of the SSMS method:[2 7]

1. Universal character: overall elemental coverage
2. High and uniform sensitivity: detection limit near 1 ppba throughout the periodic system
3. Simple sample preparation and measuring process: same pretreatment and measuring conditions for all elements
4. Relatively simple spectrum: relatively easy identification (qualitative analysis) and good possibility to computerize the evaluation process
5. Two different but complementary detection methods: high resolution with photoplate, better reproducibility with electrical detection

This features clearly define the role of SSMS in pollution analytics. It is the primary method used for overall survey both in regular control of inorganic pollution in a given field and in cases of pollution occurrences of unknown nature and origin. The information presented by mass spectra is helpful in determining which other method of analysis — atomic absorption spectrometry, neutron activation analysis, or occasionally others — is to be used and which procedures are less expensive and more accurate in controlling the concentration of characteristic elements of pollution in a given case.

It is expected, therefore, that SSMS laboratories will be established in the most important pollution control centers (as noted by the author in Japan). It is also expected that the performance of the SSMS method will be improved: the precision resulting from more strict control of sparking parameters, the accuracy of a higher level of standardization. Electrical detection methods will be refined as well, resulting in still lower detection limits. Computer evaluation and control will also improve performance characteristics and shorten the time now required for survey analysis.

REFERENCES

1. **Dempster, A. J.,** Ion sources for mass spectroscopy, *Rev. Sci. Instrum.,* 7, 46, 1936.
2. **Taylor, C. E.,** Survey Analysis for Trace Elements, presented at the 23rd Annual Conf. on Mass Spectrometry and Allied Topics, Houston, Texas, May 25-30, 1975.
3. **Ahearn, A. J.,** *Trace Analysis by Mass Spectrometry,* Academic Press, New York, 1972.
4. **Dietze, H. J.,** *Massenspektroskopische Spurenanalyse* (Mass Spectroscopic Trace Analysis), Akademische Verlagsgesellschaft, Geest et Portig K.-G., Leipzig, 1975.
5. **Chupakhin, S. M., Kruchkova, O. I., and Ramendik, G. I.,** *Analiticheskie Vozmozhnosti Iskrovoi Mass-spektrometrii* (Analytical Capabilities of Spark Source Mass Spectrometry), Atomizdat, Moscow, 1972.
6. **Chupakhin, S. M., Ramendik, G. I., and Kruchkova, O. I.,** Metod posloinovo mass-spektralnovo analiza (Method of mass spectrometric layer analysis), *Zh. Anal. Khim.,* 24, 352, 1969.
7. **Franzen, J. and Schuy, K. D.,** Advances in precision of mass spectroscopic spark source analysis of conducting materials, *Z. Anal. Chem.,* 225, 295, 1967.
8. **McCrea, J. M.,** Characteristics of ion sensitive emulsions for mass spectroscopy, I-II, *Appl. Spectrosc.,* 20, 181, 1966; 21, 305, 1967.
9. **Franzen, J., Maurer, K.-H., and Schuy, K. D.,** Über den Ionennachweis mit Photoplatten (Photoplate detection of ions), *Z. Naturforsch.,* 21a, 37, 1966.
10. **Kennicott, P. R.,** An internal image discriminating developer for mass spectrograph plates, *Anal. Chem.,* 38, 633, 1966.
11. **Cavard, A.,** Study of the Action of Positive Ions on Ion-Sensitive Emulsions Used in Mass Spectrography, Ph.D. thesis, Report CEA-R-3759, Centre d'Etudes Nucléaires de Grenoble. Rep.
12. **Conzemius, R. J. and Svec, H. J.,** Electrical measurement of mass resolved ion beams, in *Trace Analysis by Mass Spectrometry,* Ahearn, A. J., Ed., Academic Press, New York, 1972, chap. 5.

13. Harrington, W. L., Skogerboe, R. K., and Morrison, G. H., Determination of trace amounts of carbon, oxygen and nitrogen in metals by SSMS, *Anal. Chem.,* 38, 821, 1966.

14. Magee, C. W. and Harrison, W. W., Automatic gap control unit for spark source mass spectrometry, *Anal. Chem.,* 45, 220, 1973.

14a. Viczián, M., (Mining Research Institute, MS Laboratory, Budapest, Hungary), personal communication.

15. Svec, H. J. and Conzemius, R. J., Advances in mass spectrometry, in *Advances in Mass Spectrometry,* Vol. 4, Institute of Petroleum, London, 1968, 457.

16. Morrison, G. H., Colby, B. N., and Roth J. R., On-line computer-controlled electrical detection in SSMS, *Anal. Chem.,* 44, 1203, 1972.

17. Bingham, R. A. and Elliott, R. M., Accuracy of analysis by electrical detection in spark source mass spectrometry, *Anal. Chem.,* 43, 43, 1971.

18. Hintenberger, H., Report of the Max Planck Institut für Chemie (Mainz), Dep. Mass Spectroscopy and Isotope Cosmology, *Naturwissenschaften,* 59, 559, 1972.

19. Schuy, K. D. and Franzen, J., Evaluation of photographic spectra in mass spectroscopy, *Z. Anal. Chem.,* 225, 260, 1967.

20. Mai, H., Contribution to the investigation of background effects in spark source mass spectrography, in *Advances in Mass Spectrometry,* Vol. 3, Institute of Petroleum, London, 1966, 163.

21. Ahearn, A. J. and Paulsen, P. J., Report of the workshop on spark source mass spectrometry at the NBS, *Anal. Chem.,* 40, 75A, 1968.

22. Kennicott, P. R., An internal image discriminating developer for mass spectrograph plates, *Anal. Chem.,* 38, 633, 1966.

23. Jaworski, J. F. and Morrison, G. H., Sensitivity calibration in SSMS, *Anal. Chem.,* 46, 2080, 1974.

24. Franzen, J. and Schuy, K. D., Improved precision of ion sensitive photoplates by a computer program, *Z. Naturforsch.,* 21a, 1479, 1966.

25. Taylor, C. E. and Taylor, W. J., Multielement analysis of environmental samples by spark source mass spectrometry, EPA-660/2-74-001, Environmental Protection Technology Series, 1974.

26. Brown, R. and Vossen, P. G. T., Spark Source Mass Spectrometric survey analysis of air pollution particulates, *Anal. Chem.,* 42, 1820, 1970.

27. Cornides, I., Application of SSMS for high sensitivity survey analysis of minerals and rocks, *Miner. Slovaca,* 5, 489, 1973.

Chapter 7
CHARACTERIZATION OF ATMOSPHERIC AEROSOLS BY NEUTRON ACTIVATION ANALYSIS

R. C. Ragaini

TABLE OF CONTENTS

I. INTRODUCTION

Junge[1] and Israël,[2] among others, have stressed the importance of measuring the composition of aerosols, especially as a function of their size, in order to evaluate (1) the sources of atmospheric aerosols, (2) the atmospheric transport and removal reactions of aerosols, and (3) the aerosol effects on man and the environment. The results of these studies are highly critical, since they can lead directly to policy decisions dealing with source emission standards and ambient air standards.

Neutron activation analysis (NAA) can play a key role in the study of atmospheric aerosols because of its capability of detecting a large number of the inorganic constituents. As is shown in Table 1, NAA can be used to detect windblown dust aerosols (Al and rare earths), marine aerosols (Na, Cl), and anthropogenic aerosols (V, Cd, Zn, As, etc.).[3]

Neutron activation analysis was first used in 1936; it experienced a great development in the 1940s with the construction of reactors and again in the late 1960s with the development of high-resolution, solid-state, Ge(Li) γ-ray detectors, and fast pulse analysis systems. In recent years, more intense neutron sources and better neutron generators have expanded the applications of NAA. Developments in pneumatic sample transfer systems, rapid radiochemical procedures following neutron activation (RNAA), techniques for instrumental neutron activation analysis (INAA), and computerization of the analysis calculations have made NAA a useful, rapid, highly accurate, sensitive (often better than 1 ng), multi-element tech-

TABLE 1

Key Chemical Constituents of Atmospheric Aerosols[3]

Element or molecular constituent	Remarks and possible origins
C (total and organics)	Primary, natural and anthropogenic sources secondary, photochem (?)
N(NO_3^-, NH_4^+, amino and pyridino)	Sources – oxidation of NO_X, NH_3, fuel additives
Na[a]	Mainly sea salt
Al[a]	Mainly soil, some possible anthropogenic
Si	Mainly soil
S(SO_4^-, SO_3^-, S $\cdots$)	Mainly secondary production from SO_2 oxidation
Cl[a]	Mainly sea salt, but some anthropogenic
K[a]	Mainly natural (?)
Ca[a]	Cement production
Ti[a]	Anthropogenic and natural
V[a]	Power plant, fuel oil
Cr[a]	Anthropogenic
Mn[a]	Anthropogenic
Fe[a]	Anthropogenic and natural
Ni[a]	Anthropogenic
Cu[a]	Anthropogenic
Zn[a]	Tire dust, smelting, fuel additives
As[a]	Combustion, metal production and processing
Se[a]	Combustion
Br[a]	Auto exhaust
Cd[a]	Metal production and processing
I[a]	Sea salt and (?)
Ba[a]	Diesel exhaust and lubrication oil atomization
Pb	Auto exhaust, industrial processing
H_2O	Liquid water content is a potentially important inert ingredient in visibility question
RE[a] (rare earths)	Mainly soil

[a] Elements detected by NAA.

From Ragaini, R. C., Ralston, H. R., Garvis, D., Kaifer, R., Report UCRL-51850, Lawrence Livermore Laboratory, Livermore, Cal., June 1975.

nique. However, NAA does not have the ability to distinguish the chemical form of the element. This characteristic can be a disadvantage or an advantage, depending on the application.

The applications of NAA to the determination of trace elements in aerosols have been widely recognized and utilized. Many of the initial studies were published in 1964 to 1965 by Winchester and co-workers,[4-8] who used RNAA techniques to study the halogens, Na, V, and Cu in the atmosphere. Between 1965 and 1970, a small number of articles on NAA of aerosols were published. Keane and Fisher[9] and Brar et al.[10] used neutron activation followed by instrumental NaI(Tl) γ-ray spectrometry to study several elements present in air particles. Dudey et al.[11] observed activation products from 23 elements in marine aerosol samples after using ion-exchange radiochemical separations for Na removal and subsequent counting with Ge(Li) detectors. Several groups[12-14] have reviewed the application of the INAA technique with Ge(Li) detectors to the study of atmospheric aerosols. As a multi-element analytical technique, INAA has both extreme sensitivity and selectivity, which are ideal characteristics for the nondestructive analysis of atmospheric aerosols. Up to 33 elements have been observed and measured in aerosols by using INAA and selecting optimum irradiation and counting schedule for measuring activation products with half-lives ranging from 2.3 min to 5.3 yr.[12-14] Since 1970, NAA of atmospheric particles has been carried out by many groups, resulting in numerous publications. This article does not constitute a complete survey of the literature; however, the cited publications are representative.

II. BASIC PRINCIPLES OF NEUTRON ACTIVATION ANALYSIS

Since many books and monographs on NAA have been published,[15-23] only a brief review is presented here. Activation analysis is an analytical technique for determining the concentrations of inorganic elements in a sample. It describes the exposure of the sample to a flux of activating species: neutrons, charged particles, or photons. NAA is specifically the quantitative measurement of radioactive species produced by neutron-induced reactions. The technique can be nondestructive and entirely instrumental (INAA) where the γ-emitting radionuclide products are detected or "counted" with a γ-ray detector (Figure 1). For INAA, β-rays are not detected because β-rays from a given nuclide are not monoenergetic, which does not permit any nuclide identification. It can also involve detection-enhancing multi-element or single-element radiochemical separations after the neutron bombardment (RNAA) followed by β- or γ-ray counting. Since more radionuclides emit γ-rays of characteristic energies, γ-ray spectroscopy is the preferred detection method for both RNAA and INAA. Most aerosol studies have been done with reactor thermal (<0.04 eV) and fission spectrum (100 eV to 15 MeV) neutrons followed by γ-ray counting. A smaller number of aerosol studies have been carried out using fast (14

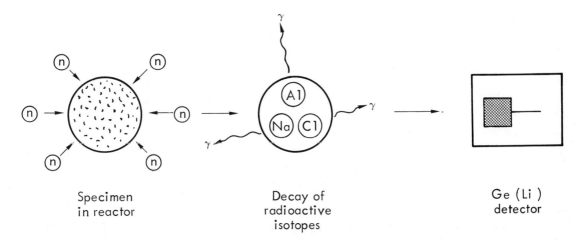

Specimen in reactor   Decay of radioactive isotopes   Ge (Li) detector

FIGURE 1.   Schematic diagram of instrumental neutron activation analysis (INAA) procedure.

MeV) neutrons from generators, for elements insensitive to thermal NAA such as Si.[24-25]

Thermal neutrons interact with stable target atomic nuclei predominantly by (n,γ) reactions, where the nucleus captures a thermal neutron, and the resulting compound nucleus de-excites by instantaneous emission of prompt γ-rays. All stable nuclides (except ^{4}He) undergo the (n,γ) reaction. The characteristic thermal neutron capture probabilities (or cross sections), and hence the sensitivity, vary greatly from nucleus to nucleus, from 10^{-3} to 10^5 b (1 barn = 10^{-24} cm^2). The de-excitation of the compound nucleus yields a product nuclide that may or may not be radioactive. When the product is radioactive, as is the case for most elements, and it decays by emission of γ-rays, these γ-rays can be detected for NAA. Fast neutrons interact with nuclei through (n,p), (n,α), (n,n') and $(n,2n)$ reactions which typically have much lower cross sections (<3 b) than thermal-neutron capture reactions.[18] The low reaction cross sections and the low fluxes (10^{11} n/cm^2/sec) available in reactors and from neutron generators restrict their application to elements for which: (1) no suitable (n,γ) product exists; (2) the (n,γ) cross sections are low compared to the fast-neutron reaction cross sections.

Many radionuclides can be formed in a neutron irradiation, and these must be separated in some manner for quantitative analysis. Separation can take place using radiochemical separations in the RNAA approach. With the INAA procedure, separation is effected by taking advantage of the differences in γ-ray energy and half-lives, and using high-resolution Ge(Li) spectrometers to resolve the γ-ray photopeaks. The radioisotopes are identified by the energies and half-lives of the γ-rays. The disintegration rate is then computed from the area of the γ-ray peaks, and the elemental concentration is calculated by comparing the activity of the sample to the activities of known elemental standards.

Figure 2 shows a general approach for carrying out activation analysis.[22] Three considerations must be properly addressed to carry out NAA:

1. Sample activation: type and flux of neutrons, irradiation time, cooling time, counting time, preparation of standards and/or flux monitors, evaluation of interferences.
2. Analysis procedures: nondestructively (INAA) or destructively (RNAA).

3. Methods of activity measurement: detection of β or γ radioactivity, type of γ-ray detector, detection characteristics of γ-ray detector, calibration of detector, counting schedule.

A. Fundamental Calculations

The quantification of the radionuclide provides a measurement of the total concentration of the parent element. The amount of a given neutron activation product is directly proportional to the amount of its parent isotope. The activity (A_i) of a given radionuclide induced in a sample can be expressed as:[26]

$$A_i(t) = \frac{f_i W N_o \sigma_i \phi}{M_i} \, [1 - \exp(-\lambda_i T)] \, \exp(-\lambda_i t) \qquad (1)$$

where W = weight of element in grams; $A_i(t)$ = decay rate of the ith isotope at a time (t) following the end of the irradiation (disintegrations per second); N_o = Avogadro's number (6.023×10^{23} atoms per g-atom); σ_i = microscopic absorption cross section for the ith isotopic reaction of interest (cm^2); M_i = atomic mass of the ith isotope (g/g-atom); f_i = fractional abundance of the ith isotope; ϕ = thermal neutron flux (n/cm^2-sec); λ_i = decay constant for the ith isotope (sec^{-1}); T = irradiation time (sec); and t = time since the end of the irradiation (sec). From Equation 1 we can see that the induced activity of the sample (A_i) does not continue to increase indefinitely with extended irradiation but is limited by the loss of the isotope due to radioactive decay. The maximum activity one can induce in a sample is given by the relationship:

$$S_i = \frac{f_i W N_o \sigma_i \phi}{M_i} \qquad (2)$$

which is the same as Equation 1, with $t = 0$ and $T = \infty$. From Equation 1 we can see that 99.2% of the saturation activity (S_i) is reached in about seven half-lives. For maximum sensitivity then, the sample would be irradiated for about seven half-lives and counted as soon after the end of the irradiation as possible.

In practice, however, the samples usually contain many elements and several activated isotopes of those elements. The activation-counting scheme must be chosen to optimize the activities of the elements of interest. A complete analysis usually involves making two or more separate irradiations for each sample and applying various counting

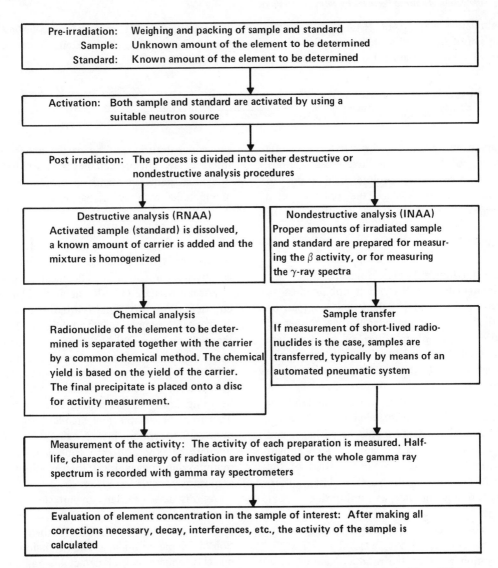

FIGURE 2.   Schematic diagram of a general process of neutron activation analysis. (From Tölgyessy, J. and Varga, S., *Nuclear Analytical Chemistry*, Vol. 3, University Park Press, Baltimore, 1974, 57. With permission.)

schedules, depending on the type of sample and the element sought in the analysis. The irradiation schedules and counting schemes used in aerosol analyses are discussed in Section V.

The elemental concentration in a sample can be calculated from Equation 1 using two methods, the absolute method or the relative method. In the absolute method, it is always necessary to determine the absolute activity of the samples. Therefore, the absolute flux and the absolute disintegration rate must be known. In using reactor fluxes, the ratio of epithermal neutron flux to the thermal neutron flux must be known for the irradiation position. When activating the sample in a position with an appreciable epithermal neutron flux, interfering reactions, which form the same daughter, can disturb the determination. For example, ^{24}Na can be formed by the reactions ^{23}Na(n,γ) ^{24}Na ^{24}Mg(n,p) ^{24}Na; ^{27}Al(n,α) ^{24}Na. During the irradiation, the flux of activating particles should be measured with flux monitors. This method is more complicated than relative activation analysis, which is the method used by most groups.

The relative method of NAA is based on a comparison of the activity of the sample with the activity of a known standard produced under the identical irradiation, cooling, and counting conditions. The weight of the element can then be expressed as:

$$\left(\begin{array}{c}\text{weight of}\\\text{element}\end{array}\right)_{\text{sample}} = \left(\begin{array}{c}\text{weight of}\\\text{element}\end{array}\right)_{\text{standard}} \times \frac{(\text{Activity of Element})_{\text{sample}}}{(\text{Activity of Element})_{\text{standard}}} \qquad (3)$$

This approach eliminates the need for careful monitoring of the neutron flux, cross sections, activation time, and absolute calibrations of counting instruments, since the only required measurements are the relative counting of the activation products in the sample and in the standard.

B. Elemental Standards/Flux Monitors

In multi-elemental NAA, it is not practical to use separate individual elemental comparison standards. Composite standards that consist of compatible chemical mixtures of elements are usually used. Five to 10 elements are usually combined in a standard solution at optimum concentrations to provide adequate counting rates without interelement interferences. Standards for each element should be included in each irradiation. Elemental standards should be packaged and irradiated identically as the samples. They should be prepared in a matrix similar to that of the sample to minimize neutron self-shielding, different irradiation geometries, and interfering nuclear reactions.

The use of single or multiple flux monitors should be considered very carefully, since the disintegration rates of the product radionuclides will depend not only on the neutron flux but also on the neutron-energy spectrum, i.e., the ratio of thermal to epithermal neutron fluxes. Decorte et al.[27] used a triple flux monitor method with ^{60}Cu, ^{198}Au, and ^{114m}In to take into account the epithermal/thermal variations. This approach is valid when the counting calibrations are absolute.

Standard reference materials (SRM) of known elemental concentrations can be used as activation standards. The National Bureau of Standards (NBS) supplies biological SRMs: orchard leaves, tuna meal, and bovine liver, and environmental SRMs: coal, fly ash, fuel oil, and gasoline. The U.S. Geological Survey supplies a wide variety of standard rock reference materials[28] whose elemental concentrations are not as well documented as those of the NBS SRMs. However, they are useful reference materials for analysis of soils. If the concentrations of the particular elements of interest are not detectable in the SRM, then individual elemental standards should be prepared and included with the irradiations.

C. Sources of Error

Valkovic[22] discusses in great detail the sources of error in NAA. Sources of error can be classified as physical or chemical. The most predominant physical errors are (1) self-shielding, (2) homogeneity of neutron flux, and (3) interferences.

In a large sample or in a sample that has a high thermal neutron absorption cross section, the neutron flux in the interior of the sample is attenuated by neutron scattering and absorption, "self-shielding." This can be minimized or avoided by using small samples or dilute solutions and by preparing the sample and standard to show the same self-absorption. In order that the sample and standard be exposed to the same neutron flux (to minimize inhomogeneities), they should be located as close as possible.

As discussed earlier, an interference occurs when the daughter radionuclide of interest is also present from some reaction other than the activation of the parent nuclide. The (n, fission) reaction on U produces several detectable radionuclides, such as ^{140}La, which are identical to thermal-neutron-capture radionuclides (^{139}La (n,γ) ^{140}La). Also, (n,p) and (n,α) reactions can produce the same type of interferences. For example, the radionuclide ^{60}Co can be formed three ways: ^{59}Co (n,γ) ^{60}Co; ^{60}Ni(n,p) ^{60}Co; and ^{63}Cu(n,α) ^{60}Co.

D. Neutron-Capture γ-ray Activation Analysis

Neutron-capture γ-ray activation analysis is based on the instantaneous γ-ray decay of excited nuclear states produced by the capture of a thermal neutron. The irradiation of the sample and the measurement of its activity are performed simultaneously. The energy of the capture γ-rays ranges from several kiloelectronvolts to approximately 10 MeV, producing a very complex spectrum. External reactor thermal neutron beams used for this purpose range from 10^7 to 10^9 n/cm^2/sec. This technique can be utilized to measure those elements that have high neutron-capture cross-sections, but which are not observed in conventional NAA where neutron capture could lead to stable nuclides, or extremely short half-life nuclides, or nuclides which emit no intense decay γ-rays. Lombard and Isenhour[28] have studied the analytical capabilities of neutron-capture γ-ray spectroscopy. They measured 18 elements that could be observed in quantities below 500 μg using an external reactor beam of 1.1×10^7 n/cm^2/sec and a 20-cm^3 Ge(Li) detector. They determined detection sensitivities of 1 ppm for Be and 2 ppm for Cd based on a 100 min exposure.

III. NEUTRON SOURCES

Neutron sources can be classified into three groups: (1) Nuclear reactors: neutrons produced by prompt and delayed neutron emission resulting from fission of uranium, (2) Neutron generators: neutrons from (p,n) and (d,n) reactions with proton and deuteron beams from accelerators; and (3) Isotopic sources: (a) neutrons from spontaneous fission of ^{252}Cf nuclei, (b) neutrons from photonuclear reactions (γ,n), (c) neutrons from alpha-neutron reactions (α,n). Each neutron source can be characterized by the energy and intensity of neutrons emitted. The neutron flux, which is the number of neutrons passing through a unit area in unit time at the irradiation site (n/cm^2/sec), is the most important characteristic of the source for NAA, since a higher flux over the sample allows higher sensitivity.

A. Nuclear Reactors

The most widely used neutron source for activation analysis at the trace-element level is the research nuclear reactor. Modern research reactors have power levels ranging from 10 KW to 10 MW with thermal neutron fluxes from 10^{11} to 10^{15} n/cm^2/sec.

Neutrons released from uranium fission have a continuous energy spectrum from 0 to 25 MeV, with an average of 2.0 MeV. Neutrons that experience no moderation after emission are termed fission neutrons and represent a small percentage of the reactor neutrons. The fission neutron flux decreases rapidly with distance away from the fuel rods due to thermalization. Thermal neutrons are those that have reached a state of thermal equilibrium from collisions with moderator atoms and have energies <0.4 eV, with a most probable energy of 0.025 eV. Epithermal neutrons are those that have undergone moderation, but are not thermalized, and they have energies >0.4 eV. The resultant neutron energy spectrum in a reactor ranges from 0.001 eV to 15 MeV. In most research reactors, the fast neutron fluxes are an order of magnitude lower than the thermal neutron fluxes.

Reactor neutron fluxes have both vertical and horizontal gradients that must be taken into account in any quantitative analysis. Horizontal flux gradients are minimized by rotation of the standards and samples. Vertical flux gradients are measured by vertical placements of flux monitors or standards throughout the sample stack.

Becker and LaFleur[30] have discussed the techniques used for characterizing a nuclear reactor for NAA. They describe thermal neutron flux determinations, neutron energy distribution measurements using cadmium ratios and threshold foil detectors, and the determination of excess sample pressures generated during irradiation at the NBS nuclear reactor. During long irradiations (hours), physical and chemical reactions can take place. The most dangerous process is the release of gas. This is usually taken into account by the use of double encapsulation, and leaving dead air space within the primary irradiation container.

High neutron fluxes can be obtained in triga reactors by pulsing the reactor. Using the appropriate neutron velocity moderators, typical peak thermal neutron fluxes greater than 10^{15} n/cm^2/sec in 20-μsec wide pulses with 100 pulses per second repetition rate are achieved for an average power of 0.5 MW and a pulsing power of 250 MW.[19] Some radionuclides with very short half-lives can be detected using pulsed NAA, e.g., Pb can be determined using the ^{204}Pb (n,n'γ) ^{204}Pb reaction.

Reactors are equipped with irradiation channels into which samples to be activated are inserted, usually by means of a pneumatic transfer system or "rabbit." These rabbits can terminate in the core or in a thermal column (which contains mostly thermalized neutrons), or in a reflector. Rabbit irradiations are routinely used for activating samples for induced activity half-lives of seconds to minutes.

B. Neutron Generators

In all of the various types of generators, a

charged particle is accelerated to an appropriate energy and then strikes a target which produces neutrons through (p,n), (d,n), (^{3}He,n), or (γ,n) reactions. Fast neutron activation is mostly limited to irradiation of samples by 14 MeV neutrons because of neutron source availability. Monoenergetic neutrons of about 14 MeV are produced by: ^{3}H (d,n) ^{4}He, Q = 17.58 MeV, which is the most useful neutron-producing reaction. Deuterons in the energy range 100 to 200 keV are produced by Cockcroft-Walton® deuteron accelerators. Such machines are capable of producing 10^{12} neutrons per second of 14 MeV from a tritium target with a 1-mA deuteron beam current with useful 14 MeV neutron fluxes of up to 5 × 10^{10} neutrons/cm^2/sec.[18]

C. Isotopic Sources

A third source of neutrons is the isotopic source in which the emitted radiation of the radioactive isotope bombards a target material to produce neutrons. The photoneutron sources and alpha-neutron sources such as ^{124}Sb-Be (γ,n) and Pu-Be (α,n) do not have sufficient neutron intensity for trace-element analysis. The most commonly used source is the spontaneous fission source ^{252}Cf. Approximately 10^9 neutrons per second are emitted by 1 mg of ^{252}Cf, with an average neutron energy between 1 and 2 MeV. With subcritical multiplier blankets of U or Pu, fluxes of 10^{12} n/cm^2/sec per gram of ^{252}Cf are attainable. The use of ^{252}Cf has been reviewed by Perkins et al.[32]

IV. GAMMA-RAY SPECTROSCOPY AND ANALYSIS

The majority of NAA aerosol studies have been carried out on activation products which are γ-ray emitters. The two types of detectors used for γ-ray spectroscopy have been the NaI(Tl) scintillation crystals and the solid-state Ge(Li) diodes. Up to 10 years ago, the only available detectors were NaI(Tl) detectors, which are capable of resolving only a few components in complex mixtures of radionuclides due to their inherently low energy resolution capability (6 to 7%). However, the introduction of Ge(Li) diodes, with their excellent energy resolution (0.15 to 0.22%), has revolutionized NAA in general and INAA in particular.

A. Ge(Li) Spectroscopy

The use of Ge(Li) detectors in NAA allows the resolution of nearly all the γ-rays of neutron activation products created by irradiation of environmental samples, provided their elemental concentrations are sufficiently high for detection. The γ-ray energies of all neutron activation products are now accurately known to within 0.1 keV.[31] Since γ-rays can be resolved and their energies accurately measured (± 0.3 keV) with Ge(Li) detectors, nearly all γ-ray emitters can be identified. Even though Ge(Li) detectors have much lower γ-ray detection efficiency relative to NaI(Tl) crystals (1 to 20%), the Ge(Li) detector efficiency is adequate to detect most activation products of atmospheric particles when using high flux nuclear reactors. Figures 3A through E show the many isotopes and elements observable in a 24-hr high-volume air filter irradiated in a nuclear reactor and counted with a Ge(Li) detector.[33]

The operations and applications of Ge(Li) detector systems have been described in the literature.[21,34,35] A complete Ge(Li) spectrometer system would consist of a Ge(Li) diode, analog electronics including a preamplifier and amplifier, and digital electronics including a 4096- or 8192-channel analyzer (or a minicomputer) accompanied by input/output peripheral devices as shown in Figure 4. A Ge(Li) detector as part of an automatic sample-changing system at Lawrence Livermore Laboratory (LLL) in Livermore, California is shown in Figure 5.[3] More sophisticated systems utilize anticoincidence shielding with NaI(Tl) detectors for background and Compton-event suppression.[36]

B. NaI(Tl) Spectroscopy

The NaI(Tl) crystals are approximately 5 to 100 times more efficient than Ge(Li) detectors. They are particularly useful for measuring trace amounts of radionuclides that have been radiochemically separated and purified. Such radionuclides would otherwise be undetected by Ge(Li) detectors because of interferences with other radionuclides or because of undetectable amounts of activity. The most common NaI(Tl) crystal configurations are well designs, or solid cylinders. As with Ge(Li) systems, the background and Compton events can be reduced on NaI(Tl) systems by employing anticoincidence shielding with large annular NaI(Tl) crystals or with large plastic phosphors.[37]

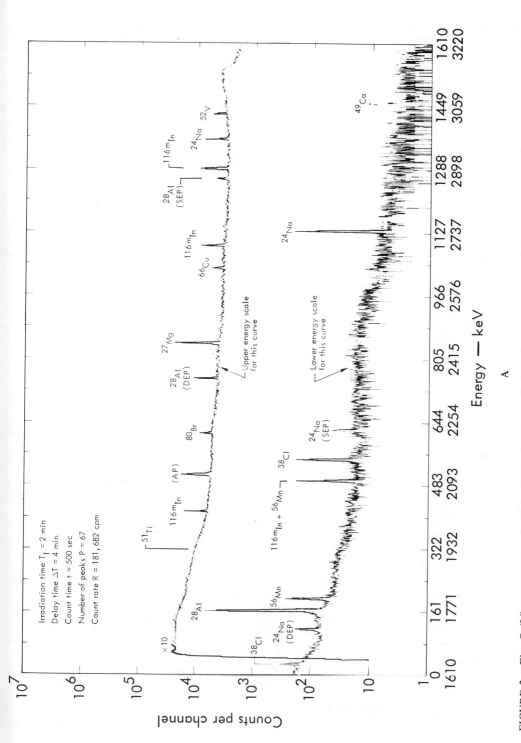

FIGURE 3. Five Ge(Li) γ-ray spectra of a high-volume air filter run for 24 hr near a Pb smelter in Kellogg, Idaho. The five spectra were taken after two different irradiations and five different counting conditions as identified in the spectra. The mass loading was 226 μg/m³; the mass on the filter was 216 mg. SEP, single escape peak; DEP, double escape peak; AP, annihilation peak. (From Ragaini, R. C., Heft, R. E., and Garvis, D., Report UCRL-52092, Lawrence Livermore Laboratory, Livermore, Cal., July 1976.)

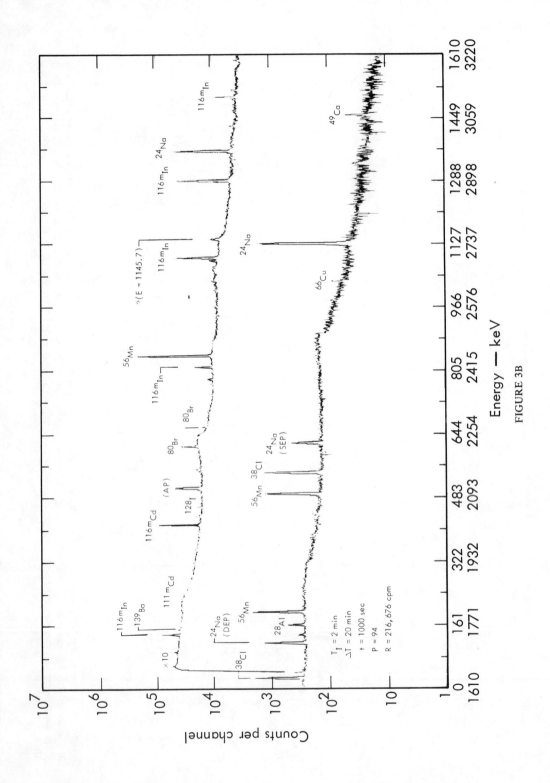

Energy — keV

FIGURE 3B

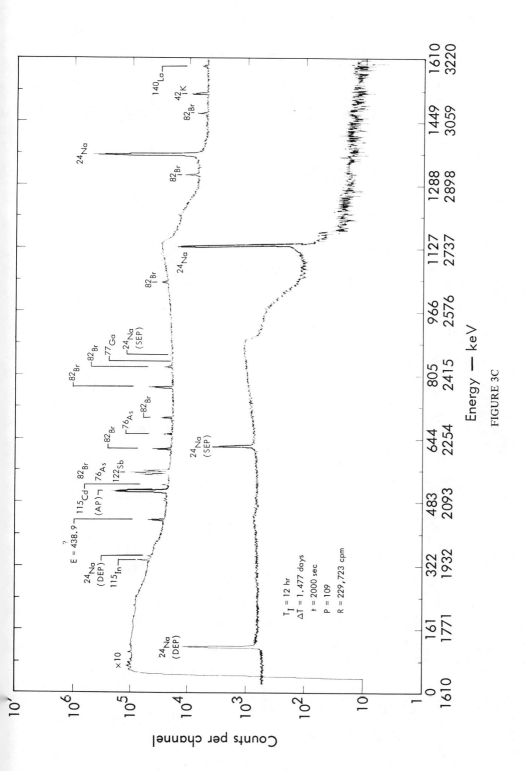

FIGURE 3C

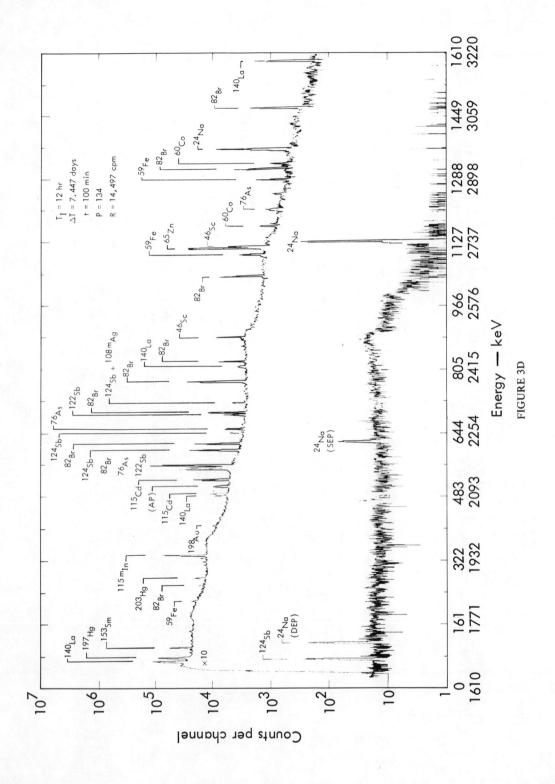

Energy — keV

FIGURE 3D

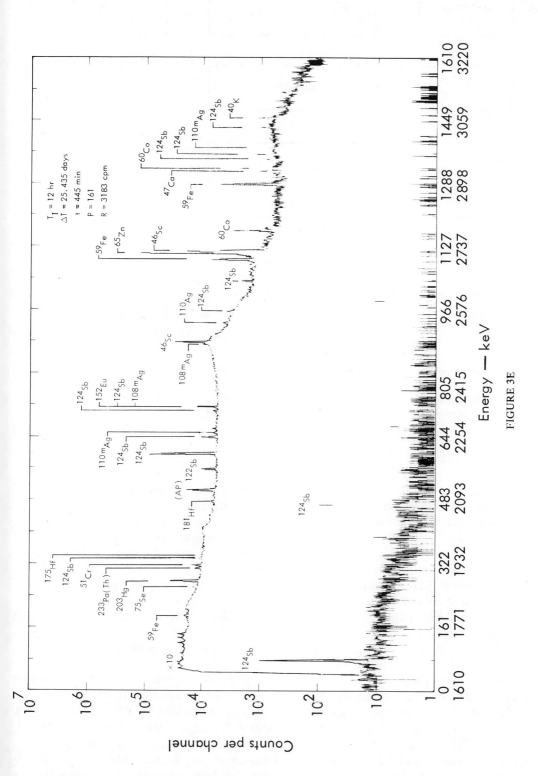

FIGURE 3E

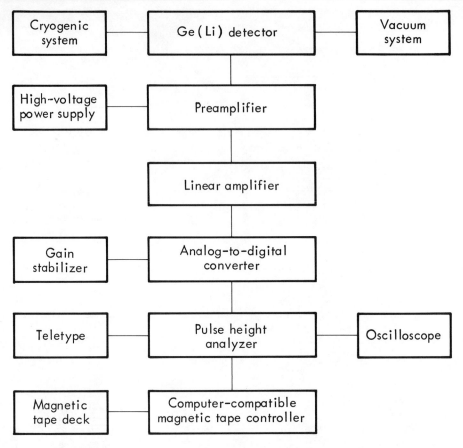

FIGURE 4.   Block diagram of Ge(Li) detector system used for γ-ray spectroscopy.

C. Spectral Analysis

The calculation of the disintegration rate and the elemental abundance from the measured γ-ray spectra involves several steps:

1. Location of γ-ray peaks in the spectra
2. Determination of the peak energies
3. Measurement of the peak areas
4. Calculation of the disintegration rate
5. Isotopic identification of the peak
6. Calculation of the elemental abundance

For analyses of many samples or for multi-elemental analysis, the least time-consuming and most rapid approach is the use of computer analysis codes as a data reduction system for γ-ray spectra. Published computer-based data reduction systems for Ge(Li) spectra have been summarized and reviewed by Hoste et al.[21] All systems generally involve seven steps in the course of the data reduction:

1. Preliminary data treatment (smoothing, derivatives, etc.)
2. Location of photopeaks
3. Determination of the peak limits
4. Application of statistical and peak-shape tests
5. Evaluation of exact peak centroid location
6. Determination of net peak area and standard elevation
7. Estimation of detection limit

One of the most powerful and sophisticated computer codes written for analysis of Ge(Li) γ-ray spectra is GAMANAL.[38] A secondary code, NADAC,[33] reduces the output of GAMANAL (disintegrations per second per radionuclide) to micrograms per element. Figure 6 shows the schematic block diagram of the steps in the GAMANAL and NADAC codes.

The GAMANAL code is a general-purpose

FIGURE 5. Automatic sample changing system utilizing a Ge(Li) detector at Lawrence Livermore Laboratory, Livermore, Cal.

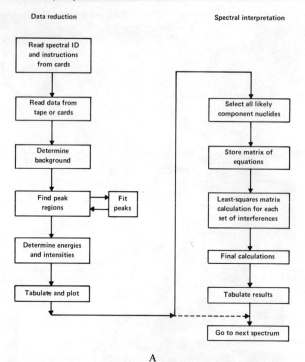

FIGURE 6.   (A) Schematic block diagram of the GAMANAL and (B) NADAC computer programs. (From Ragaini, R. C., Ralston, H. R., Garvis, D., and Kaifer, R., Report UCRL-51850, Lawrence Livermore Laboratory, Livermore, Cal., June 1975.)

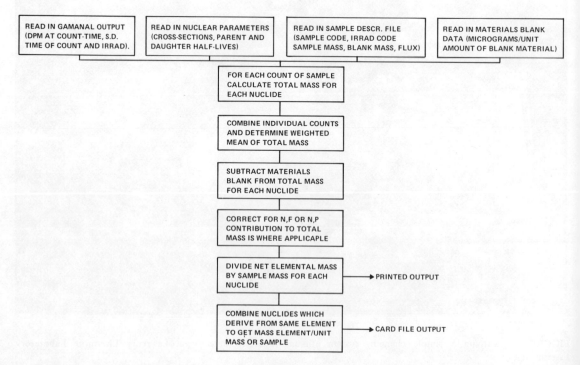

B

computer program for data reduction and interpretation of gamma spectra. It examines the pulse-height data for background and peak regions, fits these peaks with the proper shape functions, and corrects for the effects of geometry, attenuation, and detector efficiency in evaluating the photon emission rate. The program then searches an extremely detailed library of decay-scheme information[38] and makes tentative assignments for each of the observed peaks. A matrix of equations is formed so that the intensity of each peak is described as a linear addition of the identified nuclides. The quantitative value, as well as the degree of interference, is the result of a least-squares solution of this set of equations. Unlikely components are also eliminated in this process.

For each nuclide in a sample, NADAC calculates the expected disintegration rate per microgram of element at counting time. The total micrograms of element for each nuclide detected is calculated by dividing the disintegration rate found by GAMANAL by the expected rate calculated by NADAC for the irradiation, cooling, and counting times for the sample. The weighted mean for the total micrograms of each element is corrected (1) for interfering production of the measured nuclide by an alternate reaction (e.g., ^{27}Al (n,p) ^{27}Mg interferes with ^{26}Mg (n,γ) ^{27}Mg), and (2) for contribution to the observed total micrograms by mounting or sample containment materials that were irradiated and counted with the sample. In addition to interference and blank corrections, NADAC also provides for correctly calculating total micrograms when:

1. Counting time is long compared to the half-life of the nuclide.
2. Sample was subjected to multiple or interrupted irradiations.
3. Nuclide counted is the daughter of the nuclide produced, including the case where the daughter nuclide has some independent production.

The physical parameters used in the NADAC calculation are the half-lives of the parent and daughter, the correction for blanks, the rate of production of nuclide per microgram of element, and the rate or production of nuclide per microgram of element giving the interfering reaction. The rate of production of nuclide by reactions

yielding (n,p) or (n, fission) interferences is determined by irradiating a known quantity of the element producing the interference in each region of the reactor used.

V. ANALYTICAL PROCEDURES

A. Collection Procedures

Atmospheric particulate matter is usually collected by pumping air through filters or through cascade impactors. The material used for the filter should have low, reproducible, blank concentrations of the inorganic elements of interest. The filter should also have a high flow-through rate together with a high collection efficiency for small particles. Dams et al.[39] evaluated 10 filter materials for atmospheric particulate sampling and INAA using Ge(Li) detectors. They concluded that cellulose fiber filters, such as Whatman® 41, were the best choice considering low blanks, particle capture efficiency, and ease of handling. However, Whatman 41 filters do have certain drawbacks. They have minimum efficiencies for particles below 0.2 μm in diameter, and they load up quickly. A 10-cm filter allows a flow rate of > 25 m^3/hr, but cannot be run for 24 hr without plugging.

Other types of filters have also been used. Nuclepore® filters have been widely used in many aerosol trace-element studies. They are smooth polycarbonate films which have straight, near-circular holes as pores in the filter. Nuclepore filters have been evaluated and compared with membrane and cellulose filters.[40] Lui and Lee[41] concluded that the membrane filter Fluoropore® was considerably more efficient than a Nuclepore filter of the same or comparable pore size for collection of submicron aerosols. Investigators should make a very careful analysis of the characteristics of the available filters in order to judge the suitability for a particular experiment.

In an impactor, air is pumped through a series of orifices (either single or multiple) of decreasing size. Baffles (or stages) downstream from each orifice collect particles of high inertia which cannot stay in the airflow bending around the baffle. The velocity of the air stream increases as the air is forced through successively smaller orifices thereby impacting smaller particles on each succeeding impactor stage. Typically, impactors collect particles from $\gtrsim$ 8 μm down to ~0.2 μm and usually contain a backup filter to collect the submicron particles. Neutron activation analysis is normally carried out on particles collected on high-purity polyethylene, Mylar®, or polycarbonate sheets placed on the stages. However, most low flow-rate impactors (1 to 2 m³/hr) only collect 2 to 5 mg of atmospheric particulate matter distributed over several collection stages for a 24-hr sampling in an environment with 100 $\mu g/m^3$ atmospheric loading. For the study of trace elements, i.e., elements with concentrations of ≈ 10 ng/m³, where only 0.2 to 0.5 μg of the element is collected, NAA is one of the few useful analysis techniques because of its high sensitivity.

The initial use of cascade impactors and INAA was reported by Duce et al.[42] who studied midocean Cl particle size distributions; Nifong and Winchester[43] who studied aerosols in the Chicago, northwestern Indiana, and southern Michigan regions; and Rahn,[44] who studied midcontinental natural aerosols in remote areas of the northern U.S. and Canada.

The use of particular cascade impactors requires careful consideration of several operational factors; flow rate, number of stages, particle cut-off diameters, particle bounce-off and re-entrainment effects, particle loss on the impactor walls, area and geometry of the particle deposits. Gordon et al.[45] have discussed these problems and have compared several types of impactors (Andersen®, Lundgren®, and Scientific Advances®) by using INAA.

Figure 7 shows typical impactor particle-size distributions of Br, Cl, Na, and Al collected at the Harbor Freeway, Los Angeles with a 4-stage Lundgren rotating-drum impactor.[3]

B. Sample Preparation and Handling

Sample preparation and handling are of great importance in trace element work. Clean techniques must be employed at every step prior to irradiation. After the sample is activated, care can be relaxed since the sample is easily differentiated from any contamination other than radioactivity that might be introduced. The insensitivity to handling procedures after irradiation is one of the most attractive features of activation analysis, consequently preirradiation handling should be minimized.

If possible, the sample should be transferred from its irradiation container into a counting container to eliminate the problem of subtracting the trace-element content of the irradiation con-

tainer. If the sample is to be counted in its irradiation container or with a binder, the trace-element content must be determined from a statistically significant number of control containers of binder samples. The same clean handling techniques must be applied to the container or binder.

Two common irradiation container materials are polyethylene and quartz. For determination of nonvolatile elements, high-purity polyethylene is often used for low-temperature facilities. However, for volatile samples or for determination of volatile elements, such as Hg, which can diffuse through polyethylene containers, quartz containers must be used.[23]

In practice, the activated samples and standards are usually transferred from the irradiation containers into standard counting geometries. Geometry must be considered in two aspects: irradiation and counting. In some cases, the same geometry will apply to both, namely when a disk shape, which is ideal for counting, can be irradiated. When the sample cannot be irradiated in disk form, it may be possible to approximate such a configuration during counting by placing a powdered sample in a flat polyethylene envelope.

Techniques have been developed for preparing most dry samples in a disk shape for irradiation with a hydraulic press and appropriate dies.[33] The disks are made in a variety of diameters, ranging from a ¼ to 1 in. If binder is necessary, a known amount is added to the weighed sample. Powdered polyethylene and Avicel® are two commercially available binders. Polyethylene requires heating to form a wafer but does not require much pressure. Avicel requires no heat, but more pressure. Both materials have acceptable blanks for most analyses.

Atmospheric particles are generally irradiated along with the collection media. The blank for the collection media must be determined as for other materials. The size of samples varies with the type of sample, the matrix, and information desired. Experimental conditions often determine sample size. Air sampling is limited by the concentrations of aerosol, sampling time, and capacity of the collection medium. Typical sample sizes range from 10 mg to 10 g.

C. INAA Irradiation, Counting Procedures, and Sensitivities

In INAA studies, two different irradiations are routinely used: first, a short irradiation (seconds

HARBOR FREEWAY, SEPT. 20, 1972

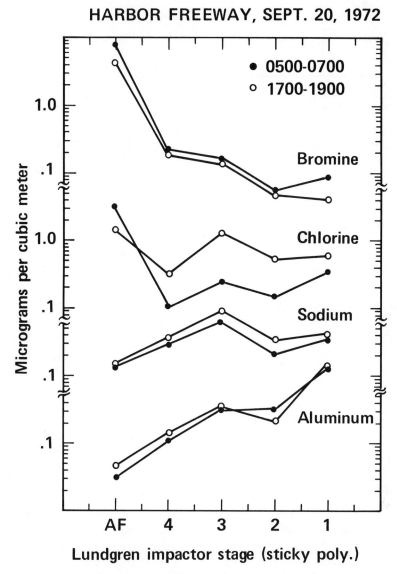

FIGURE 7. Impactor distributions of Br, Cl, Na, Al collected with a Lundgren rotating drum impactor at the Harbor Freeway in Los Angeles. (From Ragaini, R. C., Ralston, H. R., Garvis, D., and Kaifer, R., Report UCRL-51850, Lawrence Livermore Laboratory, Livermore, Cal., June 1975.)

to minutes) is used to produce radionuclides with half-lives between 2 min and several hours. Gamma-ray spectra of the sample are then taken after a several minute cooling period. After about 15 days, the same sample is irradiated again for several hours to produce radionuclides of long half-lives. Several gamma-ray spectra are taken over a period of weeks to emphasize radionuclides with half-lives ranging from hours to months. Figures 3A through E show five γ-ray spectra of a 24-hr high-volume air filter: two spectra (3A and B) taken after a short irradiation (2 min) and three spectra (3C through E) taken after a long irradia-

tion (12 hr) in a thermal neutron reactor flux of 5 × 10^{12} n/cm^2/sec.[33] Table 2 shows the irradiation and counting schemes used in Reference 13 for INAA with a 30-cc Ge(Li) detector of up to 33 elements in atmospheric particles. Table 3 lists the nuclear data relating to the measurement of air filter samples by INAA.[46]

Detection sensitivities for multi-element INAA of atmospheric particles are not fixed, but are dependent on the composition of the particles and on the collection, irradiation, and counting procedures. Estimates of the minimum detectable

TABLE 2

Irradiation Time and Counting Procedure for INAA of Atmospheric Particles

Irradiation time[a]	Cooling time	Count time[b] (sec)	Elements observed
5 min	3 min	400	Al, S, Ca, Ti, V, Cu
	15 min	1000	Na, Mg, Cl, Mn, Br, In, I
2–5 hr	20–30 hr	2000	K, Cu, Zn, Br, As, Ga, Sb, La, Sm, Eu, W, Au
	20–30 day	4000	Sc, Cr, Fe, Co, Ni, Zn, Se, Ag, Sb, Ce, Hg, Th

[a]In fluxes 2×10^{12} n/cm²/sec.
[b]On a 30 cc. Ge(Li) detector.

Data from Dams, R., Robbins, J. A., Rahn, K. A., and Winchester, J. W., *Anal. Chem.*, 42, 256, 1970.

concentrations of 37 elements in urban aerosols as measured by INAA are given in Table 4.[13] These are based on a 24-hr sampling period at a sampling rate of 1 liter/min/cm² with 25-mm diameter circular polystyrene filters using the irradiation/counting schedule shown in Table 2. The INAA method is sensitive enough to measure most of the metal and nonmetal pollutants of interest. With the short irradiation procedure Al, Mg, Ti, V, Cu, Cl, Na, Mn, Br, I, Ba, and In can be measured. With the long irradiation procedure one can also measure As, W, Ga, K, Mo, Sm, Au, Hg, La, Sb, Zn, Fe, Cr, Co, Se, Cd, Ag, Ce, Cs, Eu, Sc, Th, Ni, Ta, Hf, Rb. Lead and Sn are never observed; S, Cd, Ag, Cu, and Hg are measured occasionally.

Activation analysis with 14-MeV neutrons is used mainly for the determination of light elements for which thermal NAA has relatively poor sensitivity. Van Grieken and Dams[25] have determined silicon in natural and pollution aerosols by INAA using a SAMES® Type J (150 kV, 1.5 mA) 14-MeV neutron generator and counting the ^{28}Al activity from the ^{28}Si(n,p)^{28}Al reaction in a 5- by 5- in NaI(Tl) well-type crystal. With an irradiation, cool, and count periods of 50, 75, and 120 sec, respectively, at an average neutron generator beam intensity of 500 μA (flux $\cong 2 \times 10^9$ n/sec/cm²), they reported a detection limit of 2.0 μg Si. This translates to a sensitivity of 5 μg Si/m³ for a 30 min sampling time using Whatman 41 filters and a high volume pump (20 m³/hr). For a typical 24-hr sampling time in urban air, approximately 50 mg of suspended particles were collected from 400 m³ of air on a 10-cm circular filter. The sensitivity

could be increased to 0.05 μg/m³ for sampling in remote areas by lengthening the sampling time.

Persiani[24] has reported INAA of Si, Al, Cl, and Hg using 14-MeV neutrons from a 200-kV Cockcroft-Walton accelerator TMC Model 211. The experimental conditions with a 14-MeV neutron flux of 9×10^9 n/cm²/sec are summarized in Table 5.

Janssens et al.[47] have determined U, Sb, In, Br, and Co in atmospheric aerosols using epithermal neutron activation and a low-energy photon detector. A 24-hr air sampling was carried out by passing about 400 m³ of air through an 11-cm diameter Whatman 41 cellulose filter and collecting about 40 mg of airborne dust. Pellets of 12.7×3 mm were pressed from one half of a filter. The flux monitor and sample were wrapped in Mylar foils and placed in a Cd box to absorb the thermal neutron flux, and a pneumatic rabbit containing the Cd box was irradiated for 0.5 min in a nuclear reactor. Irradiations were performed in a thermal flux of 2.6×10^{12} n/cm²/sec, an epithermal flux of 1.1×10^{11} n/cm²/sec, and a fast flux of 6.5×10^{11} n/cm²/sec. A high-resolution low-energy photon detector (LEP) was used to count the low-energy γ-rays. Table 6 contains the relevant information on the irradiation and counting procedures. With this approach the detection limit for U is reported as 5 to 10 ng, and is limited by the filter blank content.

Silver iodide is a commonly used agent in cloud-seeding experiments. With Ag air concentrations extremely low (0.1 ng/m³), sensitive analytical techniques are very important for the

TABLE 3

Nuclear Data Relating to the Measurement of the Elements in Air Filter Samples by Neutron Activation Analysis

Irradiation Parameters $\Big\{$ Irradiation Interval – 0.5–1.0 min
Neutron Flux – 1 × 10^{13} Neutrons per Square Centimeter per Second

Element	Target isotope	Isotopic abundance (%)	Product nuclide	Half-life	Thermal neutron cross section	Best γ for measurement (KeV)	Number of γ's per 1000 decays	Associated γ-rays KeV (γ's/1000 disintegrations)	Possible interfering nuclide in diode measurement	Possible interfering nuclear reactions producing nuclides of interest
Al	^{27}Al	100	^{28}Al	231 min	0.24 b	1779	1000	None	^{151}Nd(1776)	^{28}Si(n,p)^{28}Al ^{31}P(n,α)^{28}Al
Br	^{79}Br	50.5	^{80}Br	17.6 min	8.5 b	617	70	666(10)	^{108}Ag(614)	^{80}Kr(n,p)^{80}Br
Cl	^{37}Cl	24.5	^{38}Cl	37.3 min	0.4 b	1643	380	2170(470)	None	^{38}Ar(n,p)^{38}Cl ^{41}K(n,α)^{38}Cl
Mg	^{26}Mg	11.2	^{27}Mg	9.45 min	0.027 b	844 1014	700 300	180(7)	^{101}Mo(840) ^{151}Nd(841) ^{87}Kr(846) ^{56}Mn(847) ^{101}Mo(1012) ^{151}Nd(1016) ^{125}Sn(1017) ^{188}Re(1018)	^{27}Al(n,p)^{27}Mg ^{30}Si(n,α)^{27}Mg
Mn	^{55}Mn	100	^{56}Mn	2.58 hr	13.3 b	847 1811	990 290	2110(150)	^{27}Mg(844) None	^{56}Fe(n,p)^{56}Mn ^{59}Co(n,α)^{56}Mn
Na	^{23}Na	100	^{24}Na	15.0 hr	0.53 b	1369	1000	2754(1000)	^{125}Sn(1369)^{117}Cd(1373) ^{188}Re(1368)	^{24}Mg(n,p)^{24}Na ^{27}Al(n,α)^{24}Na
Ti	^{50}Ti	5.3	^{51}Ti	5.79 min	0.14 b	320	950	605(15) 928(50)	^{199}Pt(317) ^{151}Nd(320) ^{105}Ru(317)^{182m}Ta(318)	^{51}V(n,p)^{51}Ti ^{54}Cr(n,α)^{51}Ti
V	^{51}V	99.8	^{52}V	3.76 min	4.9 b	1434	1000	None	^{139}Ba(1430)	^{52}Cr(n,p)^{52}V ^{55}Mn(n,α)^{52}V
As	^{75}As	100	^{76}As	26.5 hr	4.5 b	559	430	1220(50) 1440(7)	^{193}Os(558) ^{114m}In(558) ^{82}Br(554) ^{76}As(562) ^{122}Sb(564)	^{76}Se(n,p)^{76}As ^{79}Br(n,α)^{76}As
Br	^{81}Br	49.48	^{82}Br	35.9 hr	3 b	657 777	60 830	1789(3) 2100(3) 554(660) 619(410) 698(270) 828(250) 1044(290) 1317(260) 1475(170)	^{110m}Ag(657) ^{187}W(772) ^{76}As(775) ^{152}Eu(779) ^{97}Mo(777) ^{131m}Te(774)	^{82}Kr(n,p)^{82}Br ^{85}Rb(n,α)^{82}Br
Co	^{59}Co	100	^{60}Co	5.24 year	37 b	1173 1332	1000 1000	None	^{188}Re(1176) ^{160}Tb(1178) ^{79}Kr(1332)	^{60}Ni(n,p)^{60}Co ^{63}Cu(n,α)^{60}Co
Cr	^{50}Cr	4.31	^{51}Cr	27.8 day	17 b	320	90	None	^{192}Ir(317) ^{177m}Lu(319) ^{147}Nd(319) ^{193}Os(321) ^{177}Lu(321) ^{177}Lu(321)	^{54}Fe(n,α)^{51}Cr
Eu	^{151}Eu	47.77	^{152}Eu	12.7 year	5900 b	1408	220	122(370) 245(80) 344(270) 779(140) 965(150) 1087(120) 1113(140)	None	^{152}Gd(n,p)^{152}Eu
Fe	^{58}Fe	0.31	^{59}Fe	45 day	1.1 b	1099 1292	560 440	143(8) 192(28)	^{160}Tb(1103) ^{182}Ta(1289) ^{115m}Cd(1290) ^{152}Eu(1292)	^{59}Co(n,p)^{59}Fe ^{62}Ni(n,α)^{59}Fe
Hf	^{180}Hf	35.22	^{181}Hf	43 day	10 b	482	810	133(480) 346(130)	^{192}Ir(485) ^{140}La(487)	^{181}Ta(n,p)^{181}Hf ^{184}W(n,α)^{181}Hf
Hg	^{202}Hg	29.8	^{203}Hg	46.9 day	4 b	279	770	None	^{197m}Hg(279) ^{75}Se(280) ^{193}Os(280)	^{203}Tl(n,p)^{203}Hg ^{206}Pb(n,α)^{203}Hg
La	^{139}La	99.91	^{140}La	40.3 hr	8.9 b	1596	960	329(200) 487(400) 815(190) 923(100) 2530(90)	^{72}Ga(1596) ^{154}Eu(1596)	^{140}Ce(n,p)^{140}La

113

TABLE 3 (continued)

Nuclear Data Relating to the Measurement of the Elements in Air Filter Samples by Neutron Activation Analysis

Irradiation Parameters $\Big\{$ Irradiation Interval — 6—8 hrs / Neutron Flux — 5×10^{12} Neutrons per Square Centimeter per Second

Element	Target isotope	Isotopic abundance (%)	Product nuclide	Half-life	Thermal neutron cross section	Best γ for measurement (KeV)	Number of γ's per 1000 decays	Associated γ-rays KeV (γ's/1000 disintegrations)	Possible interfering nuclide in diode measurement	Possible interfering nuclear reactions producing nuclides of interest
Na	^{23}Na	100	^{24}Na	15 hr	0.53 b	1369	1000	2754(1000)	^{134}Cs(1365)	^{24}Mg(n,p)^{24}Na ^{27}Al(n,α)^{24}Na
Sb	^{123}Sb	42.75	^{124}Sb	60.2 day	3.3 b	1691	500	602(980) 723(98) 2091(69)	None	^{124}Te(n,p)^{124}Sb ^{127}I(n,α)^{124}Sb
Sc	^{45}Sc	100	^{46}Sc	83.8 day	13 b	889 1120	1000 1000	None	^{110m}Ag(884) ^{192}Ir(884) ^{65}Zn(1115) ^{182}Ta(1121) ^{131m}Te(1125)	^{46}Ti(n,p)^{46}Sc
Se	^{74}Se	0.87	^{75}Se	120 day	30 b	265 280	600 250	97(33) 121(170) 136(570) 401(120)	^{169}Yb(261) ^{115}Cd(263) ^{182}Ta(264) ^{203}Hg(279) ^{133}Ba(296) ^{192}Ir(283)	^{78}Kr(n,α)^{75}Se
Sm	^{152}Sm	26.6	^{153}Sm	47.1 hr	210 b	103	280	70(54)	^{182}Ta(100)	^{153}Eu(n,p)^{153}Sm ^{156}Gd(n,α)^{153}Sm
Ta	^{181}Ta	99.99	^{182}Ta	115 day	21 b	68 1221	420 270	100(140) 152(70) 222(80) 1122(340) 1189(160) 1231(130)	^{169}Yb(63) ^{121}Te(65) ^{182}Ta(65) ^{75}Sc(66) None	^{182}W(n,p)^{182}Ta ^{185}Re(n,α)^{182}Ta
Tb	^{159}Tb	100	^{160}Tb	72 day	46 b	879	310	87(120) 299(300) 966(310) 1178(150)	^{185}Os(879)	^{160}Dy(n,p)^{160}Tb
Th	^{232}Th	100	^{233}Pa	27.0 day	7.4 b	312	270	300(50)	^{192}Ir(317)	None
Zn	^{64}Zn	48.89	^{65}Zn	243 day	0.46 b	1116	506	511(500β^{+})	^{129m}Te(1112) ^{152}Eu(1212) ^{182}Ta(1113) ^{160}Tb(1115) ^{46}Sc(1126)	None

From Rancitelli, L. A., Cooper, J. A., and Perkins, R. W., Paper IAEA-SM-175/24, in *Comparative Studies of Food and Environmental Contamination*, International Atomic Energy Agency, Vienna, Austria, 1974, 438. With permission.

TABLE 4

Sensitivities for Determination of Trace Elements in Aerosols by INAA

Element	Detection limit[b] (μg)	Minimum detectable concentration in urban air ($\mu g/m^3$), 24-hr sample[b]
Al	0.04	0.008
S	25.0	5.0
Ca	1.0	0.2
Ti	0.2	0.04
V	0.001	0.002
Cu	0.1	0.02
Na	0.2	0.04
Mg	3.0	0.6
Cl	0.5	0.1
Mn	0.003	0.0006
Br	0.02	0.004
In	0.0002	0.00004
I	0.1	0.02
K	0.075	0.0075
Cu	0.05	0.005
Zn	0.2	0.02
Br	0.025	0.0025
As	0.04	0.004
Ga	0.01	0.001
Sb	0.03	0.003
La	0.002	0.0002
Sm	0.00005	0.000005
Eu	0.0001	0.00001
W	0.005	0.0005
Au	0.001	0.0001
Sc	0.003	0.000004
Cr	0.02	0.00025
Fe	1.5	0.02
Co	0.002	0.000025
Ni	1.5	0.02
Zn	0.1	0.001
Se	0.01	0.0001
Ag	0.1	0.001
Sb	0.08	0.001
Ce	0.02	0.00025
Hg	0.01	0.0001
Th	0.003	0.00004

[a] Irradiation and counting conditions listed in Tables 2 and 3.
[b] Air sampling rate of 12 l/min/cm^2.

determination of Ag as part of these experiments. Albini et al.[48] have published a method for nondestructive determination of silver traces in atmospheric particles collected on cellulose filters (Whatman 41) based on thermal neutron activation followed by β-counting of the ^{110}Ag 24.4-sec activity. A reactor thermal neutron flux of 3.5 $\times$ 10^{11} n/cm^2/sec was used for irradiation. The counting apparatus consisted of two thin-window 2π β-counters. Irradiation time was 2 min, cooling time was 40 sec. The decay of the samples was followed with a multiscaler for 2000 sec. The decay curves were resolved by a computer code. The only appreciable interfering radionuclide with a half-life similar to that of ^{110}Ag was 27.1-sec ^{19}O formed by the ^{18}O (n,γ)^{19}O reaction. The determination limit for a 10% confidence limit was estimated at 10^{-9} g. The determination limit using the method of counting the 657.5-keV γ-ray of ^{110}Ag with a Ge(Li) detector was also reported as 6 $\times$ 10^{-8} g. This method is applicable to the determination of Ag in cloud-seeding experiments, where the samples are generally collected in nonindustrial areas, and where the interfering element concentrations are reasonably low.

D. NAA of Aerosols Utilizing Radiochemical Procedures

Although the use of INAA has increased tremendously in recent years, some elements cannot be detected without the use of radiochemical separations. Elements may not be detectable because their activation products may be obscured by other trace elements, or by the activation of major elements. In either case, radiochemical separations may be necessary.

In environmental samples, the ^{24}Na activities of irradiated samples are sometimes high enough to mask other nuclides of similar or shorter half-lives.[23] Girardi and Sabbioni[51] have shown that hydrated antimony pentoxide (HAP) is highly selective for removing ^{24}Na from 8 M H$_2$SO$_4$.

Radiochemical separations may be carried out for individual elements or for groups of elements. Detailed procedures on individual elements may be found in the Nuclear Science Series *The Radiochemistry of the Elements*.[49] Group separations are suitable for use with the high-resolution Ge(Li) spectrometer systems. The separation techniques can include precipitation, distillation, inorganic separators, ion exchange, chromatography, paper and thin-layer chromatography, and solvent extraction.[50]

The determination of Hg in air with high sensitivity (1 ng/m^3) is important in air pollution

TABLE 5

Sensitivities for Trace Elements in Aerosols with 14-MeV Neutrons

		Irradiation					
Element	Isotope	Flux (n/cm²/sec)	Time (min)	Delay time (min)	Count time (min)	Sensitivity (μg)	Ref.
Si	^{28}Al	2×10^9	0.83	1.25	2	3	25
Si	^{28}Al	9×10^9	10	0.5	5	5	24
Al	^{27}Mg	9×10^9	10	10	10	20	24
Cl	^{37}S	9×10^9	10	0.5	10	20	24
Hg	^{199m}Hg	9×10^9	10	20—40	20	5	24

TABLE 6

Nuclear Data for Trace Elements in Aerosols with Epithermal Neutrons

Element	Isotope	$T_{1/2}$	$I\gamma/\sigma th^a$	γ-Ray energy (KeV)
Br	^{80m}Br	4.5 hr	15	37.0
	^{82m}Br	6.1 m	17.4	46.0
Co	^{60m}Co	10.5 m	1.92	58.6
Sb	^{122m}Sb	4.2 m	32.7	61.6
		4.2 m	32.7	76.3
U	^{239}U	23.5 m	97	74.0
In	^{116m}In	54 m	13.6	138.3

Note: Epithermal flux = 1.1×10^{11} n/cm²/sec; irradiation time = 5 min; cooling time = 3 min; and count time = 6 min.

[a]Ratio of resonance integral to thermal neutron activation cross section.

From Janssens, M., Desmet, B., Dams, R., and Hoste, J., *J. Radioanal. Chem.*, 26, 305, 1975. With permission.

studies. Total gaseous Hg in air has been determined by concentrating Hg on purified activated charcoal followed by NAA.[52] With a flow rate below 5 l/min, 1.5 to 2 m³ of prefiltered air was sampled. The charcoal was irradiated in a thermal neutron flux of 5×10^{12} n/cm²/sec for 12 hr in a quartz capsule. After irradiation, the Hg was separated from the radioactive matrix by evaporation and absorption on a second charcoal absorbent. After 24 hr cooling, the 0.770 MeV γ-ray of ^{197m}Hg was counted with a well-type NaI(Tl) crystal. Concentrations of Hg down to 0.5 ng/m³ have been determined using this method.[52] The loss of Hg by diffusion during irradiation of Hg (II) solutions in polyethylene containers has been

previously discussed.[23] Several explanations have been proposed for this effect. This problem is avoided by irradiation in quartz vials where there is no Hg loss.

The concentrations of Cd and Cu in atmospheric aerosols are usually too low to measure by INAA. Post irradiation radiochemical separations of the air filters are necessary to improve the detection sensitivities. Since the aerosol particles are mostly resuspended soils by weight, the required radiochemical separations are similar to those used for RNAA of soils. The filters with their collected particles are dissolved by digestion in H_2SO_4-HF or HNO_3-HF. Once the samples are in solution, the specific elemental separation procedures may be carried out.[23]

In INAA procedures, Ni is determined by the ^{58}Ni(n,p) ^{58}Co (T = 71 day) reaction using the 810 keV γ-ray of ^{58}Co. However, this method is not very sensitive, with upper limits of 0.02 μg/m³.[13] Das et al.[53] have proposed a RNAA procedure using the ^{64}Ni (n,γ) ^{65}Ni(T = 2.56 hr) reaction. Aerosols were collected by filtering 250 to 2500 m³ of air in 24 hr through a 22- by 13-cm Microsorban® filter yielding about a 150 mg sample, of which 2.5 mg was used for the Ni analysis. The sample was irradiated for 30 min in a thermal flux of 5×10^3 n/sec/cm². The filter paper was wet ashed in HNO_3 and the Ni was extracted by liquid-to-liquid extraction with chloroform (as the dimethylglyoxime complex) from a slightly alkaline 50% ammonium citrate (or Tartrate) solution. After back-extraction with HCL the solution was passed through a Dowex® anion-exchange column to eliminate any traces of ^{61}Cu. Using this procedure the sensitivity is 0.02 μg Ni and 0.005 μg Ni/m³ sampling 1000 m³ of air.

Schramel et al.[54] applied a chemical separation to determine As, Br, Cd, Co, Cr, Cu, Hg, La, Sb, Se, Sn, and Zn in atmospheric particles. The samples were irradiated for 24 hr in a thermal flux of 1×10^{13} n/cm^2/sec, and then were wet ashed with a mixture of H_2SO_4, HNO_3, and H_2O_2. Volatile bromides of As, Hg, Sb, Se, and Sn were distilled by the addition of HBr. The volatile bromide group can be futher separated in Hg, Sb, Sn and As, Se groups by anion exchange of the halogen complexes formed in HCL and HBr. After removal of silica as SiF_4, the undistilled trace elements in the H_2SO_4 solution, Cd, Co, Cr, Cu, La, and Zn were separated from alkalies, alkaline earths, and phosphorous by selective absorption on a Bio Rad-Chelex® 100 ion-exchange resin. After the separations each of four groups were counted for 15 min with a Ge(Li) detector: (1) Hg, Sb, Sn; (2) As, Se; (3) Br; (4) Cd, Co, Cr, Cu, La, and Zn.

The atmospheric halogens are part of the marine air cycle, and they are significant atmospheric constituents in areas where automobile exhausts predominate due to the combustion emission of volatile lead halides. Halogens may also play a role in artificial ice nuclei formation. Moyers and Duce[55] have reported a procedure for the determination of gaseous Br and I using NAA involving the absorption of gaseous I and Br on purified activated charcoal. At a sampling rate of 1 to 1.5 m^3/hr, air was pulled through an electrostatic precipitator to remove particles and across a 1.5-g bed of purified activated charcoal. Sampling times were 4 to 6 hr for nonurban marine air and 2 to 3 hr for urban polluted air. The charcoal was irradiated in a thermal flux of 5×10^{12} n/cm^2/sec for 20 min. After irradiation, the I and Br were removed from the charcoal by digestion with NaOH and purified by using selective oxidizing and reducing agents, solvent extraction, and precipitation as AgX. The amounts of I and Br were then determined by β-counting on a low background, anticoincidence, gas-flow proportional counter. The sensitivities were 0.12 ng of Br and 0.15 ng of I.

VI. ATMOSPHERIC ENVIRONMENTAL STUDIES USING NAA

The focus of atmospheric environmental research has been to understand the impact of pollution on our health and our environment.

Since these impacts are related to the chemical composition of aerosols, NAA groups have been able to contribute in this area because of the accuracy and sensitivity of NAA trace element analyses. The NAA studies can be generally classified in five groups:

1. Pollution source studies
2. Urban aerosol studies
3. Background aerosol studies
4. Tracer studies
5. Atmospheric chemistry studies

A. Pollution Source Studies

An example of a pollution source study is a study of northwest Indiana aerosols.[56] Twenty-five filter samples were run for 24 hr near and around the northwest Indiana industrial area at urban and nonurban sites. Each filter was analyzed by INAA for 30 elements, and area-wide distribution patterns were established. At urban sites the Fe, Mn, Zn, Sb, Cr, W, Co, Sc, La, Ce, Th, Ca, Mg concentrations were higher than the concentration at nonurban sites. This increase was apparently associated with the steel-producing industries. Selenium, on the other hand, was evenly distributed over the same area. Selenium is geochemically associated with S and is a fossil-fuel combustion effluent.

Because of the potential biomedical and environmental hazards associated with release of potentially toxic substances during coal combustion, several groups using INAA techniques have sought to characterize particulate emissions from coal-fired power plants.[57-60] Results of these studies indicate that elemental concentrations are enriched inversely with particle diameter through a vaporization-surface condensation mechanism and through preferential penetration of smaller particles through emission-control devices. In addition, considerations of mass balances indicate that emission of several species (including Hg, As, Se, Br, I) have a significant vapor-phase component at stack temperatures.

Elemental particulate concentrations are compared by normalizing to a nonvolatile reference element, such as Al, Si, or Fe. This is defined as the elemental ratio ER where

$$ER = (X/Al)_{sample} \qquad (4)$$

These ratios are independent of dilution by air. A more revealing ratio is the comparison of the sample elemental ratio to the same ratio in a reference material. This is defined as the enrichment factor (EF) where

$$EF = \frac{ER_{sample}}{ER_{reference}} = \frac{(X/Al)_{sample}}{(X/Al)_{reference}} \qquad (5)$$

If the reference material has an elemental composition similar to that of the source under study, then the elemental EF values should equal one. Enrichment factor values much greater or much less than one indicate enrichment or depletion processes, respectively, taking place in the source.

Studies were carried out at a western U.S. coal-fired power plant using parallel electrostatic precipitator (ESP) and wet scrubber control devices.[58,59] In scrubber emissions, 95% of the mass of insoluble species was associated with submicron particles. Particulate emissions from the ESP unit were associated with much larger particles. Preliminary results indicate that concentrations of As, Se, Zn, Sb, Mo, Ga, W, V, U, and Ba on fly ash were enriched relative to coal. The enrichments for these elements ranged from ~1.5 to ~3.0 for stack emissions, with similar enrichments in the plume.

Figures 8 and 9 show enrichment factors for Sc and As, respectively, for particulate in-stack samples at 5 to 10 min, and at 10–20 min in the plume. Scandium represents the behavior of other nonvolatile elements Na, K, Ca, and the rare earths, which are predominantly associated with large particles emitted from the precipitator units. Enrichment factors for these elements tended to be near 1.0 in ESP and plume samples. In addition, there appeared to be a decrease in the enrichment factor of particles less than 3 μm in diameter as the distance from the plant increased.

Enrichment factor vs. particle size curves for As, shown in Figure 9 were similar to those of Ba, Se, Sb, and Ga. These elements were highly enriched on smaller particles leaving the ESP; however, enrichment factors for the fine particles decreased with increasing distance from the plant. This decrease in enrichment factor can probably best be explained in terms of heterogenous coagulation. For elements that vaporize during combustion and later condense on the surface of particles, enrichments are greatest on the smallest particles. A decrease in the small-particle enrichment factor results from preferential removal by large particles of the smallest, most highly enriched particles present in the fine-particle distribution. Mass

median diameters for virtually all of the elements decreased with increasing distance from the plant, primarily as a result of sedimentation of the larger particles.

In summary, it appears that the major transformations observed were physical in nature. For liquid aerosols emitted from scrubber units, the major observable change appears to have been evaporation of moisture to form submicron aerosols highly enriched in several metals. For aerosols emitted from precipitator units, size distributions and enrichment factors of trace elements appear to have been modified primarily by coagulation and sedimentation rather than by processes involving evaporation or condensation of volatile species. Based on filter experiments, major changes in the enrichment factor for total suspended particulate matter were apparently due primarily to sedimentation of large particles, almost immediately after release into the atmosphere. Furthermore, changes in the enrichment factors observed in total suspended particles were relatively small.

Other pollution-source studies using NAA include studies of smelters,[62,63] agricultural field burning,[64] jet aircraft emissions,[65] and motor vehicle studies.[66]

B. Urban Aerosol Studies
A clear understanding of urban pollution is necessary for controlling local air quality and for determining the impact of urban pollution plumes on nonurban areas. Recent studies of urban pollutant sources have been coupled with studies on nonurban receptor sites. Urban studies using INAA include Heidelberg,[67] Boston,[68] Livermore, California,[69] San Francisco,[70] Cleveland,[71,72] Chicago,[10,11] Cambridge, Massachusetts,[12] Toronto,[76] Fairbanks,[73] St. Louis,[74] Ann Arbor,[75] and Ghent.[77] Hopke et al.[78] applied multivariate analysis to the concentrations of 18 elements determined by INAA to identify sources of selected elements on the Boston aerosol.

C. Background Aerosol Studies
Rahn[44] has measured the aerosol compositions at five remote sites in central and western Canada for a 3-month period using cascade impactors and filter samples and INAA. Even though INAA is a sensitive method, 1000 to 1500 m³ of air had to be sampled because of the extremely clean air. Aluminum was found at all stations at similar

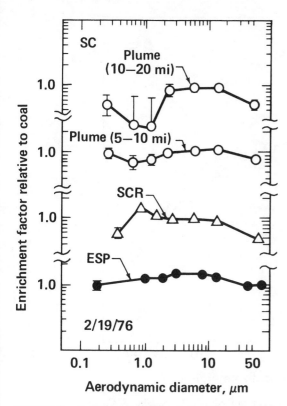

FIGURE 8. Enrichment factor as a function of particle size for scandium in fly ash collected in stack and downwind from a coal-fired power plant using an ESP emission control system. (From Ondov, J. M., Ragaini, R. C., Biermann, A. H., Choquette, C. E., Gordon, G. E., and Zoller, W. H., Report UCRL-78825, Lawrence Livermore Laboratory, Livermore, Cal., May 1977, 655.)

concentrations and on large particles, due to its origin in resuspended soil dust. However, As, Se, In, and Hg, Cu, Zn, and Sb were found concentrated on small particles, suggesting some vapor-to-particle conversion either at the source of during atmospheric transport. However, neither anthropogenic nor natural sources for these elements have yet been demonstrated. Similar results have been found at the South Pole,[79,80] pointing to possible global atmospheric environmental problems.

D. Transport/Tracer Studies

Atmospheric transport research depends on tracer studies. The direct injection of tracer material such as fluorescent dyes or radioactive gases has been used to trace the dispersal of atmospheric pollutants. Of course, the trace elements that occur in source effluents can be used as tracers. However, for meteorological model testing, difficulties of emission rate variations and of ubiquitous industrial elemental emission patterns can be overcome by using artificial activable trace elements. Indium chloride has been used as an aerosol tracer for precipitation scavenging investigations.[81,82] Indium was determined with RNAA by means of detection of 54-min ^{116m}In with a sensitivity of about 0.1 ng, which is below natural background of 1 to 2 ng/l rain water samples.

In a recent tracer study, chlorides or oxides of one of the rare earths La, Ce, Sm, and Dy were simultaneously injected into each of three Albany, Oregon industrial plants to study meteorological dispersion models.[83] The relative concentrations of these elements were determined at various downwind sampling locations. These elements were chosen since their background levels are negligible; they are easily detected by INAA (sensitivities as low as 10^{-12} g); they should show similar chemical and physical behavior; and the 20 to 40 g release amounts are nontoxic to man and the environment. These tracers were used to test the Gaussian plume model in the generalized and trapped plume forms.

E. Atmospheric Chemistry Studies

The potential for anthropogenic operations to cause large-scale atmospheric chemical effects has focused attention on studies of chemical and physical transformations of natural and man-made pollutants. Pillay et al.[84] have determined that NAA of Se in the atmosphere can be used as an indicator of the levels of total S pollutants. Additionally, the Se/S ratio measurements can be used to distinguish the S pollution caused by coal combustion products from that of crude oil combustion. The advantage of this technique is that NAA for Se can be performed by filtering 10 to 20 m^3 of air, while much larger volumes are required for sulfur analysis.

Since Pb, Br, and Cl have been shown to originate from automobile exhausts, there have been many studies of these elements in urban[85-87] and marine[88] environments. Bromine and Cl are usually analyzed by NAA and Pb is analyzed by other techniques. Because of its high sensitivity, INAA has also been used to study V concentrations in urban and nonurban aerosols.[89,90] Zoller et al.[90] have concluded that V

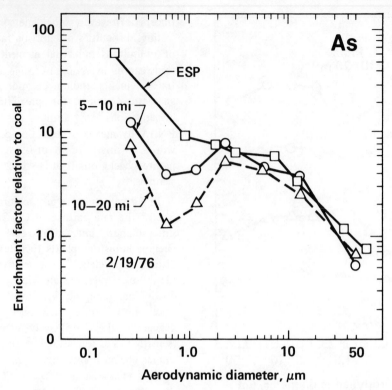

FIGURE 9.   Enrichment factor for arsenic. (From Ondov, J. M., Ragaini, R. C.,
Biermann, A. H., Choquette, C. E., Gordon, G. E., and Zoller, W. H., Report
UCRL-78825, Lawrence Livermore Laboratory, Livermore, Cal., May 1977, 656.)

can be used as a useful indicator of the anthropogenic
activities, particularly in the northeastern U.S.,
where residual fuel combustion is the only significant source of atmospheric V. Duce and Hoffman[91] have further concluded that most of the V
present in the northern hemisphere westerlies over
the Atlantic and Pacific Oceans is from anthropogenic sources.

REFERENCES

1.   Junge, C. E., *Air Chemistry and Radioactivity,* Academic Press, New York, 1963.
2.   Israël, H. and Israël, G. W., *Trace Elements in the Atmosphere,* Ann Arbor Science, Ann Arbor, Mich., 1974.
3.   Ragaini, R. C., Ralston, H. R., Garvis, D., and Kaifer, R., Trace Elements in California Aerosols. Part 1. Instrumental
     neutron activation analysis techniques, Report UCRL-51850, Lawrence Livermore Laboratory, Livermore, Cal.,
     June 1975.
4.   Duce, R. A. and Winchester, J. W., Determination of iodine, bromine, and chlorine in atmospheric samples by
     neutron activation analysis, *Radiochim. Acta,* 4, 100, 1965.
5.   Duce, R. A., Winchester, J. W., and Van Nahl, T. W., Iodine, bromine, and chlorine in the Hawaiian marine
     atmosphere, *J. Geophys. Res.* 70, 1775, 1965.
6.   Duce, R. A., Winchester, J. W., and Van Nahl, T. W., Iodine, bromine, and chlorine in winter aerosols and snow
     from Barrow, Alaska, *Tellus,* 238, 1966.
7.   Lininger, R. L., Duce, R. A., Winchester, J. W., and Matson, W. R., Chlorine, bromine, iodine, and lead in aerosols
         from Cambridge, Massachusetts, *J. Geophys. Res.*, 71, 2457, 1966.
8.   Winchester, J. W. and Duce, R. A., Coherence of iodine and bromine in the atmosphere of Hawaii, northern Alaska,
     and Massachusetts, *Tellus,* 18, 281, 1966.

9.   Keane, J. R. and Fisher, E. M. R., Analysis of trace elements in airborne particulates by neutron activation and gamma-ray spectrometry, *Atmos. Environ.*, 2, 603, 1968.
10. Brar, S. S., Nelson, D. M., Kanabrocki, E. L., Moore, C. E., Burnham, C. D., and Hattori, D. M., Thermal neutron activation analysis of airborne particulate matter in Chicago Metropolitan Area, in *Modern Trends in Activation Analysis*, Vol. 1, Devoe, J. R., Ed., National Bureau of Standards Special Publication No. 312, U.S. Government Printing House, Washington, 1969, 43.
11. Dudey, N. D., Ross, L. E., and Noshkin, V. E., Thermal neutron activation analysis of airborne particulate matter in Chicago Metropolitan Areas, in *Modern Trends in Activation Analysis*, Vol. 1, Devoe, J. R., Ed., National Bureau of Standards Special Publication No. 312, U.S. Government Printing House, Washington, 1969, 55.
12. Zoller, W. H. and Gordon, G. E., Instrumental neutron activation analysis of atmospheric pollutants using Ge(Li) γ-ray detectors, *Anal. Chem.*, 42, 256, 1970.
13. Dams, R., Robbins, J. A., Rahn, K. A., and Winchester, J. W., Nondestructive neutron activation analysis of air pollution particulates, *Anal. Chem.*, 42, 861, 1970.
14. Pillay, K. K. S., Thomas, C. C., Jr., and Hyche, C. M., Neutron activation analysis of inorganic constituents of airborne particulates, *Nucl. Technol.*, 10, 224, 1971.
15. Rakovic, M., *Activation Analysis*, Iliffe Books, London, 1970.
16. Kruger, P., *Principles of Activation Analysis*, John Wiley & Sons, New York, 1971.
17. DeSoete, D., Gijbels, R., and Hoste, J., *Neutron Activation Analysis*, John Wiley & Sons, New York, 1971.
18. Nargolwalla, S. S. and Przybylowicz, E. P., *Activation Analysis with Neutron Generators*, John Wiley & Sons, New York, 1973.
19. Tölgyessy, J. and Varga, S., *Nuclear Analytical Chemistry*, Vol. 1–3, University Park Press, Baltimore, 1974.
20. Amiel, S. Ed., *Nondestructive Activation Analysis*, Elsevier, Amsterdam, 1977.
21. Hoste, J., Op de Beeck, J., Gijbels, R., Adams, F., Van Den Winkel, P., and De Soete, D., *Instrumental and Radiochemical Activation Analysis*, CRC Press, Cleveland, 1971.
22. Valkovic, V., *Trace Element Analysis*, Taylor and Francis, London, 1975.
23. Robertson, D. E. and Carpenter, R., Neutron Activation Techniques for the Measurement of Trace Metals in Environmental Samples, NAS-NS-3114, National Technical Information Science Department of Commerce, Springfield, Mass., 1974.
24. Persiani, C., The detection of selected air pollutants using a laboratory neutron generator, *J. Am. Ind. Hyg. Assoc.*, p. 573, September 1971.
25. Van Grieken, R. and Dams, R., Determination of silicon in natural and pollution aerosols by 14-MeV neutron activation analysis, *Anal. Chim. Acta*, 63, 369, 1973.
26. Lenihan, J. M. and Thomson, S. J., Eds., *Advances in Activation Analysis*, Vol. 1, Academic Press, New York, 1969.
27. Decorte, F., Speecke, A., and Hoste, J., Reactor neutron activation analysis by a triple comparator method, *J. Radioanal. Chem.*, 3, 205, 1969.
28. Lombard, S. M. and Isenhour, T. L., Neutron capture gamma-ray activation analysis using lithium drifted germanium semiconductor detectors, *Anal. Chem.*, 40, 1990, 1968.
29. Flanagan, F. J., U.S. Geological Survey Standards. II. First Compilation of Data for the new U.S.G.S. Rocks, *Geochim. Cosmochim. Acta*, 33, 81, 1969.
30. Becker, D. A. and LaFleur, P. D., Characterization of a nuclear reactor for neutron activation analysis, *J. Radioanal. Chem.*, 19, 149, 1974.
31. Pagden, I. M. H., Pearson, G. J., and Bewers, J. M., An isotope catalogue for instrumental activation analysis, *J. Radioanal. Chem.*, 9, 101, 1972.
32. Perkins, R. W., Rancitelli, L. A., Cooper, J. A., and Brown, R. E., Laboratory and environmental mineral analysis using a Californium-252 neutron source, *Nucl. Appl. Technol.*, 9, 861, 1970.
33. Ragaini, R. C., Heft, R. E., and Garvis, D., Neutron activation analysis at the Livermore pool-type reactor for the environmental research program, Report UCRL-52092, Lawrence Livermore Laboratory, Livermore, Cal., July 1976.
34. Keil, G and Bernt, H., Eds., Proceedings of the second symposium on semiconductor detectors for nuclear radiation, *Nucl. Instrum. Methods*, 101, 1, 1972.
35. Cooper, J. A., Applied Ge(Li) gamma-ray spectroscopy, in *Contemporary Activation Analysis*, Ryan, V. A., Ed., Marcel Dekker, New York, 1973.
36. Cooper, J. A. and Perkins, R. W., A versatile Ge(Li)-NaI(Tl) coincidence-anticoincidence gamma-ray spectrometer for environmental and biological problems, *Nucl. Instrum. Methods*, 99, 125, 1972.
37. Wogman, N. A., Robertson, D. E., and Perkins, R. W., A large well crystal with anticoincidence shielding, *Health Phys.*, 13, 767, 1967.
38. Gunnink, R. and Niday, J. B., The GAMANAL Program, Report UCRL-51061, Lawrence Livermore Laboratory, Parts 1–3, Livermore, Cal., 1972.
39. Dams, R., Rahn, K. A., and Winchester, J. W., Evaluation of filter materials and impaction surfaces for nondestructive neutron activation analysis of aerosols, *Environ. Sci. Technol.*, 6, 441, 1972.
40. Stafford, R. G. and Ettinger, H. J., Filter efficiency vs. particle size and velocity, *Atmos. Environ.*, 6, 353, 1972.
41. Liu, B. Y. H. and Lee, K. W., Efficiency of membrane and nuclepore filters for submicrometer aerosols, *Environ. Sci. Technol.*, 10, 345, 1976.

42. **Duce, R. S., Woodcock, A. H., and Moyers, J. L.,** Variation of ion ratios with sye among particles in tropical oceanic air, *Tellus,* 19, 369, 1967.

43. **Nifong, G. D. and Winchester, J. W.,** Particle-size Distributions of Trace Elements in Pollution Aerosols, Report COO-1705B, University of Michigan, Ann Arbor, Mich., 1970.

44. **Rahn, K. A.,** Sources of Trace Elements in Aerosols – An Approach to Clean Air, Ph.D. thesis, University of Michigan, Report COO-1705-9, 1971.

45. **Gordon, G. E., Gladney, E. S., Ondov, J. M., Conry, T. J., and Zoller, W. H.,** Intercomparison of Several Types of Cascade Impactors, presented at American Chemical Society Division of Environmental Chemistry, Los Angeles, March 1974.

46. **Rancitelli, L. A. and Tanner, T. M.,** Multielement Characterization of Atmosphere Pollutants by X-ray Fluorescence Analysis and Instrumental Neutron Activation Analysis, Proc. Symp. Development of Nuclear Techniques, International Atomic Energy Agency, Vienna, Austria, 1976.

47. **Janssens, M., Desmet, B., Dams, R., and Hoste, J.,** Determination of uranium, antimony, indium, bromine, and cobalt in atmospheric aerosols using epithermal neutron activation and a low-energy photon detector, *J. Radioanal. Chem.,* 26, 305, 1975.

48. **Albini, A., Cesana, A., and Terrani, M.,** Determination of silver traces in air by neutron activation analysis, *J. Radioanal. Chem.,* 34, 185, 1976.

49. Nuclear Science Series, *The Radiochemistry of the Elements,* National Academy of Sciences, Washington, D.C., 1974.

50. **Newton, G. W. A.,** *Radiochemistry,* Vol. 2, The Chemical Society, Burlington House, London, 1975.

51. **Girardi, F. and Sabbioni, E.,** Selective removal of radio-sodium from neutron activated materials by retention on hydrated antimony pentoxide, *J. Radioanal. Chem.,* 1, 168, 1968.

52. **Van der Soot, H. A. and Das, H. A.,** Determination of mercury in air by neutron activation analysis, *Anal. Chim. Acta,* 70, 439, 1974.

53. **Das, H. A., Evendijk, J. E., and De Jong, J. P. M.,** The determination of nickel in air dust by neutron activation analysis, *J. Radioanal. Chem.,* 13, 413, 1973.

54. **Schramel, P., Samsahl, K., and Pavlu, J.,** Determination of 12 selected microelements in air particles by neutron activation analysis, *J. Radioanal. Chem.,* 19, 329, 1974.

55. **Moyers, J. L. and Duce, R. A.,** The collection and determination of atmospheric gaseous bromine and iodine, *Anal. Chim. Acta,* 69, 117, 1974.

56. **Harrison, P. R., Rahn, K. A., Dams, R., Robbins, J. A., Winchester, J. W., Brar, S. S., and Nelson, D. M.,** Area-wide trace metal concentrations measured by multi-element neutron activation analysis: a one-day study in northwest Indiana, *J. Air Pollut. Control Assoc.,* 21, 563, 1971.

57. **Ragaini, R. C. and Ondov, J. M.,** Trace contaminants from coal-fired power plants, in *Proc. Int. Conf. Environmental Sensing Assessment,* Vol. 1, Institute of Electrical and Electronic Engineers, New York, 1976, 17.

58. **Ragaini, R. C. and Ondov, J. M.,** Trace element emissions from western U.S. coal-fired power plants, *Proc. Int. Conf. Modern Trends in Activation Analysis,* Vol. 1, March 1976, 654.

59. **Ondov, J. M., Ragaini, R. C., Biermann, A. H., Choquette, C. E., Gordon, G. E., and Zoller, W. H.,** Elemental Emissions from a Western Coal Fired Power Plant: Preliminary Report on Concurrent Plume and In-stack Sampling, American Chemical Society Meeting, Division of Environmental Chemistry, New Orleans, March 25, 1977; Lawrence Livermore Laboratory Report UCRL-78825, May 1977.

60. **Zoller, W. H., Gladney, E. S., Gordon, G. E., and Bors, J. J.,** Emissions of trace elements from coal-fired power plants, in *Proc. 8th Conf. Trace Elements in Environmental Health,* University of Missouri Press, Columbia, Mo., 1974.

61. **Klein, D. H.,** Pathways of thirty-seven trace elements through a coal-fired power plant, *Environ. Sci. Technol.,* 9, 973, 1975.

62. **Wesolowski, J. J., John, W., and Kaifer, R.,** Lead source identification by multi-element analysis of diurnal samples of ambient air, in *Trace Elements in the Environment,* Advances in Chemistry Series, No. 123, American Chemical Society, Washington, D.C., 1973, 1.

63. **Ragaini, R. C., Ralston, H. R., Roberts, N., Garvis, D., and Langhorst, A.,** Environmental trace metal contamination in Kellogg, Idaho near zinc and lead smelters, Report UCRL-77733, Lawrence Livermore Laboratory, Livermore, Cal., 1976.

64. **Shum, Y. S. and Loveland, W. D.,** Atmospheric trace element concentrations associated with agricultural field burning in the Willamette Valley of Oregon, *Atmos. Environ.,* 8, 645, 1974.

65. **Fordyce, J. S. and Sheibley, D. W.,** Estimate of contributions of jet aircraft operations to trace element concentration at or near airports, *J. Air. Pollut. Control Assoc.,* 25, 721, 1975.

66. **Ondov, J. M., Zoller, W. H. and Gordon, G. E.,** Trace Elements on Aerosols from Motor Vehicles, Paper 74, 67th Annual Meeting of the Air Pollution Control Assoc., Denver, Colo., June 1974.

67. **Bogen, J.,** Trace elements in atmospheric aerosol in the Heidelberg area, measured by instrumental neutron activation analysis, *Atmos. Environ.,* 7, 1117, 1973.

68. **Gladney, E. S., Zoller, W. H., Jones, A. G., and Gordon, G. E.,** Composition and size distribution of atmospheric particulate matter in Boston area, *Environ. Sci. Technol.,* 8, 551, 1974.

69. Rahn, K., Wesolowski, J. J., John, W., and Ralston, H. R., Diurnal variation of aerosol trace element concentrations in Livermore, California, *J. Air Pollut. Control Assoc.,* 21, 406, 1971.

70. John, W., Kaifer, R., Rahn, K., and Wesolowski, J. J., Trace element concentration in aerosols from the San Francisco Bay Area, *Atmos. Environ.,* 7, 107, 1973.

71. King, R. B., Fordyce, J. S., Antoine, A. C., Leibecki, H. F., Neustadter, H. E., and Sidik, S. M., Elemental composition of airborne particulates and source identification: an extensive one year survey, *J. Air Pollut. Control Assoc.,* 26, 1073, 1976.

72. Neustadter, H. E., Fordyce, J. S., and King, R. B., Elemental composition of airborne particulates and source identification: data analysis techniques, *J. Air Pollut. Control Assoc.,* 26, 1079, 1976.

73. Winchester, J. W., Zoller, W. H., Duce, R. A., and Benson, C. S., Lead and halogens in pollution aerosols and snow from Fairbanks, Alaska, *Atmos. Environ.,* 1, 105, 1967.

74. Tanner, T. M., Young, J. A., and Copper, J. A., Multielement analysis of St. Louis aerosols by nondestructive techniques, *Chemosphere,* 5, 211, 1974.

75. Harrison, P. R., Matson, W. R., and Winchester, J. W., Time variations of lead, copper and cadmium concentrations in aerosols in Ann Arbor, Michigan, *Atmos. Environ.,* 5, 613, 1971.

76. Paciga, J. J. and Jervis, R. E., Multielement size characterization of urban aerosols, *Environ. Sci. Technol.,* 10, 1124, 1976.

77. Heindryckx, R., Comparison of the mass-size functions of the elements in the aerosols of the Gent industrial district with data from other areas. Some physicochemical implications, *Atmos. Environ.,* 10, 65, 1976.

78. Hopke, P. K., Gladney, E. S., Gordon, G. E., Zoller, W. H., and James, A. G., The use of multivariate analysis to identify sources of selected elements in the Boston Urban aerosol, *Atmos. Environ.,* 10, 1015, 1976.

79. Zoller, W. H., Gladney, E. S., and Duce, R. A., Atmospheric concentrations and sources of trace metals at the South Pole, *Science,* 183, 198, 1974.

80. Duce, R. A., Hoffman, G. L., and Zoller, W. H., Atmospheric trace metals at remote northern and southern hemisphere sites: pollution or natural?, *Science,* 187, 59, 1975.

81. Gatz, D. F., Dingle, A. N., and Winchester, J. W., Detection of indium as an atmospheric tracer by neutron activation, *J. Appl. Meteorol.,* 8, 229, 1969.

82. Dingle, A. N., Gatz, D. F., and Winchester, J. W., A pilot experiment using indium as a tracer in a convective storm, *J. Appl. Meteorol.,* 8, 236, 1969.

83. Shum, Y. S., Loveland, W. D., and Hewson, E. W., The use of artificial activable trace elements to monitor pollutant source strengths and dispersal patterns, *J. Air Pollut. Control Assoc.,* 25, 1123, 1975.

84. Pillay, K. K. S., Thomas, C. C., Jr., and Sondel, J. A., Activation analysis of airborne selenium as a possible indicator of atmospheric sulfur pollutants, *Environ. Sci. Technol.,* 5, 74, 1971.

85. Martens, C. S., Wesolowski, J. J., Kaifer, R., and John, W., Lead and bromine particle size distributions in the San Francisco Bay area, *Atmos. Environ.,* 7, 905, 1973.

86. Paciga, J. J., Roberts, T. M., and Jervis, R. E., Particle-size distributions of lead, bromine, and chlorine in urban-industrial areas, *Environ. Sci. Technol.,* 9, 1141, 1975.

87. Moyers, J. L., Zoller, W. H., Duce, R. A., and Hoffman, G. L., Gaseous bromine and particulate lead, vanadium and bromine in a polluted atmosphere, *Environ. Sci. Technol.,* 6, 71, 1972.

88. Martens, C. S., Wesolowski, J. J., Harriss, R. C., and Kaifer, R., Chlorine loss from Puerto Rican and San Franciscan Bay area marine aerosols, *J. Geophys. Res.,* 78, 8778, 1973.

89. Martens, C. S., Wesolowski, J. J., Kaifer, R., John, W., and Harriss, R. C., Sources of vanadium in Puerto Rican and San Francisco Bay area aerosols, *Environ. Sci. Technol.,* 7, 817, 1973.

90. Zoller, W. H., Gordon, G. E., Gladney, E. S., and Jones, A. G., Sources and distribution of vanadium in the atmosphere, in *Trace Elements in the Environment,* Advances in Chemistry Series, No. 123, American Chemical Society, Washington, D.C., 1973, 31.

91. Duce, R. A. and Hoffman, G. L., Atmosphere vanadium transport to the ocean, *Atmos. Environ.,* 10, 989, 1976.

Chapter 8

CHARACTERIZATION OF INDIVIDUAL AIRBORNE PARTICLES BY LIGHT MICROSCOPY, ELECTRON AND ION PROBE MICROANALYSIS, AND ELECTRON MICROSCOPY

M. Grasserbauer

TABLE OF CONTENTS

I. DEFINITION OF ANALYTICAL GOAL

Characterization of airborne particulate matter consists not only of the determination of average elemental composition but also includes identification and quantitative determination of the compounds present and a characterization of the particles in terms of size, shape, and morphology (see Chapter 1).

Investigation of type and quantity of the different compounds is important since the physi-

ological effects of dust are dependent on the chemical species in which the elements are present.[1] Furthermore, knowledge of the compounds, rather than of the elements, aids identification of pollution sources and their contribution to pollution situations.[2] Another important aspect of this investigation is the measurement of chemical changes in an aerosol as it travels between an emission source and an immission site.

Size, shape, and morphological parameters are of special interest since they also determine the physiological effect of dust to a great extent for the following reasons: The harmful intake of airborne particles by the human lung shows a maximum in the particle size range of 0.1 to 5 μm diameter,[3] and the composition of an airborne particulate changes greatly with particle size.[4-6] Furthermore, particle size is an important parameter for source determination,[7-9] and studies of aerosol formation mechanism.[10] Particle shape influences the deposition characteristics of particles in the lung, noted in asbestos studies. The morphology (surface structure) of a particle determines its active surface and is, therefore, an important parameter for the adsorption or chemisorption of gases on particles. Adsorbed or chemisorbed gas is in activated state and, thus, can have a more harmful effect ("synergistic effect"). Morphology can also yield valuable information about the formation process of a particle, e.g., whether a vaporization and condensation process has taken place.

Chapters 9 to 12 deal with cummulative methods of obtaining information about the compounds present. Techniques such as X-ray photoelectron spectroscopy (XPS), infrared (IR) spectroscopy, X-ray diffraction (XRD), and thermoanalysis provide possibilities for the qualitative identification and, in some cases, even the quantitative determination of dust compounds — especially for C, N, S compounds, the mineral matrix or asbestos. Generally, these methods can be characterized by the following features:

1. The analytical information obtained from a large number of particles (due to the amount of sample usually analyzed, which is in the microgram to milligram range) is statistically proven and, therefore, representative of the airborne dust examined.

2. These methods provide unique information about a group of substances that figure prominently in pollution and that can hardly be characterized by other methods (especially the C, N, S compounds).

3. Application of these analytical methods is limited to a rather small number of substances and major or minor components — relative detection limit (DL) 0.1 to 10%, depending on substances, composition, and method; in the analysis of such a complex mixture, interferences frequently occur (as in IR spectroscopy). Some analytical effects that are evaluated are rather small (as in XPS); the signal often does not yield unequivocal information about the species (as in thermoanalysis or XRD).

It is obvious that identification of the different chemical species of airborne particulate matter would be easier if it were possible to separate the compounds to be characterized from the complex mixture. While this is possible by chemical treatment in some cases, generally, a separation of dust into its different compounds cannot be achieved. The only method of breaking down a complex mixture is by spatial separation of the conglomerate and analysis of the individual particles. Therefore, this method of compound identification of airborne particulate matter has to include a suitable method of sample preparation (which provides that the individual dust particles are more or less spatially separated from each other) and the application of different analytical techniques with a spatial resolving power of the order of particle size.

Single particle analysis also provides other information required for characterization of airborne particulate matter. The number of particles of a specific chemical composition (which is necessary for the quantitative determination of compounds) size, shape, and morphological features can be determined with the imaging capabilities of the techniques applied. Since particle size can vary within the range of <100 Å diameter to >100 μm, it is evident that different analytical techniques will have to be used for the characterization of individual components. Further complications are that the type and concentration of the chemical compounds vary greatly.

As a result of these difficulties encountered in the identification of such particles — especially in the submicron range — development of these analytical techniques to solve this specific analytical problem is still in progress. Therefore, this chapter will not provide instructions for analytical routine work but rather basic ideas, proposals, and practical hints for small particle analysis. Emphasis will be placed on more recent developments in the identification of particles in the submicron range. Since it is not intended to give a description of the fundamentals of the different methods applied, the reader is asked to refer to the works cited.

II. SURVEY OF ANALYTICAL METHODS AND INFORMATION

The most important methods for the characterization of individual airborne particles are listed in Table 1. These are light microscopy (LM), electron probe microanalysis (EPMA) with an electron

TABLE 1

Survey of Analytical Methods for the Characterization of Individual Airborne Particles

Analytical method	Reagent	Signal	Analytical information	Relative sensitivity	Lower limit of particle diameter
LM	Light	Reflected, transmitted light	Type of compounds (species, structure), size, shape, morphology	Only pure species can be identified	~0.5 μm
EPMA	Electrons	X-Ray spectrum	Type of elements and their concentration Number of particles of a specific composition	0.X%	~0.1 μm
		SE	Shape, size, morphology Number of particles of a specific composition		~100 Å
		BSE, AE	Shape, size, morphology Number of particles of a specific composition		~0.1 μm
IPMA	Ions (O_2^+, O^-, Ar^+)	Secondary ions	Type of elements and their concentration	ppm	0.X μm
STEM, TEM	Electrons	X-Ray spectrum	Type of elements and their concentration	Major and minor compounds	~200 Å
		Secondary electrons	Shape, size morphology		~50 Å
		Transmitted electrons	Size, shape		~10 Å
		Diffracted electrons	Structure and lattice parameters	Pure species	~200 Å
		Energy spectrum of transmitted electrons	Type of elements	Major components	~100 Å

microprobe or a scanning electron microscope (SEM), ion probe microanalysis (IPMA), transmission (TEM) and scanning transmission electron microscopy (STEM).

In LM, optical characteristics of the particles are used to gain information about the type of compounds present. By simple optical imaging, the number of particles, their size, shape, and morphology can be determined. Usually particles consisting of one specific compound can be identified. Particles consisting of mixtures of different compounds (fine agglomerations, inclusions, or adsorbed layers of other substances in the particle matrix) cause serious identification problems. The lower limit of particle diameter is in the order of approximately 0.5 μm.

In EPMA, signals that are typical of the chemical identity of the particle — such as the X-ray spectrum — and of size, shape, and morphology — such as the secondary electrons (SE), backscattered electrons (BSE), and absorbed electrons (AE) — are excited. The evaluation of the characteristic X-ray spectrum yields direct information about the elemental composition of a particle, and is, therefore, basically different from LM identification. Although the volume of excitation of the characteristic X-ray spectrum in a solid with a finely focused electron beam (diameter in the order of 100 to 1000 Å) is several cubic microns (approximately 1 to 5 μm in diameter and depth), quantitative elemental analysis of individual particles is possible by applying a modified method of quantification (ratio method, see Section III.C.2.) to a lower particle size limit of about 0.1 μm. Contrary to LM, minor components of a particle (inclusions, adsorbed layers of solids, etc.) can be detected if their concentration is higher than several tenths of a percent. This detection limit is valid for particle sizes in the order of the excited volume.

Shape, size, and morphology can be determined for a large size range — from the largest particles to those 100 Å in diameter. Due to the high resolution, which is usually equal to the beam diameter of 50 to 100 Å, and high depth of field of SE images, the SEM is inherently better for determining these particle features than the LM. By combining elemental analysis of individual particles with particle counts and area measurements, the fraction of specific compounds among the total particulate matter can be determined.

With IPMA, the type and concentration of the elements of a particle can be determined by evaluation of the secondary ion mass spectra. The most important features of this method are its high relative sensitivity, which enables the identification and determination of trace components in individual particles, and its capability of surface analysis. The lower size limit for particle identification is in the range of several tenths of a micron. A quantification procedure, similar to that used in EPMA, must be employed, since the diameter of the excited area is approximately 1 to 2 μm. The depth of signal excitation is approximately one monolayer. However, to obtain a significant number of secondary ions, much more material must be sputtered off during analysis. Therefore, IPMA is not primarily a surface analytical technique when applied to the identification of airborne particles having diameters in the range of a few microns to several tenths of a micron. This capability of the method can only be utilized when larger particles are analyzed.

TEM and STEM provide a variety of analytical signals that facilitate characterization of particles in the submicron range. As in EPMA, the X-ray spectrum obtained with an STEM yields information about the type of elements present and their concentration. The sensitivity is some what lower, due to decrease of X-ray intensity with particle size. The lower size limit of particle identification is on the order of 200 Å. STEM enables the imaging of particles by SE to a diameter as small as 50 Å. This limit can be extended further to approximately 10 Å by using transmitted electrons. Electron diffraction patterns of individual particles yield information about structure and lattice parameters and, therefore, about the chemical identity of the particles. They provide valuable additional information. The newest technique used in STEM is the evaluation of the energy spectrum of the transmitted electrons. The energy loss peaks are characteristic of the elements in the particle, and they can be used for qualitative identification. The size limit is on the order of approximately 100 Å particle diameter.

Application of each of these analytical methods is mainly determined by the size of the particles to be characterized. The size range covered by these methods is shown in Figure 1. The methods listed here have the unique capability of the chemical identification of individual particles and exact determination of size and offer the only possibility of gaining information about shape and morphology. However, they also have serious limitations: (1) To gain representative information about the particle collective, it is usually necessary to investigate a rather large number of individual particles, which can be time consuming. (2) Generally, only solid substances can be investigated, and virtually no information is obtainable for liquid or gaseous adsorption products. When the analytical signals are excited with electrons or ions, the particles heat up to a certain degree (depending on sample preparation, particle size, and thermal conductivity), which can result in evaporation (due also to the high vaccum in the sample chamber) or destruction of thermally nonstable compounds (e.g., some C,N,S compounds).

III. CHARACTERIZATION OF AIRBORNE PARTICLES

A. Sample Collection and Preparation

While the method of collecting airborne

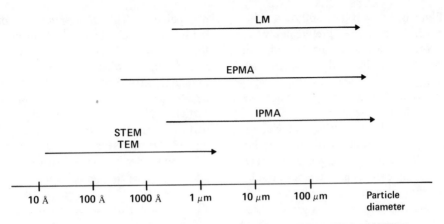

FIGURE 1. Particle size range covered by LM, EPMA, IPMA, TEM, and STEM.

particulate matter must be chosen according to the analytical goal (e.g., determination of short-term immission values, long-term fallout of dust, chemical composition of a certain particle size range) sample preparation is mainly determined by the analytical method applied. For complete characterization of individual dust particles, a spatial separation of the particles is usually necessary in order to analyze one particle without interference from neighboring particles. This is especially important for X-ray techniques when submicron particles are investigated, since lateral distribution of the electrons far exceeds particle size. It is, therefore, convenient to distinguish — as far as sample preparation is concerned — between sampling methods that yield a certain amount of spatially nonseparated dust and those that yield a spatially separated collective.

Sampling with dust jars, cascade impactors, high-volume filters, or electrostatic precipitators yields either a loose pile of dust or a more or less dense layer on a substrate (filter, impactor plate). Unless these layers are very thin (as obtained with an impactor during very short sampling times), such a sample is preferably investigated for qualitative information only. If quantitative studies are performed the particulate matter should be transferred and spread onto a suitable substrate. The properties of these substrates are determined by the analytical techniques applied. These are

1. For all methods: substrates should have smooth surface.

2. For LM: high transparency, colorless, nonstaining, chemically resistant; glass or quartz slides are generally used.

3. For EPMA and IPMA: when elemental analysis is carried out, low level of detectable elements (high-purity substrate having low atomic number), nonhygroscopic. Be plates, C blocks, or thin plastic or carbon films are appropriate. When only imaging is performed, any highly polished substrate is sufficient.

4. For TEM and STEM: transparancy for high energy electrons: Thin Formvar,® carbon, or SiO films (mechanically supported by grids) are used.

Ultrasonic removal of filter samples may be necessary especially for the fine particles. For impactors, ultrasonic suspension in a liquid may be required. Dispersion of the particles in an inert liquid, transferring a few drops on the substrate, and drying seems an appropriate method of minimizing losses but may cause agglomeration of fine particles. For subsequent SEM investigation, McCrone and Delly[11] advocate suspension of the particles in a dilute collodion or Formvar® solution. After drying, the particle is covered with a thin film, which also causes fixation of particles to the substrate. A modified technique can be utilized to prepare samples for the electron microscope.[11] In order to obtain a thin film, a drop of the solution containing the suspended particles is placed directly on a grid or on the surface of water then collected with a grid. After drying, the particulate matter on the substrate is prepared for the subsequent investigation — when LM is used, a drop of an appropriate mounting medium (preferably having a high refractive index like Aroclors®) is added, and the sample is covered with a slide.

For EPMA investigations, the particles are coated with a thin electrically and thermally conductive film (thickness 100 to 300 Å) — either of carbon, which may be preferable when subsequent X-ray analysis is performed, or with gold or gold-palladium alloy, when high resolution images are made. Sputter deposition of gold is preferable, since it avoids the shading effects, obtained with evaporation techniques, that cause problems resulting from electrostatic charging of the noncoated areas during analysis (especially silicates of larger grain sizes). Particles in the submicron range often need not be coated when the substrate is conductive. For quantitative elemental analysis of larger particles (diameter $> 5\mu m$), embedding in resin and polishing can be advantageous. If loose dust of a grain size above approximately 1 μm is examined, often the simple method of dusting the material on a Teflon® plate covered with a suspension of graphite in an organic solvent (as used for coating TV tubes) can be applied. The graphite film has a very smooth surface after the solvent has evaporated.

For IPMA investigations, larger particles may also be coated with C or Au. For smaller particles, it is sufficient that they be deposited on an electrically conductive substrate, especially when negative primary ions are used for analysis, which decreases charging.

In electron microscopy, the particles can be viewed directly, or the contrast can be enhanced by shadowing with metal oxides (e.g., SiO) or metals. Replication techniques are most useful when surface characteristics are to be studied.[11] Replication can be achieved by placing a film of particles suspended in collodion solution on a smooth substrate (glass), partially dissolving the film with ethyl acetate vapor and pressing them into a polysterene wafer. After separation of the surface, the particles stick to the collodion and the polysterene is metal shadowed and carbon coated. The polysterene is dissolved and a positive carbon replica obtained.[11,12]

In some instances, a separation of the complex conglomerate dust may be necessary, especially when collected with high-volume filters or dust jars. Among the methods listed by McCrone and Delly[11] the following are especially important:

1. Size separation by sieving: This technique is used preferably for the elimination of large particles ($> 44 \mu m$[11]), as they are often collected in dust jars.

2. Mechanical separation with micro-manipulators: This means actually picking out individual particles under the light microscope on the basis of their appearance. The technique is mainly applied when these particles have to be examined on a special substrate, as demonstrated by McHugh and Stevens[13] for IPMA of oil soot particles. McCrone and Delly[11] advocate that particles as small as 1 μm can be manipulated.

3. Density separation by floating in liquids of varying density in a centrifuge tube: It is possible to separate combustion products (e.g., magnetic spheres from pulverized coal boilers), all biological particulates, most industrial dusts, and most minerals from heavy minerals and most metals by the use of methylene iodide (ρ = 3.33 g cm^{-3}) as a separation liquid.

Despite the fact that sophisticated techniques for the preparation of a sample with spatially separated particles exist, the analyst concerned with characterization of individual particles will try to apply a collection method that yields a sample fulfilling the requirements for subsequent examination with as little manipulation as possible. The most promising method of obtaining a sample of particles spatially separated on a substrate and suitable for direct analysis of the individual particles seems to be collection with centrifuges and Nuclepore® filters.[14-16] Centrifuges allow collection of airborne particulate matter directly on a suitable substrate (thin plastic sheets, etc.) with a high resolution for grain size differentation. Nuclepore filters — thin (5 to 10 μm) polycarbonate films — also produce grain size fractionization due to their uniform pore size (0.015 μm to 8 μm) (see also Figure 3 in Chapter 1). The low airflow through the filter, as compared to other filters, is advantageous for characterization of individual particles, for it assures low loading and, therefore, spatial separation of particles on the filter within reasonable sampling times.

The material properties of the filter are such that they conform to the requirements for LM, EPMA, and IPMA. When applying these techniques, coating with C or Au before or after filtration is necessary. When particles are investigated with LM under polarized light, the birefringence of the substrate has to be considered (η = 1.623 to 1.585). Observation of the small particles captured around the pore entrances is often difficult. In order to visualize these particles, the pore outlines can be removed by exposing the filter to $CHCl_3$ or $CH_2Cl\text{-}CH_2Cl$.[14] A further disadvantage is that some small particles are collected in the pores.[17] Microchemical tests can be made directly on the filter, since polycarbonate is resistant to most acids and organic solvents (except halogenated hydrocarbons and strong bases). Furthermore, the particles can be stained directly on the substrate, since the filter is nonstaining.

The application of Nuclepore filters as substrates for SEM investigations has been treated and compared with other filters (Gelman® polyvinyl-chloride filter, Millipore®, and silver membrane filter) by Denee and Stein.[18] The authors claim that the polycarbonate filter is best suited for SEM analysis.

For TEM and STEM, surface replication is necessary because the filter is not electron transparent. SiO replicas can be prepared, according to Frank and Spurny,[17] by coating the membrane with a layer of SiO of approximately 1000 Å thickness and additionally with approximately 200 Å of Cr (at 15°). The sample is placed on a grid and the membrane dissolved with $CHCl_3$ vapor. The resulting replica is a thin-film SiO-Cr-substrate that contains the $CHCl_3$ insoluble particles. For X-ray analysis carried out in STEM, a carbon sandwich sample should be prepared. The membrane is coated with carbon before and after filtration. After dissolution of the membrane, a sample containing the particles between two carbon layers on a grid is obtained.

Spatial separation of the particles can also be achieved with the automatic sampling device described by Heard and Wiffen.[19] The authors developed an instrument for the collection of a continuous series of samples on Millipore filters. They obtain a particle density, especially suitable for electron microscopic investigation. For short-term measurements (collection time of minutes), the impactor can also be used since particle density on the plate is sufficiently low. Electron microscopic C films supported by grids are placed directly on the impactor plate, thus yielding a sample suited for electron microscopic investigation.[20] Schütz[21] used an impactor with a rotating disc. Due to the larger deposition area, a spreading of the particles can be achieved even for normal sampling duration.

Another technique for collecting spatially separated airborne particles is the exposure of suitable substrates to settling dust in the air. Small particles are collected on C films with grid support for subsequent electron microscopic investigation. A disadvantage of this technique is that dust settlement is not representative of the actual composition and size distribution of the particles in the aerosol, as smaller particles settle much more slowly (see Chapter 2). Wind erosion of the deposited particulate and rain can also disturb the sample. Sticky tapes are not recommended for sampling since the properties of the surface of the tape are not well suited for investigation with EPMA, IPMA, or electron microscopical techniques.

B. Light Microscopy

In characterization of individual airborne particles, LM has been extensively demonstrated and described.[11,22] McCrone and Delly[11] include a comprehensive survey of the use of LM for particle identification and describe the sophisticated techniques applied in detail. This chapter will include an abstract of these authors' presentation to enable the reader to judge the applicability, advantages, and disadvantages of LM as compared to other methods for particle characterization.

The chemical identification of individual airborne particles is based on certain properties that can be observed with LM and that are typical for a specific chemical composition of the particle. The most important properties are transparency, opacity (primarily for visible radiation), color, refractive index, birefringence, size, shape, and morphology of a particle. These properties can be determined for individual particles to a lower size limit of approximately 0.5 to 1 μm in diameter, which corresponds to a particle mass of approximately 10^{-9} g.

Transparency, opacity, and the colors observed with transmission or reflectance illumination yield primary information about particle identity, since many substances possess a more or less distinct color, as seen in the micrographs of McCrone and Delly.[22] The refractive index, which is related to the molecular weight of the compound, and the anisotropy of the refractive index, which appears in noncubic crystals, provide further information for identification of a particle. The refractive index of individual particles is determined by microscopic immersion methods by varying the index of the mounting liquid using the so-called Becke Test (for details see McCrone and Delly[11]), which yields better than ± 0.001 accuracy. The anisotropy of individual particles is studied under polarized light, and the refraction indices can be measured for different crystallographic directions. Under crossed polars, the birefringence properties of anisotropic crystals will cause the crystal to display a color, which depends on thickness and differences in refractive indices for the two vector components of the polarized light in the crystal. With the aid of the Michel-Levy chart, which relates interference colors observed to crystal thickness and differences in refractive indices, the birefringence of a crystal can be estimated by a skilled microscopist. Further information is provided by the location of the lower and higher refractive index with the aid of compensators (such as thin quartz plates, the quartz wedge, or the quarter-wave plate) or the observation of interference patterns obtained in conoscopic observation. According to McCrone and Delly,[11] these sophisticated techniques are helpful in determining the principal refractive index directions in a crystal and in determining the crystal system and thus, for characterizing and identifying compounds.

Another very useful technique applied in LM identification of dust particles is dispersion staining. The term dispersion staining is used to describe the color effects produced when a transparent object, immersed in a liquid having a refractive index near that of the object, is viewed under the microscope with transmitted white light. The color spectrum that appears at the edge of the particle (using annular or central stops in the back focal plane of the microscope) can be used for systematic identification of many isotropic and anisotropic compounds.[23] This technique is especially useful to distinguish between components of a complex mixture such as dust, preferably when the components differ only slightly in refractive index. Dispersion staining charts are provided by McCrone and Delly.[11,47] This technique has been used for the identification of quartz in lung tissues,[24] glass fragments,[25] settled dust,[26] and for particle counting.[27]

Size, shape, and morphology of the individual particles are prime features for identification because certain species often appear in rather narrow size ranges, depending on the specific origin of a compound. Shape and morphology of individual particles are closely related to material properties, which in turn are characteristic for a substance or group of substances: e.g., fibrous

materials (of organic or inorganic compounds) and amorphous and crystalline substances (which usually display distinct differences in particle shape). Particle shape may also suggest the crystalline structure, an important clue in identification. The question arises in what terms or numbers these particle features can be described or documented. The most simple and widely used method for individual particles is documentation by imaging as described by McCrone and Delly.[22] Quantification of these features (description of individual particles or particle collectives by a set of numbers) is a difficult task and presently the problem is not completely solved.

The most important parameters in describing particle size are linear extension values and particle area. Such linear parameters represent the actual diameter of a regular particle, the projected or a statistical diameter of an irregular particle, or length and width for rectangular particles (such as fibers). The projected diameter is the diameter of a circle equal in area to the profile of the particle. Therefore, it is actually a parameter that characterizes the area of the particle and is obtained by area measurements. A statistical diameter is obtained from linear measurements and requires random orientation of particles and a sufficient number of measured particles. It is, therefore, used as a mean value for a particle collective. Two of the most commonly used are Martin's and Feret's diameter, illustrated in Figure 2.[28]

For the description of particle size, not only individual size parameters or average values are important but also the particle size distribution. This can be determined with microscopic techniques by classification of a large number of particles according to size. The curves representing the number of particles (fraction of total number) of a certain size vs. particle size reveals valuable

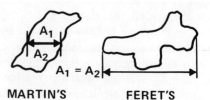

MARTIN'S   FERET'S

FIGURE 2.   Schematic representation of Martin's and Feret's diameter of particles. (From Giever, P. M., *Air Pollution*, Vol. 2, Academic Press, New York, 1968, 249. With permission.)

information about the physical characteristics (aerodynamic behavior, etc.) of a collective. Due to the large numbers necessary to obtain a statistically proven distribution curve, automation of the measurement is necessary. For automated particle size analysis, the different image analysis systems (such as the Zeiss-Mikro Videomat®,[29] Leitz-Classimat®,[30] or Imanco-Quantimet®,[31] and the Bausch and Lomb-Ommicon®[32]) that are based on the direct area measurement of the individual particles are recommended. From the individual parameters, the mean values and distribution curves are calculated. Since the images are measured with television scanning speed, this method ensures performance of particle size analysis within a reasonable time span.

Not only LM images but any image or photomicrograph can be evaluated. Thus, automated image analysis can be applied in electron microscopy where particle counting and size determinations are even more important than in techniques examining larger particle size ranges. The reason for this is that for particle sizes of approximately 0.1 μm and larger, a cumulative size determination can be achieved by applying sampling methods that employ size fractionation (such as, impactors, centrifuges, Nuclepore filters). These methods yield size distribution curves of mass or elemental concentration vs. a diameter (aerodynamic, real diameter). Since they provide rapid results and yield the distribution of the elements in different size fractions, these methods are generally preferred.

To achieve the same information about the size distribution of the elements, it would be necessary to differentiate between various compounds in particle size analysis. Since, the signal used in LM is chemically nonspecific and the different particles are sorted according to their black and white contrast, it is seldom possible to obtain size distribution curves for specific elements or chemical species. However, if this is intended, EPMA must be utilized as it enables direct chemical differentiation of the particles by evaluation of the X-ray signal. Unfortunately, this method is much slower, and thus it is used only for very specific problems.

Although the electron microscopic pictures suffer from the same disadvantage as LM images (even to a larger extent since the contrast in electron microscopic pictures cannot be related to a certain species), automated analysis of number

and size of submicron particles is routinely employed. This method not only allows rapid size classification and determination of the total amount of material (it is assumed that on the statistical average the projected diameter can be used to calculate the volume of each particle) but also a differentiation of particles that have a significantly different shape, i.e., asbestos fibers. Modern systems allow separate counting and size determination for particles of different shapes by employing appropriate computer subroutines that can distinguish, e.g., between fibers and other particles on the basis of length-to-width ratios or the ratio of interference of a particle to area.[33] Thus, they permit determination of the quantity of a certain species characterized by shape, e.g., asbestos. For further information about particle size analysis via stereology see the References 34–42.

Since the exact description of particle shape still presents a problem, especially for irregular particles, and morphology cannot be quantified, McCrone and Delly[11,22] use a set of descriptive terms that are quite perceptual and useful. Due to the complex relationship between the properties of dust particles described thus far and their chemical identity, which may cause severe problems, supplementary information about the particle by determining physical or chemical properties must be obtained. The most important physical and chemical properties which can be measured for individual particles are density, melting point, hardness, magnetic susceptibility, IR absorption, solubility in solvents, and chemical reactivity.[11] Unfortunately, some of these parameters — density, melting point, IR absorption — can only be determined for larger particles, thus excluding application of these techniques to respirable particles.

The determination of these properties involves a set of analytical procedures such as hot-stage microscopic investigations, application of microdensity gradient tubes, crush tests, observations of particle movement in a magnetic field, and others. One of the most valuable tools — unfortunately, only for larger particles (diameter > 30 to 50 μm) — is the registration of an IR absorption spectrum of individual particles with the aid of an IR microscope attachment for an IR spectrometer. As outlined by Kellner (Chapter 11), an IR pattern of more or less pure substances usually provides a specific pattern for the chemical composition.

Methods for obtaining physical properties of individual particles are described in detail by McCrone and Delly.[11]

As further supplementary tests for the chemical identity of particles, microchemical reactions can be carried out under the light microscope. The general problem lies in controlling the drop size of the reagent or the area of precipitation. Several approaches have been developed.[43-46] Although microchemical reactions may be useful for survey checks (especially important are solubility tests), they generally have been replaced by the direct techniques of local analysis utilizing physical methods as described below. Microchemical reactions are still important for detection of certain anions (such as NO_3^-, SO_4^{2-}, $S_2O_3^{2-}$) that cannot be identified with electron or ion techniques or substances that are volatile in high vacuum and under electron or ion bombardment. Chemical identification tests furthermore prove useful in increasing the chemical specificity of the identification of submicron particles under the electron microscope.

As seen from this brief survey of methods and techniques that determine optical, physical, and chemical characteristics, the LM identification of individual particles is an extremely complex task. This becomes more apparent if one keeps in mind the great variety of particles (due to many different sources) and the large size range with an accumulation of the important anthropogenic substances at particle sizes just above or even below the spatial resolution of the technique. Extensive experience and the aid of a suitable encyclopedic collection of photographs and data is required. McCrone and Delly[22,47] provide such a collection. Their book contains 609 colored photomicrographs of the most important dust particles, morphological descriptions, and ample data dealing with optical, physical, and chemical characteristics. The particles are divided into four groups: (1) wind erosion particles (biological substances, fibers, rocks, and minerals); (2) industrial dusts (abrasives and polishes, catalysts, cements, detergents and cleaners, fertilizers, food processing, metal refining and processing substances, pigments, polymers, and miscellaneous industrial substances; (3) combustion products (auto and trash burners, oil soot, coal fly ash, and incinerator dust); and (4) miscellaneous (such as brick particles, feathers, etc.)

Identification of an unknown particle follows a classification scheme that determines opacity, color, birefringence, refractive index relative to the mounting medium Aroclor® and shape. These general characteristics are expressed in a six digit binary code. This code leads to a small number of particles pictured in the *Particle Atlas.*[22] By comparison of the microscopic image with the photographs (morphology, any obvious detail) and/or determination of the refractive index, the particle should be identifiable.

To successfully apply LM techniques in particle identification, the use of McCrone and Delly's particle atlas seems to be a necessity — especially since most analytical chemists concerned with environmental analysis do not have the intensive background in mineralogical LM required. Despite the existence of such extremely valuable aids, the complexity of LM identification, which is based on the logical combination of many properties (which are normally not unequivocally typical for specific composition), is certainly a serious disadvantage of this technique. Another serious limitation is that the minimum size of identifiable particles is approximately 1 μm. This means that a large and important fraction of airborne dust cannot be characterized. A third disadvantage is encountered in morphological studies due to the low depth of field of the LM that causes difficulties in the imaging of rough surfaces. Therefore, utilization of analytical techniques that provide (1) means for a direct chemical identification of individual particles, (2) possibility of characterization of particles in the submicron range, and (3) high depth-of-field images is recommended. These techniques are EPMA, IPMA, STEM, and TEM.

C. Electron Probe Microanalysis

As described in Section II, EPMA characterizes individual airborne particles by analysis of their chemical composition and determination of size, shape, and morphology. For this task either scanning electron microscopes (SEM) or microprobes are used. Although there is no sharp distinction between these two types of instruments, differences exist as far as optimum use is concerned. The SEM equipped with an energy dispersive X-ray spectrometer (EDS) is recommended for morphological studies and qualitative and semiqualitative X-ray analysis, since high lateral resolution of imaging is of primary importance. The microprobe is emphasized for chemical identification of particles by means of quantitative elemental analysis and X-ray valence band spectroscopy. In this instance, high precision and accuracy of X-ray measurement and quantification are of main concern and are best obtained with normal beam incidence and high X-ray take-off angle.

The general technique applied for characterization of individual particles involves imaging with the SE signal and subsequent local analysis using either an EDS or wavelength-dispersive spectrometer (WDS). For survey purposes, the BSE (or absorbed electron) signal can be used, since it allows rapid differentation between compounds containing heavy elements and those containing light elements. As illustrated in Figure 3, a particle containing a heavy element (such as Fe) is distinctly different from the mineral silicate particles. By employing the BSE images, industrial metal emission products can often be rapidly identified within the matrix of dust.

To measure the X-ray signal the EDS is generally preferred, unless its limitations indicate the need for WDS. The advantages, as far as particle identification is concerned, include a rapid qualitative elemental analysis, multi-element capacity, the influences induced by surface roughness on X-ray generation and absorption are more equalized than with application of WDS (since all elements are measured from the same direction), and the distribution of elements within a large sample area can be obtained since no Rowland circle exists. WDS is preferably utilized for qualitative and semiquantitative analysis of second period elements (especially O and C), for highly accurate and sensitive quantitative analysis and the recording of X-ray valence band spectra. (For further information on the EDS and the differences between EDS and WDS also see Chapter 3 of this book).

To solve the central difficulty of determining chemical identity (species) of a dust particle, three main approaches can be used in EPMA. The first approach is based on a combination of qualitative or semiquantitative elemental analysis and morphological features that, when combined, are often characteristic for a certain chemical species. The second method of identification is a quantitative elemental analysis that determines the compound by stoichiometric calculation of the formula from elemental concentrations. The third approach utilizes evaluation of X-ray valence band spectra that may have a fine structure characteristic of a certain compound.

1. Particle Characterization by Combination of Qualitative and Semiquantitative Elemental Analysis and Morphology

The principle of this method is based on the fact that although airborne particulate matter from various emission sources can consist of a great variety of components, an individual particle normally contains only a few elements. Knowledge of the identity of these elements limits the number of possible compounds. Often, this limited number of compounds can be differentiated by distinct differences in morphology, size, or shape of the particles. The primary reason for this is that different compounds may have different crystal or material properties (hardness, etc.) or, usually, a different formation process, which strongly influences these properties.

Obvious advantages of a combination of morphology with qualitative or semiquantitative analysis are that instead of the complicated and elaborate procedure of quantitative analysis (see Section III.C.2.) a rapid and easy evaluation can be performed, and the morphological parameters are obtained in the same step of analysis with SEM. On the other hand, the evaluation of morphological features requires a pictorial atlas as described for LM. Such a collection of SEM photographs at different magnifications (illustrating a total view of the particle and details of the surface structure) and the X-ray spectra of these particles has been provided by McCrone and Delly.[49] Volume 3 of their *Particle Atlas* contains images and spectra of the same 609 various types of dust components that were characterized by LM.[48]

The atlas represents a valuable aid to the analytical chemist, for it actually forms the bridge between qualitative analysis and identity of the species and makes this simple method of identification applicable on a broad basis. It is especially important in connection with LM identification as the combination of the two techniques provides an extensive amount of information about the particle. Unfortunately, simultaneous LM and SEM investigation is only possible with a few instruments.

The atlas demonstrates the validity of the basic assumption that different compounds of similar elemental composition have different morphology. It is possible to differentiate between such similar substances: PbO (particle no. 406) forms plates

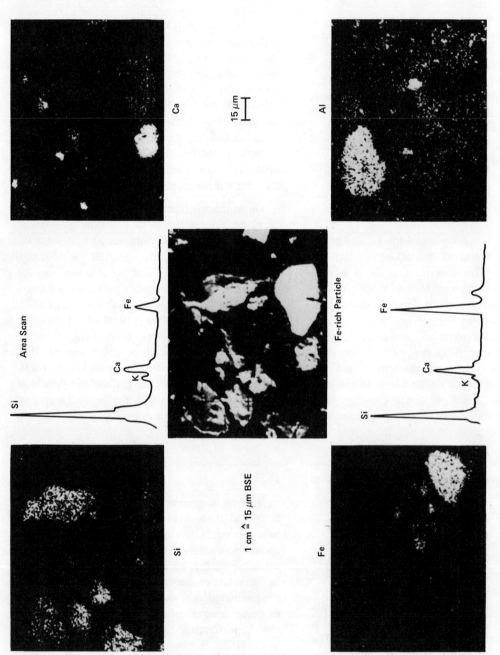

FIGURE 3.   Electron probe microanalysis of airborne particles — BSE micrograph (center), energy dispersive X-ray spectra of total area imaged and of an individual particle and distribution of different elements. (From Malissa, H. and Grasserbauer, M., *Mikrochim. Acta*, 2, 325, 1975. With permission.)

with rounded edges as large as 10 μm. These plates are aggregates of smaller particles; 2 Pb $CO_3 \cdot$ Pb$(OH)_2$ (particle no. 442) appears as small irregular plates (diameter 0.5 to 2 μm) forming fluffy, irregularly shaped aggregates; and Pb_3O_4 (particle no. 444) is made up of well-formed equant crystals (diameter $\sim$ 0.2 to 1 μm), coalescing to irregularly shaped, stringy aggregates. Differentiation is still more pronounced when LM is used as an additional tool, since the optical properties of these three substances are quite different.

One limitation of the application of this combined identification method is that the majority of the particles discussed in the atlas are large particles in the size range of several micrometers to several hundred micrometers. Very few examples of airborne particles in the submicron range, which is especially important, are given. The other limitation is connected with the unique pos-sibility of gaining direct information about the origin of the particle by comparing particle morphology observed with the photomicrographs (showing particles of specific origin) and becomes valid if a particle of a certain qualitative elemental composition has a different origin than assumed or if size, shape, or morphology are changed due to secondary mechanical (grinding, abrasion, erosion etc.) or chemical (selective solution of components, burning off of parts as with soot particles, etc.) processes. Identification problems also occur when the particle consists mainly of second period elements that normally cannot be detected with an EDS, as in urban aerosols. In this case, many investigations are confined to morphological studies.

When there are distinct differences between some important anthropogenic particles, like soot from oil firing (Figure 4) and the mineral background (Figure 5), morphology alone can be used

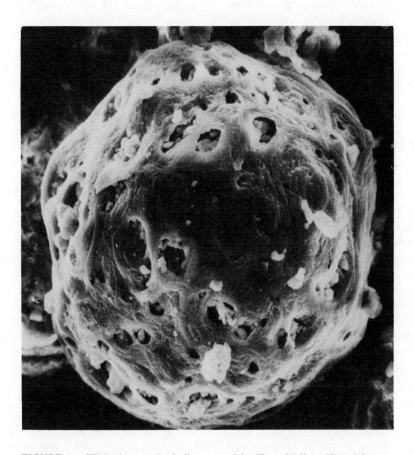

FIGURE 4. SEM micrograph of oil soot particle. (From Malissa, H. and Grasser-bauer, M., *Mikrochim. Acta*, 2, 325, 1975. With permission.)

FIGURE 5.   SEM micrograph of mineral particle. (From Malissa, H. and Grasserbauer, M., *Mikrochim. Acta*, 2, 325, 1975. With permission.)

for the determination of the fraction of soot particles. These oil soot particles can also be differentiated clearly from fly ash particles originating from coal firing. Fly ash particles, which are created by combustion, are identifiable by their spherical structure, indicating that they were in a liquid state when airborne (Figure 6). According to Henry and Blosser,[50] these particles range in size from ~ 0.2 μm to 15 μm and are the most prevalent material in samples collected in Washington, Denver, Cincinnati, Chicago, Philadelphia, and St. Louis. Cheng et al.[51] compared the particles emitted by oil- and coal-fired boilers in power plants. Morphological studies carried out with SEM showed that more than 90% of the collected particles from both sources were spherical, but distinct differences in surface structure do exist, as already described. These authors also tried to gain some information about the chemical composition of the particles – especially about the presence of trace metals that might participate in the catalytic oxidation of SO_2 to SO_3 in the atmosphere. By qualitative X-ray analysis of individual particles carried out with EDS, they found that the particulates from oil-fired burners contain S, Si, Ca, V, and Fe as major and Mg, Al, P, Cr, Mn, and Ni as minor elements. Glassy spheres from the coal-fired burner, however, contain primarily Si, K, Ca, Ti, Fe, and S and as minor elements, Al, P, Cl, Mn, and Cu. Other investigations[52] revealed that these spherical

particles collected from coal firings are mainly silicates. Oil soot particles naturally are, more or less, a carbon skeleton containing other elements such as sulfur as shown in Figure 7.

Schulz, Engdahl and Frankenberg[54] carried out morphological studies of submicron particles emitted from a pulverized coal-fired power plant. They found that practically all particles are spherical and have the typical appearance shown in Figure 6. The authors stress the importance of investigating the size and shape of a particulate when a size differentiated collection method is applied in order to control the accuracy of the cumulative size determination, which yields an aerodynamic parameter and not a direct size. They found, in the case of spherical particles with a uniform density, that the diameter determined from impactor deposition and the diameter measured in the SEM are in fairly strong agreement. Flachsbart, Stöber, and Hochrainer[55] investigated the relation between the real particle size and the aerodynamic diameter of fibers with a high length-to-width ratio (on the order of 100) and used the SEM for the measurements of real fiber length and width. They found that, for asbestos fibers collected with a centrifuge, the aerodynamic diameter is about three times actual diameter when the length-to-width ratio is in the range of 10 to 200. This means that for fibers, the diameter is dominant for the deposition characteristics and not their length.

Morphological studies also play an important role in the investigation of processes of particle formation, since SEM imaging of individual particles may provide direct clues to reaction mechanisms. Buckle and Pointon[56] investigated the growth of Cd aerosols and found that a roughening of the surface, due to growth steps, takes place when the particles fall through the supporting gas mixed with Cd vapor (Figure 8). They hint that this may be a temperature effect. Such model investigations provide basis for the understanding of a mechanism of condensation processes.

A further area of interest, where morphological investigations with SEM lead to significant results even if no chemical identification is carried out, is the study of the deposition of particulate matter on plant surfaces and the associated physiological effect. Ricks and Williams[57] investigated the stomata (pores) of the leaves of *Quercus petraea* (oak) and found that a significant portion of them is occluded by fine particles (Figure 9), thus

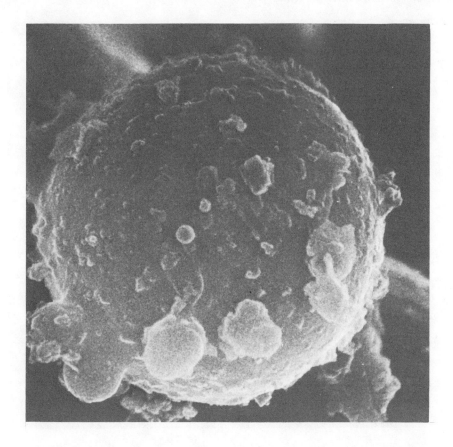

FIGURE 6.   SEM micrograph of typical particle occurring in fly ash. (Magnification ×
5000.)

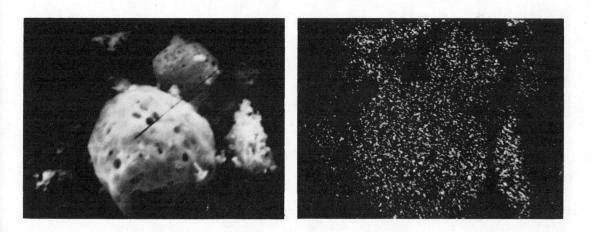

FIGURE 7.   SEM micrograph of oil soot particle and distribution of sulfur. (From Malissa, H., *Angew. Chem. Int. Ed.
Engl.*, 15, 123, 1976. With permission.)

altering gas exchange and transpiration. These
studies are especially important for understanding
the effect of solid pollutants on plant life and for

gaining more information about the filtering
capacity of plants.

Schütz[21] uses the morphology of sea salt

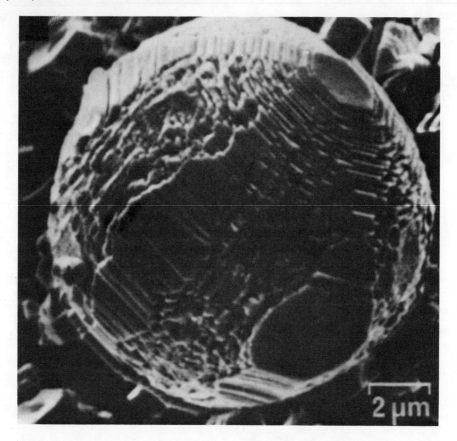

FIGURE 8.   SEM micrograph of a Cd aerosol particle condensed at room temperature. (From Buckle, E. R. and Pointon, K. C., *Atmos. Environ.*, 8, 1335, 1974. With permission.)

particles (cubic) and mineral particles (not cubic) to distinguish between these two types and to determine the size distribution of particles from ocean spray and from Sahara sand collected near the Cape Verde Islands. The author emphasizes the fact that it is not possible to identify the different mineral particles on the basis of morphology alone. The particle size limit for qualitative elemental analysis depends largely on the geometrical arrangement of the solid-state detector (solid angle of X-ray detection), the electron beam density (current per unit of beam diameter), the substrate used, and excitation conditions.

For normal SEM's (not having a variable sample-detector geometry) and metal substrates, the lower size limit for qualitative identification is on the order of 0.2 μm^{21} when low excitation energies (5 to 10 keV) are used. Using substrates that have a lower Bremsstrahlung – background (such as electron transparent films) – the minimum size for particle identification can be lowered significantly.

Maggiore and Rubin[58] used SEM equipped with a field emission gun for the identification of asbestos fibers on the basis of qualitative elemental analysis. Due to a high beam current density and a high solid angle of X-ray acceptance by the detector (variable geometry), the authors were able to analyze amosite fibers as small as 500 Å in diameter. The authors stress that the fundamental limitation of X-ray analysis of small particles is imposed by fluorescence of other parts of the specimen by backscattered electrons when the distance between the particles is in the order of 1 μm or below.

2. Particle Identification by Quantitative Elemental Analysis

a. Quantitative Analysis of Particles Larger than Excited Volume

The excited volume (volume of X-ray production) can be roughly calculated using Equation 1[59] for the depth of signal excitation (d_E) and assuming lateral electron diffusion to be approxi-

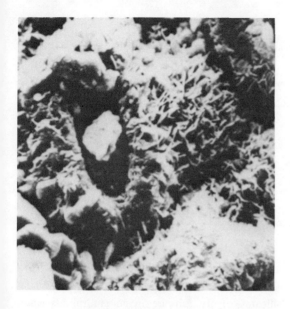

FIGURE 9. SEM micrograph of an oak leaf showing occlusion of stomatal pores by airborne particles. (From Ricks, G. W. and Williams, J. H., *Environ. Pollut.*, 6, 87, 1974. With permission.)

mately equal to excitation depth. The contribution of the electron beam diameter to the lateral spread does not have to be taken into account for modern electron microprobes since it is significantly smaller. Therefore, in the context of particle analysis, d_E means the diameter of a sphere where X-ray production takes place. According to Equation 1, d_E depends mainly on the density of the material, the energy of the measured line, and the excitation energy:

$$d_E = 0.033 (E^{1.7} - E_c^{1.7}) \cdot \frac{A}{\rho \cdot Z} \qquad (1)$$

where d_E = depth of signal excitation (μm); E = excitation energy (keV); E_c = critical excitation energy of the measured X-ray line (keV); A = atomic weight; Z = atomic number; ρ = density (g/cm^3).

Table 2 lists calculated values of d_E for some important compounds of airborne particulate matter. Generally, the excited volume ranges from ~ 0.5 to ~ 10 μm. Principally, it would be possible to decrease this value by using lower excitation energies. In this case, however X-ray intensity (I) also diminishes (I is proportional $\frac{E^{1.7}}{E_c}$) (see Figure 10)[21] which may create the problem of insufficient intensity. In every case, careful optimization of the excitation parameters is necessary.

If the particle diameter is significantly larger than d_E, quantitative elemental analysis of individual particles is rather straightforward. The normal quantitation procedure (Equation 2) can be used:

$$\frac{I_A}{I_A^x} = \frac{c_A}{c_A^x} \cdot \frac{f_Z \cdot f_A \cdot f_F}{f_Z^x \cdot f_A^x \cdot f_F^x} \qquad (2)$$

where I_A = X-ray intensity (peak) of element A in sample or standard (x); c_A = concentration of element A in sample or standard (x) (wt %); and f_Z, f_A, f_F = correction factors for atomic number, absorption and fluorescence effects in sample and standard (x). As standards defined mineral phases, metals or alloys can be taken as polished samples.

Since many of the compounds of airborne particulates belong to the group of substances that are quantitatively analyzed in mineralogy, ceramics, or metallurgy on a routine basis, the choice of appropriate standards does not usually present a problem, and correction procedures can often be avoided. EDS is recommended for rapid quantitative analysis of particles. Appropriate spectra evaluation procedures[60] allow quantitative analysis having an accuracy not significantly lower than when WDS is employed.

Precision and accuracy of quantitative elemental analysis of particles larger than approximately 10 μm in diameter depend on sample preparation. If it is possible to embed and polish individual particles,[52,61] then the accuracy is the same as for mineralogical phase analysis — approximately 1 to 5 relative percent. If the particles are analyzed as they are, the influence of surface roughness decreases the precision and accuracy due to uncontrolled electron backscattering and X-ray absorption. Very irregular morphology (holes, etc.) presents a further problem. When sponge-like particles (as shown in Figure 4) must be analyzed, an estimation of the dead volume can help to make an empirical correction of X-ray intensities. Generally, an accuracy for quantitative analysis on the order of 5 to 10 relative percent should be obtainable.

If second period elements must be determined quantitatively, however, results are less accurate due to the large matrix effects encountered. Therefore, these elements are only detected and calculated from stoichiometry, e.g., oxygen in oxides, etc. This procedure demands that the stoichiometric relation between cation and anion is known. Unfortunately, this is not always the case, since a number of metals are present in

TABLE 2

Volume of X-ray Generation (Expressed as d_E) for some Compounds as a Function of Excitation Energy Calculated with Equation 1

Analyzed element and line	Compound	d_E (μm)				
		E=5 keV	E=10 keV	E=20 keV	E=30 keV	E=40 keV
$F(K_\alpha)$	NaF	0.37	1.22	4.0	8.0	13.1
$Al(K_\alpha)$	Al_2O_3	0.22	0.81	2.7	5.4	8.9
$Si(K_\alpha)$	SiO_2	0.23	0.84	2.8	5.6	9.2
$S(K_\alpha)$	$CaSO_4$	0.22	0.82	2.8	5.6	9.2
$Fe(K_\alpha)$	Fe_2O_3	–	0.27	2.3	5.2	8.8
$Ag(L_\alpha)$	AgJ	0.1	0.56	2.0	4.3	7.0
$Pb(L_\alpha)$	PbO	–	–	0.52	1.9	3.7
$Pb(M_\alpha)$	PbO	0.1	0.39	1.36	2.7	4.5

different oxidation states due to differences in the chemistry of the formation process.

Precision and accuracy can also be significantly lower when particles are analyzed that are only slightly larger than the excited volume. In this case, the orientation of the particle (most particles have irregular shapes) can influence the X-ray intensity due to partial occurrence of the effects typical for small particles (see below).

Particle analysis as described in this paragraph is primarily utilized for identification of larger particles collected in dust jars or filters and for particles precipitated with air cleaning devices or deposited between formation and emission. Pietzner and Schiffers[52] investigated the composition of particles that were collected from the combustion gas of the firing room of power plants heated with pulverized coal. The material collected consisted mainly of fine coal particles and fly ash. The authors embedded the particles into a resin and performed quantitative elemental analysis of the individual particles after polishing. LM yielded some clues indicating possible substances but was not able to clearly identify particles. The authors found that fly ash particles that have a size of less than 50 μm (the majority of the particles) show different compositions but have a homogeneous element distribution within the particle. About 60% of the particles were silicates of the type akermanite ($Ca_2MgSi_2O_7$) or melilithe [(Ca, Na)$_2 \cdot$(Al, Mg)(Si, Al)$_2O_7$].

Since only a fraction of these compounds (30%) could be detected with X-ray diffraction analysis, it was concluded that about half of the

silicates of fly ash are in an amorphous state. Furthermore, oxides were found (MgO, CaO, MgO$\cdot$Fe$_2$O$_3$, 2CaO$\cdot$Fe$_2$O$_3$, etc.) to be present as homogeneous mixtures. Figure 11 shows the LM and EPMA images of a sample containing silicates (small round particles) and an ion-manganese oxide particle.

Contrary to the particles smaller than about 50 μm, which are homogeneous, larger particles show distinct inhomogeneities. Two types of particles can be differentiated. One type contains a core of silica and then zones of K-Al-silicates, Ca-silicates, and Ca-Mg-Al-silicates (Figure 12). The second type has a core that is practically free of silicates but consists of oxides (Fe, Al, Ti, Mg, Ca, etc.) and is covered by an outside layer of Ca-Mg-Al-K-silicates. The detailed study of the structure of such particles provides important clues to the firing process and the formation of fly ash particles. The authors discuss this mechanism and emphasize the importance of the alkali components for the reduction of melting temperatures, which enables the formation of liquid phases in the firing process.

In immission situations, larger particles, which can be identified easily, often consist of minerals from the soil, road debris, or buildings. In industrial areas, anthropogenic substances can be present in significant quantities in the larger size fraction. Dorn et al.[62] found that in the proximity of a lead smelter, more than 50% of the lead, zinc, and copper occurs in particles which are larger than 2 μm. However, in urban aerosols, anthropogenic substances are primarily present in

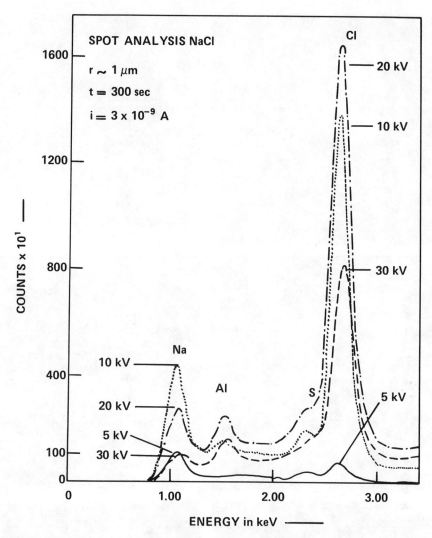

FIGURE 10.   Energy dispersive X-ray spectra of a 1 μm radius sea salt particle as a function of excitation energy. (From Schütz, L., *Berichte zue Elektronenmikroskopischen Direktabbildung von Oberflächen (BEDO)*, Vol. 9, Verlag Remy, Münster, Germany, in press.)

particle size fractions smaller than 1 μm.[5],[9] For this reason, it is necessary to develop analytical methods for the quantitative analysis of particles in the low micron and submicron range.

b. Quantitative Analysis of Particles Smaller Than Excited Volume

When particles smaller than a few microns in diameter are excited with an electron beam, the X-ray intensity is not only dependent on the concentration of the measured element in the particle but also on particle size for the following reasons:

1. The number of electrons (N_e) hitting the particle at fairly normal angle of incidence or hitting it at all decreases with particle size. Electrons striking the particle at angles smaller than 90° have a higher backscatter yield than a flat sample of the same atomic number; and electrons missing the particle are not available for X-ray production.

2. The probability that electrons leave the particle, due to side scattering or transmission before they have lost the total amount of energy exceeding the critical excitation energy, increases with descreasing particle size, thus diminishing X-ray intensity.

3. Absorption of the X-rays en route from the generation point to the particle sur-

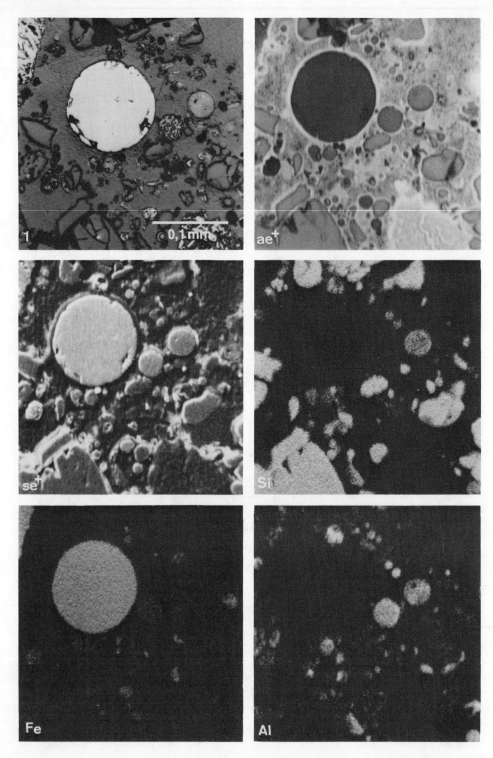

FIGURE 11. Fly ash particles (d < 50 μm) collected from firing room of a pulverized coal fired power plant. Particles were embedded and polished. 1, light microscopical image; ae,⁺ absorbed electron image; se,⁺ secondary electron image; distribution of the different elements (Si, Fe, Al, Ca, Mg, Na, K, Ti, Mn). (From Pietzner, H. and Schiffers, A., *Sonderheft der VGB,* Technische Vereingung der Grosskraftwerksbetreiber, Essen, Germany, 1972. With permission.)

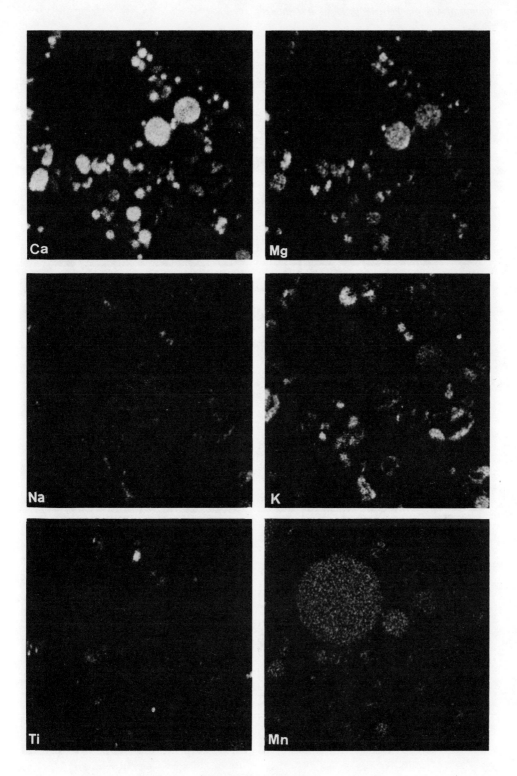

FIGURE 11. (continued)

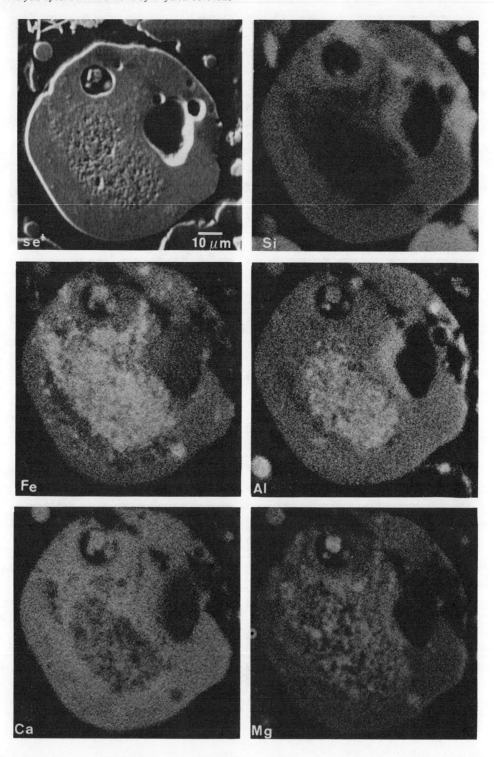

FIGURE 12.   Fly ash particle displaying an inhomogeneous structure typical for particles larger than 50 μm. (From Pietzner, H. and Schiffers, A. *Sonderheft der VGB,* Technische Vereingung der Grosskraftwerksbetreiber, Essen, Germany, 1972. With permission.)

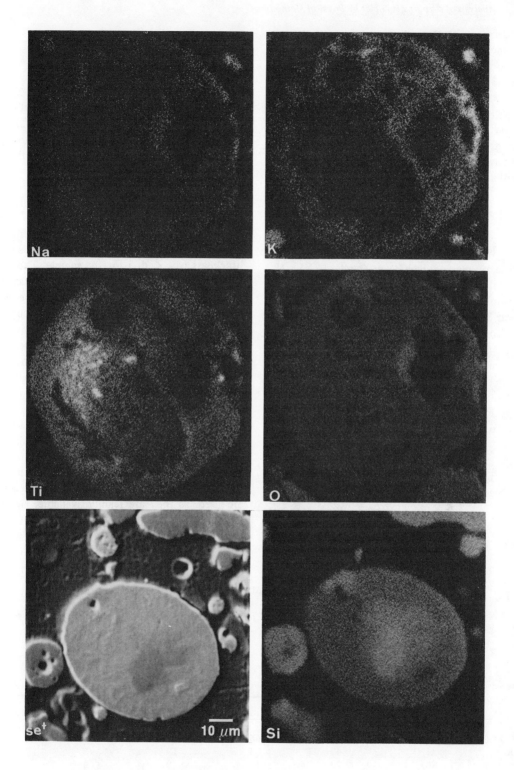

FIGURE 12. (continued)

face decreases heavily with decreasing particle size, thus increasing the measured intensity per mass unit of an element as compared to a bulk sample (Table 3).

For particles smaller than the excited volume, the relation between the measured X-ray intensity of element A in the sample and standard and its respective concentrations can be expressed by Equation 3:

$$\frac{I_A}{I_A^X} = \frac{c_A}{c_A^X} \cdot \frac{N_e}{N_e^X} \cdot \frac{f_{Z(A)} \cdot f_{A(A)} \cdot f_{F(A)}}{f_{z(A)}^x \cdot f_{A(A)}^x \cdot f_{F(A)}^x} \tag{3}$$

where N_e = number of electrons hitting the particle or measured phase in standard (X). N_e takes into account the influence of particle size on the number of electrons hitting the sample. If the chosen beam diameter is larger than the particle, then N_e is directly proportional to the projected particle area. A further advantage of this technique is minimal influence of surface morphology. $f_{z(A)}$ has the same meaning as in bulk analysis — that is, correction of the differences in backscattering and energy loss of the electrons between sample and standard. $fP_{A(A)}$ — the absorption correction factor — is the ratio of the X-ray intensity at the surface of a particle and the intensity generated under the same conditions within a bulk sample as defined by Equations 4 and 5.

$$f_{A(A)} = \frac{_0\int^{d_z} \phi_A(\rho z) \cdot e^{-\left(\frac{\mu}{\rho}\right)_A \cdot \rho \cdot x} d(\rho z)}{_0\int^{d_E} \phi_A(\rho z) d(\rho z)} \tag{4}$$

$$P_{A(A)}^x = \frac{_0\int^{d_E} \phi_A^x (\rho \cdot z) \cdot e^{-\left(\frac{\mu}{\rho}\right)_A \cdot \rho \cdot x} d(\rho \cdot z)}{_0\int^{d_E} \phi_A^x (\rho z) d(\rho z)} \tag{5}$$

where $\phi (\rho \cdot z)$ = depth (Z) distribution function of generated X-ray intensity and $e^{-\left(\frac{\mu}{\rho}\right) \cdot \rho \cdot x}$ = absorption term where $\frac{\mu}{\rho}$ = mass absorption coefficient; ρ = density; x = X-ray path length from point of generation to surface; d_z = particle thickness; and d_E = maximum depth of signal excitation. The absorption term, therefore, is not only dependent on composition but also on particle size, since the integral is taken from 0 to d_E for bulk samples but only to the actual particle thickness ($d_Z < d_E$) for the sample. The absorption correction factor contains the quantitative relation between particle size and X-ray generation and absorption. $f_{F(A)}$, the fluorescence correction factor, can normally be neglected, since these effects are usually less than 5 relative percent.[63] The dependence of X-ray intensity on particle diameter can be accounted for by using particulate material as a standard that has identical composition, particle size, and morphology. Unfortunately, such standards are hardly available. Therefore, these effects have to be taken into account by using suitable quantification methods for the X-ray intensity of the particles; thus, bulk standards (e.g., polished minerals and metals) can be used, as they are readily available. Generally, two basic approaches are especially important:

Firstly, estimation of the actual particle volume and correction of the measured intensity by

TABLE 3

Influence of Particle Size on Absorption of X-ray Intensity (d_E values for 20 keV)
Absorption of X-ray Intensity (%)

Analyzed element and line	Compound	Bulk sample $x=d_{E/2}$	Particles ($x = d_z/2$)			
			$x=1\ \mu m$	$x=0.5\ \mu m$	$x=0.2\ \mu m$	$x=0.1\ \mu m$
$F(K_\alpha)$	NaF	89	66	45	20	10
$Al(K_\alpha)$	Al_2O_3	43	34	19	8	4
$Si(K_\alpha)$	SiO_2	23	17	9	4	2
$S(K_\alpha)$	$CaSO_4$	14	11	6	2	1
$Fe(K_\alpha)$	Fe_2O_3	3	2	1	0.5	0.3
$Ag(L_\alpha)$	AgJ	29	29	16	7	3
$Pb(L_\alpha)$	PbO	2	6	3	1.2	0.6
$Pb(M_\alpha)$	PbO	60	49	29	13	7

normalizing to the volume excited in a bulk sample of the same or similar composition is determined by utilizing suitable calibration curves.[61] These calibration curves contain the measured X-ray intensity of one or more elements of a compound vs. the particle diameter (see Figure 13). Since the particle diameter can easily be measured in a microprobe or SEM and an approximate calculation of the volume is therefore possible, the expected X-ray loss can be determined from the curve and taken into account. According to Bayard[61] the accuracy of quantitative analysis applying this procedure is on the order of ±15 relative percent. Fiori et al.[64] proposed to use Monte Carlo techniques, which consist of a theoretical simulation of individual electron paths to calculate the X-ray output of particles of defined geometry as a function of size. First results indicate that such calibration curves can be calculated with high accuracy (Figure 13).

Secondly, intensity ratios of two elements in the sample and standard are employed. Bayard,[61] and Armstrong and Buseck[63] proposed to measure and evaluate the ratio of the X-ray intensities of two elements (A,B) in the sample and standard. Thus, the number of electrons hitting the particle is eliminated. The relation between intensity and concentration can be expressed by Equation 6 (the fluorescence correction factor is no longer considered):

$$\frac{\dfrac{I_A}{I_B}}{\dfrac{I_A^x}{I_B^x}} = \frac{\dfrac{c_A}{c_B}}{\dfrac{c_A^x}{c_A^x}} \cdot \frac{\dfrac{f_{Z(A)}}{f_{Z(B)}}}{\dfrac{f_{Z(A)}^x}{f_{Z(B)}^x}} \cdot \frac{\dfrac{f_{A(A)}}{f_{A(B)}}}{\dfrac{f_{A(A)}^x}{f_{A(B)}^x}} \tag{6}$$

If the backscattering characteristics, the depth distribution function of the generated X-ray spectrum, and the absorption properties are equal or nearly equal for element A and B, then the ratios of the atomic number and absorption correction factors for A and B become unity in the sample and standard. The ratios of the measured intensities then can be used directly to calculate the concentration ratio of these two elements in the particle as seen in Equation 7:

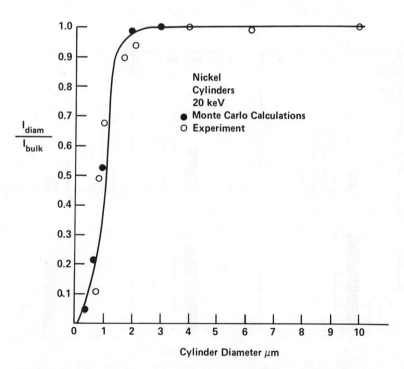

FIGURE 13. Dependance of measured and calculated X-ray intensity of nickel cylinders as function of cylinder diameter. (From Fiori, C. E., Heinrich, K. F. J., Myklebust, R. L., and Darr, M. M., *Nat. Bur. Stand. (U. S.) Spec. Publ.*, 422, 1283, 1976. With permission.)

$$\frac{c_A}{c_B} = \frac{\dfrac{I_A}{I_B}}{\dfrac{Ix_A}{I^x_B}} \cdot \frac{c^x_A}{c^x_B} \tag{7}$$

This means that by application of the ratio method, the dependence of the analytical result on particle diameter is eliminated if these conditions are fulfilled. This is generally true when the critical excitation energies (E_c) and mass absorption coefficients of the measured radiation of A and B are similar. In praxis these conditions can often be fulfilled by choosing appropriate element pairs or X-ray lines. However, in many other cases, especially when very small particles are analyzed, the effects of which are still more pronounced (as in STEM analysis which basically utilizes the same quantification procedure), this simple method will not produce sufficiently accurate results.

Armstrong and Buseck[63] developed a suitable mathematical procedure for taking into account the differences in electron backscattering, depth distribution function, and absorption term for the two elements in the particle and standard (particle model). The proposed evaluation model consists basically of a calculation of the atomic number correction factors for particle and standard (to correct for the differences in backscattering and stopping power), the generated intensity, and its absorption by integration of the function in Equation 4 between 0 and particle thickness; it takes into account the dependence of the X-ray path length on particle size, thus eliminating the influence of electron transmission and X-ray absorption on the analytical result. The application of these iterative calculations yields a corrected concentration ratio of the two elements.

Armstrong and Buseck[63] tested the model with a selection of silicate, oxide, and sulfide particles ranging in size from 0.5 to 20 μm. The experimentally determined and the calculated intensity ratio of Mn to Si in $MnSiO_3$ as a function of particle size are in strong agreement. The improvement in the accuracy of the analysis of Si, Al, and Ca in anorthite ($CaAl_2Si_2O_8$) is demonstrated in Figure 14. The authors point out that by application of the correction procedure, an average accuracy of ± 5 to 8 relative percent can be achieved. Generally, it can be stated that the ratio method is the quantitative elemental analysis

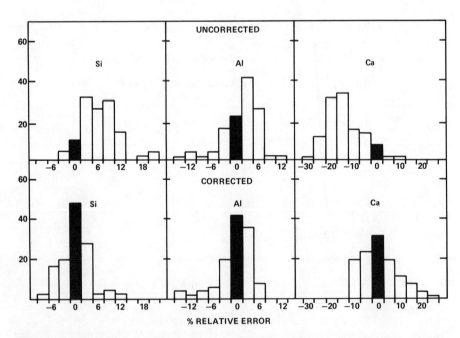

FIGURE 14.   Relative error in elemental analysis of Si, Al, and Ca in 122 particles of anorthite (size range 0.5 to 40 μm) encountered with and without correction procedure. (Reprinted with permission from Armstrong, J. T. and Buseck, P. R., *Anal. Chem.*, 47, 2178, 1975. Copyright by the American Chemical Society.)

procedure most widely applied to small particles. The concentration ratios obtained can be used to determine the compound present by comparing these ratios to the appropriate ratios of the compounds suspected. It is also possible to use these ratios to calculate the elemental composition directly if the stoichiometry with the elements that cannot be measured is known, e.g., if the assumption is valid that these elements are present as oxides in a certain stoichiometric relation (Al_2O_3, SiO_2, etc.). The elemental concentrations cannot be calculated when only one element can be measured (e.g., Fe_2O_3, $FeCO_3$, etc.) nor can a stoichiometric assumption be made, when the oxidation state of the element is unknown.

The possibility of the determination of the chemical formula of submicron particles depends on the accuracy of the elemental analysis, which (in the ideal case where no systematic influence changes X-ray intensity) is determined by counting statistics of the measured X-ray line. Counting precision determines the range of concentration of an element that is statistically proven. Since the absolute X-ray intensity decreases greatly with particle diameter, the relative standard deviation increases with decreasing particle size, thus enlarging the concentration interval in which the true concentration is contained. The interval is determined by $I \pm 2\sigma$ (for 95% confidence). Heidel and Desborough[65] demonstrated that using normal excitation conditions (15 keV excitation energy, 30 nA beam current measured in a Faraday cage, 10 sec counting time) for the analysis of 0.25-μm particles of different lead compounds with a commercial microprobe analyzer (ARL EMX-SM®, beam diameter $\sim$ 1 μm), the confidence interval of analysis is often much greater than the differences in lead content of the various lead particles. Figure 15 shows that counting precision of the Pb Mα intensities causes a significant overlap of the calculated lead concentration range for $(PbO)_2 \cdot PbCl_2$, $PbO_2 \cdot PbBrCl$, PbOHCl, $PbCl_2$, and PbBrCl, thus prohibiting an accurate identification of the different molecular species.

Insufficient X-ray intensity, and therefore the decrease of analytical precision, is also the size limiting factor. The minimum particle size for which a qualitative analysis is possible is given by the statistical law that the net signal has to be at least twice the standard deviation of the background.

For quantitative evaluation, the required accuracy (determined by the differences in the elemental concentrations in the compounds which can be present) sets the size limit, since the particle has to be large enough to deliver a net signal of sufficient precision. Therefore, it is difficult to give a general size limit for quantitative elemental analysis utilizing an electron microprobe analyzer. Landstrom and Kohler[66] report that particles as small as 0.1 μm can be successfully analyzed.

Important applications of EPMA for the chemical identification of submicron particles have been cited by Bayard[61] who identified lead components in automobile exhaust and emissions from cement plants; Henry and Blosser,[50] who investigated urban aerosols; Landstrom and Kohler,[66] who studied atmospheric samples and automobile emissions; and the Advanced Metals Research Corporation,[67] Durham, North Carolina. The latter report contains a thorough study of the effect of excitation conditions on X-ray analysis of particles ranging in size from 3 to 0.5 μm and presents results of the analysis of Los Angeles smog particles.

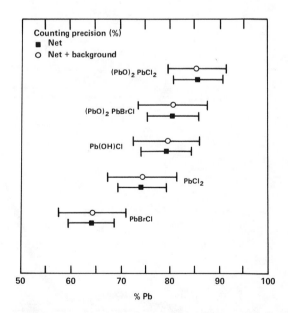

FIGURE 15. Confidence interval (2σ) of the quantitative analysis of lead in different compounds. Particle diameter is 0.25 μm. (From Heidel, R. H. and Desborough, G. A., *Environ. Pollut.*, 8, 185, 1975. With permission.)

3. Particle Identification by Evaluation of the Fine Structure of X-ray Valence Band Spectra

X-Ray valence band spectra of a solid originate from electron transitions between the valence band of a solid and an inner orbital. They are broad bands that may have a pronounced fine structure determined by chemical bonding of the element, whose X-ray spectrum is recorded. Therefore, X-ray valence band spectra not only contain information about the elements present in a solid but also, in many instances, about the compound, since the fine structure may be uniquely characteristic for a specific compound.[68-75] The most important valence band spectra that can be easily recorded with commercial electron microprobe analysers are the K bands of the second period elements, the K_β band of third period elements and the $L_{\alpha,\beta}$ bands of fourth period elements.

Figure 16 shows as an example the oxygen K band of Fe_2O_3 and $Fe_2(SO_4)_3$. The differences in peak wavelength, free width at half maximum (FWHM) of the bands, and peak symmetry are so pronounced that a distinction between those two compounds can be made on the basis of the fine structure.

The identification of the compound of an individual airborne particle must include a qualitative elemental analysis as a first step (performed with EDS) and the registration of the valence band spectra to obtain information about the species. The basic procedure is demonstrated in Figure 17, which contains a scheme for the identification of copper compounds. Qualitative elemental analysis will identify copper alone or copper and some other element (e.g., S or Fe) as the major component. Evaluation of the copper valence band

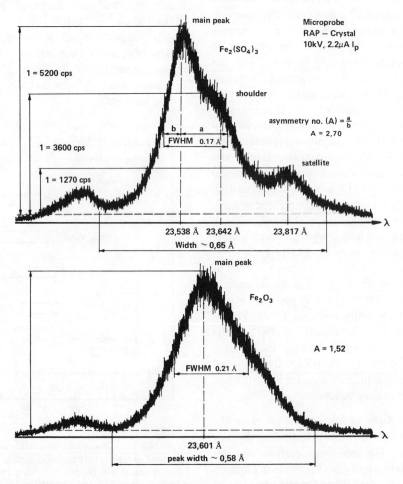

FIGURE 16. X-ray valence band spectrum of oxygen in Fe_2O_3 and $Fe_2(SO_4)_3$. (From Grasserbauer, M., *Z. Anal. Chem.*, 273, 401, 1975. With permission.)

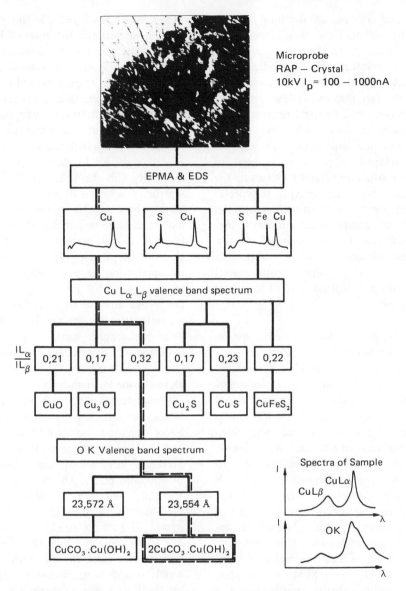

FIGURE 17. Scheme of the identification of copper compounds by qualitative elemental analysis and X-ray valence band spectroscopy. (From Grasserbauer, M., *Mikrochim. Acta*, 2, 55, 1975. With permission.)

spectrum yields information about the compound, e.g., for those substances which contain only Cu as an element detectable with EDS the L_β/L_α intensity ratio will allow differentiation between CuO, Cu_2O and the basic carbonates. These two compounds can be identified by distinct differences in the wavelength of the oxygen K peak. There are many more examples of the application of X-ray valence band spectra for identification of the chemical species, e.g., compounds of Fe, Mn, Cr, Si, Pb, and S have been identified.

The advantage of utilizing X-ray valence band spectra for the chemical identification of airborne particles is evidenced by the fact that the signal measured is directly typical of the compound; therefore, a full quantitative elemental analysis with stoichiometric calculation can be avoided. This compound-specific information can also be obtained for particles smaller than the excited volume, although the changes induced on the spectral fine features by the exponential decrease of absorption with diminishing particle size still have to be studied. The particle size limit depends primarily on the substance and ranges from 0.5 to 1 μm.

The basic disadvantages of the application of X-ray valence band spectra are the low count rates obtained and the fact that rather few systems of compounds, as compared to the large number of existing substances, can be identified. However, at the present time the capabilities are not yet completely developed. Work is in progress to enhance the fine structure (thereby increasing the number of identifiable compounds) of the spectra by mathematical treatment.[77] In its present state, valence band spectroscopy is an interesting supplementary technique for the chemical identification of some airborne substances — especially the characterization of different transition metal oxides for which the oxidation state of the metal can be determined.

4. Limitations of the Utilization of EPMA for the Chemical Identification of Airborne Particles

Although EPMA can be considered to be the most important method for the characterization of individual airborne particles of a size above several tenths of a micrometer, this method has limitations that make it necessary to use other techniques for a wide range of problems. Among these limitations is the destruction of particles due to heating in analysis. The nonconductive C, N, and S compounds can be destroyed under the electron beam. Heating can be minimized by using low beam currents. However, this reduces X-ray intensities and, therefore, demands highly sensitive detection devices, e.g., a high solid angle of X-ray acceptance of the EDS. Some SEM's are equipped with a variable sample detector arrangement. For analysis with low beam currents, the detector is moved closer to the sample. Extremely high detection sensitivities are obtained with STEM, in which the detector can be positioned very close to the sample.

X-Ray intensity is usually not sufficient to identify particles smaller than several tenths of a micron. To overcome this limit, the sensitivity of the detection system has to be increased, the background of the substrate reduced, and the beam diameter has to be reduced without great loss of beam intensity (in order to be able to focus the beam completely on a particle with a diameter well below 0.1 μm). Again, this can be achieved by application of a STEM, which can be equipped

with a highly sensitive EDS. This allows very small beam diameters (on the order of 100 Å) at beam currents that provide sufficient X-ray intensity by means of a special construction of the beam focusing system (highly excited objective lens) or the use of field emission guns. The use of electron transparent substrates reduces background to a small fraction of the bremsspectrum intensity encountered on a bulk element substrate.

Although EPMA has a very high absolute sensitivity (on the order of 10^{-15} g) its relative sensitivity, which is in the range of 0.1%, does not permit the detection of trace elements within an individual airborne particle. The relative sensitivity is only a few percent for the second period elements in particles. This limit can be overcome by application of a technique that has a high relative sensitivity for the whole element range. Ion probe microanalysis (IPMA) provides these characteristics because the signal yield is high, very little background signal is generated, and the detection sensitivity of the instruments is high.

D. Ion Probe Microanalysis

For investigation of an airborne particle with IPMA, the particle must lie on a substrate that has a low secondary ion yield in the region of interest and a low trace element level. McHugh and Stevens[13] advocate the use of high purity carbon or tantalum plates. The particles to be examined are transferred from the collection substrate with the aid of a micromanipulator. Negative primary ions ($^{16}O^-$) are preferred since the loss of particles due to charging is minimized. The lateral resolution of elemental analysis with an ion probe is equal to the beam diameter, which is approximately 1 to 2 μm for modern instruments under routine analysis conditions.

Qualitative analysis can be performed automatically with computer routines that evaluate the mass spectrum, which can look rather complicated even for simple substances due to the high sensitivity and the generation of a whole series of molecular ions. These complex molecular ions can be considered a background that limits the DL to about 100 ppm on the average. McHugh and Stevens[13] analyzed a series of oil soot particles and were able to identify 21 elements in the particles — occurring in concentrations ranging from approximately 10 ppm to more than 10%. The results of the analysis are shown in Figure 18. Comparing these results with those of EPMA, it

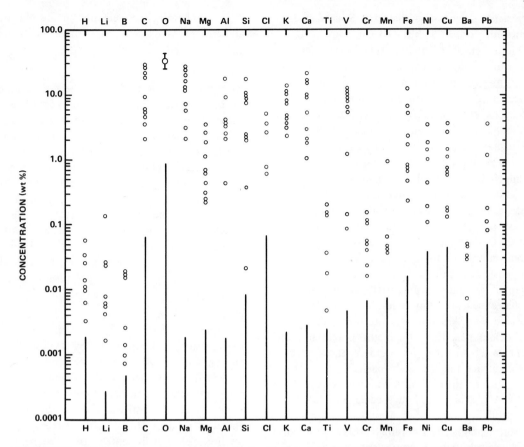

FIGURE 18. The ion microprobe analysis results for 10 oil soot particles ranging in size from 2 to 8 µm. The heights of the bars drawn in for each element represent the minimum detectable level in an oil soot matrix. (Reprinted with permission from McHugh, J. A. and Stevens, J. F., *Anal. Chem.,* 44, 2187, 1972. Copyright by the American Chemical Society.)

can be stated that with IPMA the additional elements of H, Li, Be, and B can be analyzed and the detection limit is about 1 to 3 magnitudes lower (depending on the element). Currently, the results obtained with IPMA must be considered only semiquantitative. Quantification is carried out with elemental sensitivity factors that must be determined experimentally from standard samples. Since no standard of similar composition was available in the case of oil soot particles, McHugh and Stevens[13] used refractory oxides, ceramics, and minerals to determine these factors. Sensitivity factors, however, are strongly dependent on the matrix composition. Therefore, the average accuracy of the analysis of the oil soot particles is only on the order of ±35% (with some elements being significantly worse). The development of suitable quantification procedures for IPMA is still in progress, thus there is considerable hope of improving this powerful method in this respect.

The use of IPMA (eventually supplemented by another surface analysis technique that has a high spatial resolution such as Auger electron spectroscopy) also enables the analyst to study the surface predominance of potentially toxic elements in airborne particles. Davison et al.[78] and Kaakinen et al.[79] attribute the phenomenon of surface enrichment of certain components to the condensation process of a species which has been previously volatilized in a high temperature combustion zone. Linton et al.[80] conclude from this that bulk analysis (which means in this case that the total volume of an individual particle is analyzed) leads to an underestimation of the potential environmental impact of airborne particles. This occurs because the surface regions have enhanced concentrations of elements, some of which are toxic, and the smallest particles display a higher concentration of such elements due to the larger surface area to volume ratios. The surface enrichment of potentially toxic metals is of special importance, since it is the particle surface that comes into contact with the human body.

Linton et al.[80] studied such surface enrichments of certain elements on fly ash particles from coal firings using bituminous coal. Samples were collected in the stack, thus providing large particle specimens (45 to 180 μm in diameter) that facilitated the investigation of individual particles. To assure accurate and extensive results, the authors used a multi-technique approach applying mainly IPMA, solvent leaching for surface characterization, and spark source mass spectrometery (SSMS) for bulk analysis. Solvent leaching of the fly ash particles with dimethyl sulfoxide (DMSO) and H_2O determined if elements were concentrated on the surface and the solubility of these elements, an important parameter for judging potential solubility in a natural environment. The leachates, leached and unleached particles were analyzed by SSMS and concentrations balanced; it was determined that the major potential trace elements Pb, Tl, Cr, Mn, and V show surface predominance and are highly leachable.

The authors then submitted individual particles to IPMA.[80] This technique is capable of determining depth profiles of a significant number of elements practically simultaneously, thus the depth distribution of important elements were studied. Although the depth resolution is somewhat constrained due to the rough surface of the particles, the depth profiles (Figure 19) indicate that there is a strong surface enrichment of Pb, Tl, Cr, Mn, and V within a zone of a few hundred Angstroms. Lead even reaches a surface concentration of 4%.

Linton et al.[80] also present an extensive discussion of their work. They report that on the basis of their studies with solvent leaching and IPMA it is possible to divide the constituents of fly ash into four categories:

1. The major elements Si, Al, and Ti, which show no surface predominance and low leachability

2. The volatile constituents of the original coal sample S, Li, Na, K, and Fe, which are surface enriched as sulfates and show moderate to high leachability

3. The trace elements (bulk concentration in fly ash <1000 ppm) Pb, Tl, Cr, Mn, and V, which show a strong surface enrichment and are moderately to highly leachable thus being of heavy environmental impact

4. The alkaline elements Ca and Mg, which are present as oxides, not surface predominant, and moderately leachable

As stated by the authors,[80] the use of modern surface analysis techniques, such as IPMA, greatly enhances the knowledge of the physico-chemical behavior of airborne particles. This information is the basis of a more accurate assessment of the physiological influence of particulates, aids a thorough investigation of gas reactions taking place at particle surface, which can be assumed to be influenced by the catalytic activity of the elements enriched at the surface. In this respect, it can be expected that these surface analysis techniques — especially those capable of single particle analysis — will be used extensively in pollution research.

Another reason that IPMA will play an important role in the chemical identification of airborne particles is that it opens up the possibilities of source identification by analysis of trace elements in individual particles. Andersen[81] was already able to identify the origin of asbestos fibers by determining the Be content of individual fibers. It can be estimated that other industrial emission products (e.g., particles from steel production, etc.) will contain trace elements that might allow an unequivocal source identification. Determination of the average trace content is often of little use since in immission measurements many different sources yield contributions.

E. Electron Microscopy

Electron microscopy performed with a TEM or STEM is the most important technique for the characterization of airborne particles in the submicron range. As already outlined in Section II, utilization of a STEM provides the advantages of determining X-ray spectra, electron diffraction patterns, and electron energy-loss spectra of individual submicron particles and permits investigation of size, shape, and morphology by high resolution transmitted electron (TE) or SE images. Although SEM can also be used to investigate submicron particles, the additional information provided by STEM, and which is definitely needed for unequivocal identification of a particle, makes this method far superior to SEM. A TEM, which is able to obtain selected area electron diffraction (SAED) patterns, is certainly a valuable tool, especially for studies of size and shape; however, it is inferior to an STEM equipped with an EDS.

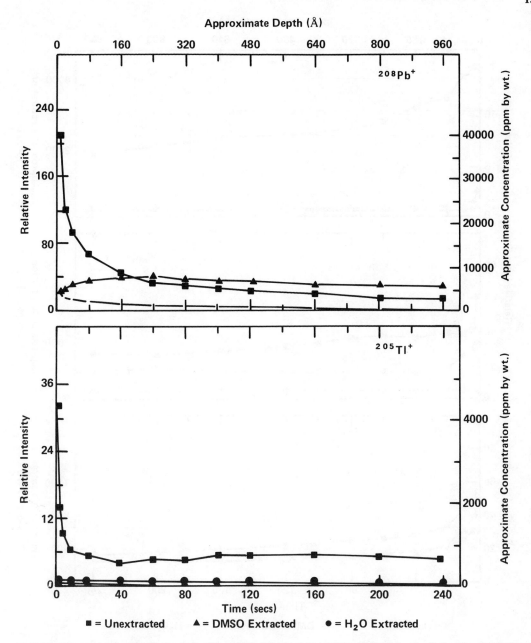

Approximate Depth (Å)

²⁰⁸Pb⁺

²⁰⁵Tl⁺

Time (secs)

■ = Unextracted   ▲ = DMSO Extracted   ● = H₂O Extracted

FIGURE 19. Ion microprobe depth profiles of Pb, Tl, Cr, Mn, and V obtained for leached and unleached fly ash particles. Sputter rate ∼ 4A/sec. (From Linton, R. W., Williams, P., Evans, C. A., and Natusch, D. F. S., *Anal. Chem.*, 49, 1514, 1977. Copyright by the American Chemical Society.)

1. Size, Shape, and Morphology of Submicron Particles

For study of size, shape, and morphology, the SE or TE signal can be used. Imaging in the TE mode has the advantage of (a) higher resolution (approximately 5 Å), (b) much higher contrast of the image, and (c) the possibility of observing the inner structure.

The higher resolution and higher contrast of TE imaging allows correct determination of size and shape, even of particles in the 100-Å region. Furthermore, it is possible to determine correct particle numbers in this 100-Å region since the particles can easily be differentiated from the substrate. This is especially important for asbestos fibers in particulates, since these often have a

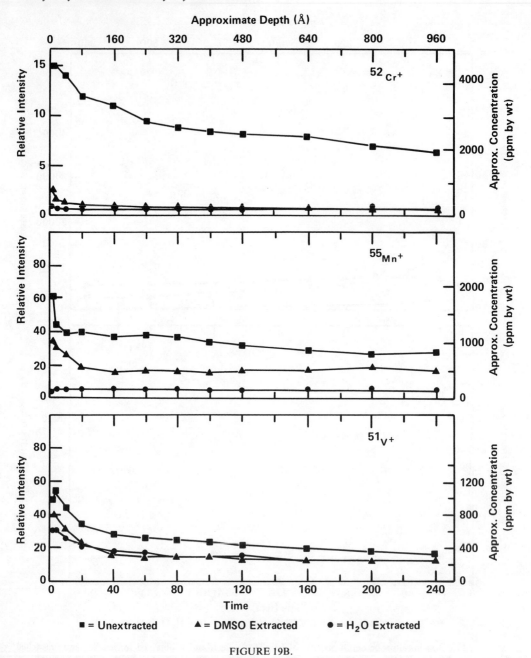

Approximate Depth (Å)

■ = Unextracted   ▲ = DMSO Extracted   ● = H₂O Extracted

FIGURE 19B.

diameter of only a few hundred Angstroms. Beaman and File[82] found that in TEM investigations of samples (talc) containing asbestos, superior imaging capabilities of small particles yielded an asbestos content 500 times higher than evaluations of SEM pictures. These authors also compare the TEM capabilities as far as detection of amphibole fibers (diameter 0.3 μm) with LM is concerned (phase contrast at 400 times and oil immersion at 1000 times magnification). They found that TEM concentrations on the average were 25 times higher.

Due to the high contrast of the image, automatic particle size analysis systems (see Section III.B.) can be used for TEM images but seldom for SE images. Evaluation of TEM images is the most important aspect of size characterization of very small particles. Renoux et al.[20] illustrated that by combining a suitable method of sampling (eight stage Andersen cascade impactor with electron microscope substrate on impactor plates and 0.4 μm Nuclepore filter) with TEM investigation of the particulates and their replicas. It is possible to obtain size distribution curves for aerosols ranging from 35 Å to 2 μm in diameter.

The possibility of observing the inner structure is of interest since such observation may yield clues on particle growth mechanisms and particle identity. Such an inner structure can usually be observed for chrysotile fibers, which are actually hollow tubes (see also Figure 28).

Shape parameters (especially the aspect ratio,[82]) which is the length-to-width ratio) play an important role in the characterization of submicron particles, since they allow a certain distinction between fibrous substances (aspect ratio >3 according to Beaman and File[82]) and particles. Unfortunately, these shape parameters alone are not sufficient to distinguish between asbestos and other materials, since quite a few substances can have a similar aspect ratio (plant fibers, fiber glass, nonasbestos silicates, synthetics, etc.). Therefore, it is practically impossible to identify a substance by its TEM appearance alone. This is also due to the fact that the often more distinctive surface morphology of a particle cannot be observed in the TEM mode. The SEM mode, as provided in a STEM, also yields more accurate information about the three-dimensional appearance of a crystal, while the TEM image is only a two dimensional silhouette. Figure 20 illustrates an indium oxide smoke particle as observed in different orientations in SEM and TEM.[83] To determine the important information here (namely, that the particle is crystalized as an octahedron), it would be necessary to investigate many particles of different orientation in the TE mode.

2. Chemical Identification of Submicron Particles with X-ray Spectra

The high beam density achieved by STEM, the solid-state detector array of the sample, and the use of electron transparent substrates permit determination of X-ray spectra of individual particles or fibers in the submicron range. These spectra are of sufficient intensity to allow quantitative analysis with the precision required for differentiation of most of the substances occurring in an aerosol. As an example, Figure 21 is a TEM micrograph of chrysotile fibers having a diameter of approximately 500 to 1000 Å and the ED spectrum obtained in 100 sec counting time. The important

elements Mg, Si, and Fe have sufficient intensity for evaluation. The copper peaks originate from the grid, which is excited by scattered electrons. The sulfur peak also results from excitation of a neighbor particle. The use of nylon grids (which can contain TiO_2 and traces of Cl[58]) diminishes this problem.

Quantitative evaluation of these X-ray spectra can principally be carried out in the same way as described in Section III.C.2.b. for EPMA of small particles. The comparison of intensity ratios determined for the particles with the intensity ratios obtained from standards seems to be preferred, although there is the problem of the eventual strong dependence of these ratios on particle size. This dependence is especially pronounced when the mass absorption coefficients of the element pair considered display larger differences. Beaman and File[82] determined these functions for a number of silicates, as shown in Figure 22. These authors also calculated the effect of particle size on X-ray intensity, for the particle dimensions normally investigated in STEM, using a ZAF correction procedure. They found that the correction factors (which indicate the difference in X-ray intensity between a small particle and a bulk standard) are on the order of 3 to 10 for Mg, Si, and Al in chrysotile fibers having a diameter of 300 to 400 Å but only 1.1 for Fe. Therefore, the quantitative analysis of such small particles is often quite inaccurate, requiring further assessment of an analytical result by evaluation of other signals. However, when judging the analytical criteria of the X-ray spectroscopic identification of such small particles, one must keep in mind that this is the only technique that yields direct semiquantitative chemical information about a particle; therefore, it is the most valuable method, which will certainly be developed further in the near future.

Suzuki, et al.[85] employed a somewhat modified version of determining elemental intensity ratios of quantitative analysis for particles approximately 0.5 to 1 μm in diameter. They calculated the intensity ratio of an element to the total net intensity in the spectrum for sample and standard. By comparing these normalized intensities in sample and standard, the concentration of the measured element can be determined. This ratio method has the same limitations as the other methods of determining X-ray intensity ratios. The authors carried out quantitative analysis of some

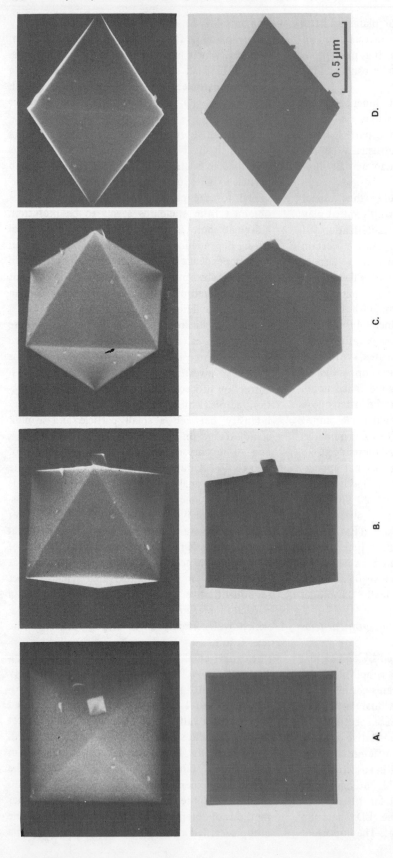

FIGURE 20.   SEM (top row) and TEM images of a single particle of In_2O_3. The particle in (A), (B), (C), and (D) is oriented with the [001], [210], [111], and [110] direction parallel to the incident beam, respectively. (From Shiojiri, M., Kaito, C., Yotsumoto, H., Aita S., *JEOL Jpn. Electron Opt. Lab. News*, 14e, 2, 2, 1976. With permission.)

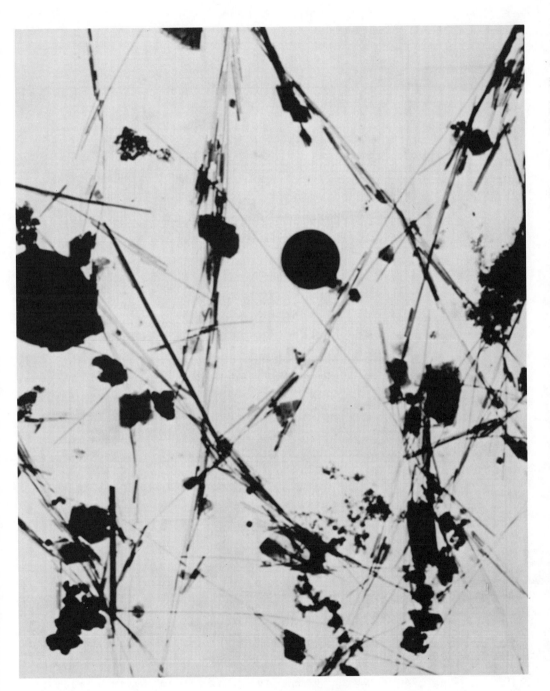

FIGURE 21.   TEM image of chrysotile asbestos (Magnification × 18,000) and ED spectrum of a single fiber. (From AEI Scientific Apparatus, Publ. 2050-27 1000 EdA 0976 Elmsford, New York, 1976. With permission.)

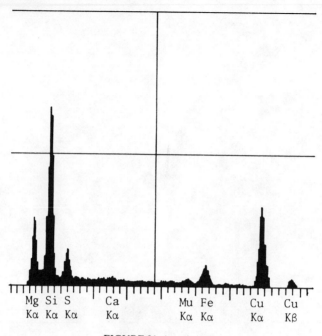

FIGURE 21. (continued)

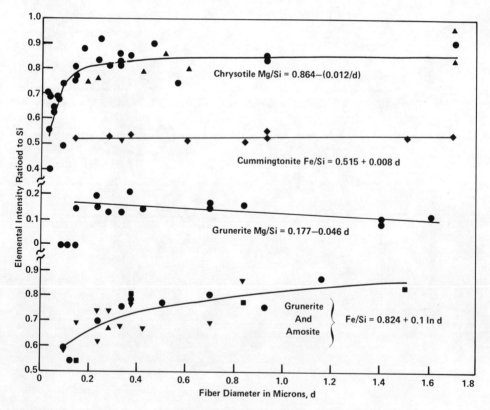

FIGURE 22.   The intensity of Mg and Fe ratioed to Si plotted as a function of fiber diameter (d) for several mineral standards (80 keV excitation energy). (Reprinted with permission from Beaman, D. R. and File, D. M., *Anal. Chem.*, 48, 101, 1976. Copyright by the American Chemical Society.)

asbestos minerals (chrysotile, amosite, and crocidolite) and kaolin (halloysite) and compared the composition obtained with the results of wet chemical analysis. Figure 23 illustrates the energy-dispersive X-ray spectra obtained from submicron particles of these substances. Table 4 lists the results obtained. The composition calculated from

the X-ray spectra compares quite well with chemical analysis.

3. Chemical Identification of Submicron Particles with Electron Diffraction Patterns

Size, shape, and morphology, as determined by TEM or STEM, are normally not sufficient data

Asbestos and Halloysite

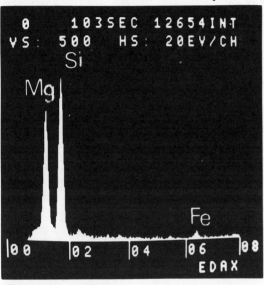

Chrysotile

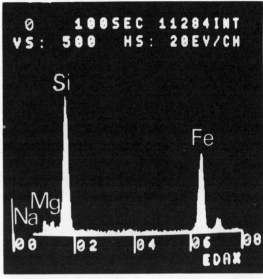

Crocidolite

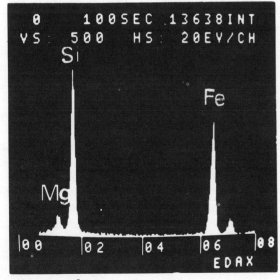

Amosite

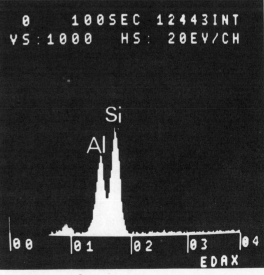

Halloysite

FIGURE 23. X-ray spectra of asbestos minerals and halloysite as obtained in a STEM with EDS at 40 keV excitation energy. (From Suzuki, M., Aita, S., and Hayeshi, H., *JEOL Jpn. Electron Opt. Lab. News*, 130, 1, 1975. With permission.)

TABLE 4

Comparison of X-ray Analysis of Individual Asbestos and Halloysite Particles with Chemical Analysis (STEM, 40 keV Excitation Energy)

| | ASBESTOS | | | | MINERALS | | KAOLIN | |
| | Chrysotile (Cassier) | | Amosite (Penge) | | Crocidolite (Kurumian) | | Halloysite (Kusatsu) | |
Composition	Chemical analysis (%)	EDS (%)	Chemical analysis (%)	EDS (%)	Chemical analysis (%)	EDS (%)	Chemical analysis (%)	EDS (%)
SiO_2	40.85	44.94	51.35	51.21	50.21	58.77	43.64	46.39
TiO_2	–		0.04		0.03			
Al_2O_3	0.95		1.45		0.49		36.41	36.26
Fe_2O_3	2.24 ⎫		3.00 ⎫		17.81 ⎫		1.33	
		2.46		35.27		30.43		
FeO	0.15 ⎭		32.17 ⎭		18.46 ⎭		0.32	
MnO	0.07		0.32		0.05			
MgO	41.38	36.93	5.95	8.7	3.35	3.52	0.20	
CaO	0.11		0.52		0.91		0.09	
K_2O	0.01		0.32		0.09		None	
Na_2O	0.03		0.10		5.63	4.26	None	
$H_2O\ (+)$	12.77		4.49		3.04		14.80	
$H_2O\ (-)$	0.98		0.23		0.39		2.90	
	Cr_2O_3							
P_2O_5	0.10							
	99.64		99.94		100.46		99.69	
Total								
	85.79	84.33	95.22	95.18	97.03	96.98	81.99	82.65

From Suzuki, M., Aita, S., and Hayashi, H., *JEOL Jpn. Electron Opt. Lab.*, 13e, 1, 1975. With permission.

for identification of airborne particles. More information about the particle must be gained — either from the X-ray spectra (as described in Section V.B.) or, if these are not accessible, from electron diffraction patterns. Electron diffraction of individual particles (selected area electron diffraction — SAED) can be carried out in practically all TEM and STEM systems, and thus it is a rather widely used technique.

The diffraction pattern obtained can be either a regular array of points, if the particle is a single crystal, or a few concentric rings, if the object is polycrystalline (shown schematically in Figure 24). The diffraction pattern of a single crystal represents a projection of a single plane in the reciprocal crystal lattice. By measuring this pattern it is possible to derive the constants of the reciprocal lattice, which in turn are related to the unit cell parameters (for further details, see McCrone and Delly[11]). Of special importance is the geometry of the array of points, which may indicate the crystal system, e.g., Figure 25 illustrates the hexagonal array of diffraction points

obtained from a hexagonal Be plate.[87] Often these patterns are used more or less as fingerprints of a specific substance.

The evaluation of the ring diagrams can be carried out in a straightforward manner using Bragg's equation. Lattice spacings of the particles are determined by measuring the diffraction ring diameters (for details see McCrone and Delly[11]). From these values a substance can be identified by the use of the American Society for the Testing of Materials (ASTM) cards. A typical ring diffraction pattern is shown in Figure 26. In this case, microcrystalline aggregates contain a sufficient number of fine crystal with random orientation to yield a ring pattern of montmorillonite.

Many airborne particles, however, do not yield a pure ring pattern or a pure point array; however, a mixed pattern is observed. These spotty lines are obtained from particles consisting of a small number of crystals as shown in Figure 27. Asbestos fibers — especially chrysotile — show a characteristic array of points which are extended to lines (Figure 28). Amorphous substances can

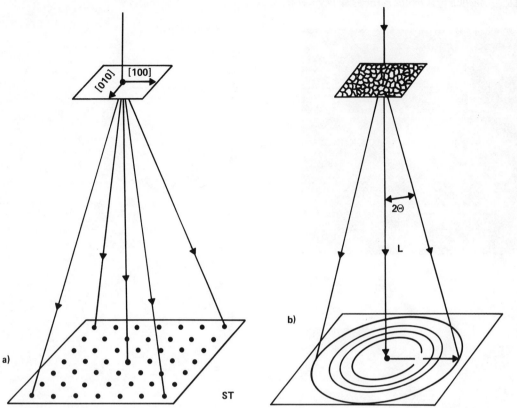

FIGURE 24. Electron diffraction patterns of a single crystal (a) and a polycrystalline particle (b). (From Radczewski, O. E., *Staub,* 22(8), 313, 1962. With permission.)

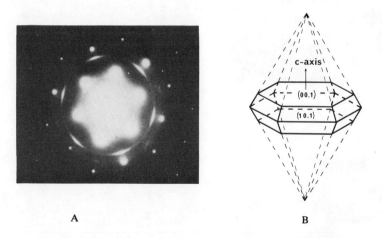

A   B

FIGURE 25. (A) SAED pattern of hexagonal Be plate. The incident beam is normal to the plate. The spots are due to Be metal, arcs are due to BeO. (B) clinographic projection of a Be crystal. (From Kimoto, K., and Nishida, I., *Jpn. J. Appl. Phys.,* 6, 1047, 1967. With permission.)

easily be differentiated from crystalline substances, because they do not show a distinct pattern (Figure 29).

SAED usually is used to provide supplementary information about a particle when the size, shape, and morphological parameters are not sufficient to yield the identity of a substance, e.g., as shown by McCrone and Delly;[11] they differentiated between

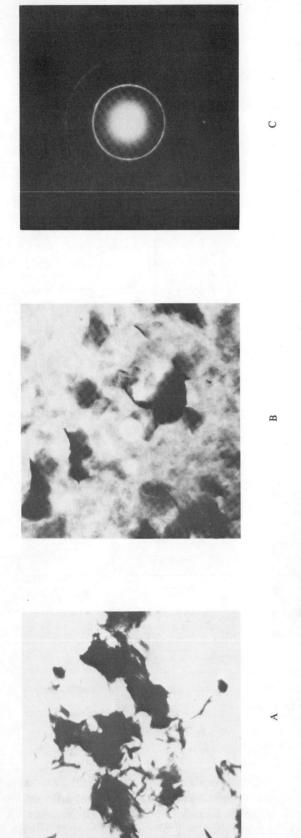

A

B

C

FIGURE 26. (A) TEM micrograph, magnification × 5,000; (B) TEM micrograph, × 25,000; and (C) SAED pattern of microcrystalline aggregates of montomorillonite (Mg, Ca) O · Al_2O_3 · $5SiO_2$ · ηH_2O. (From *The Particle Atlas*, Vol. 3, McCrone, W. C. and Delly, J. G., Ed., Ann Arbor Science, 1973. With permission.)

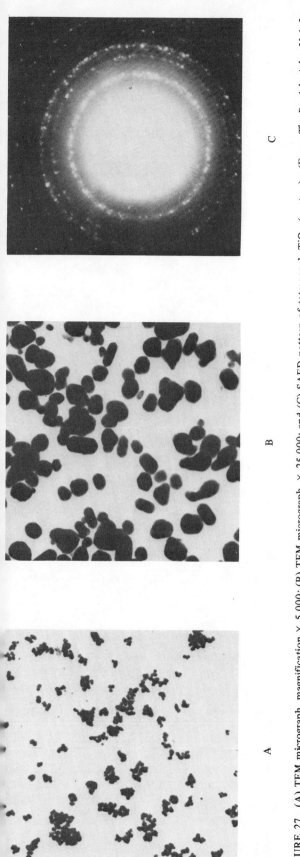

A B C

FIGURE 27. (A) TEM micrograph, magnification × 5,000; (B) TEM micrograph, × 25,000; and (C) SAED pattern of tetragonal TiO$_2$ (anatase). (From *The Particle Atlas*, Vol. 3, McCrone, W. C. and Delly, J. G., Ann Arbor Science, 1973. With permission.)

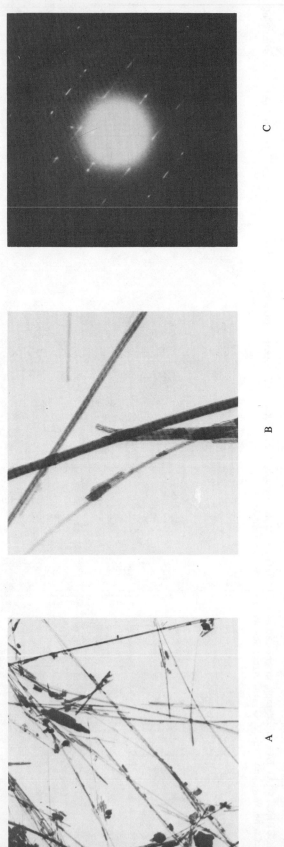

FIGURE 28. (A) TEM micrograph, magnification × 5,000; (B) TEM micrograph, × 25,000; and (C) SAED pattern of chrysotile asbestos. (From *The Particle Atlas,* Vol. 3, McCrone, W. C. and Delly, J. G., Ann Arbor Science, 1973. With permission.)

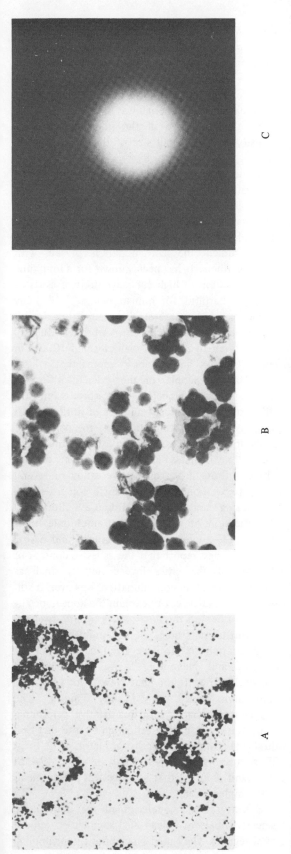

A   B   C

FIGURE 29. (A) TEM micrograph, magnification × 5,000; (B) TEM micrograph, × 25,000; and (C) SAED pattern of amorphous Fe₂O₃ . (From *The Particle Atlas*, Vol. 3, McCrone, W. C., and Delly, J. G., Anr Arbor Science, 1973. With permission.)

two substances that have a similar appearance on the TEM micrograph but show a completely different diffraction pattern. Identification on the basis of the combination of electron micrographs and electron diffraction data remains a rather complex task, carried out mainly to differentiate asbestos fibers from other fibrous materials, such as halloysite, since the major asbestos contaminant, chrysotile, displays a very distinct SAED pattern. For most other substances, this combined information is not sufficient for identification, thus, additional knowledge about a particle must be determined. The most important information is certainly the qualitative elemental composition, which can be obtained with EDS.

In the analytical praxis of the identification of submicron particles, all of the information obtained is usually combined to overcome the limitations associated with each of the signals used. Beaman and File[82] described such a combination of the different signals obtainable in STEM for the identification and quantitative determination of asbestos fibers among other particles. The first step involves a selection of the particles to be investigated further (particles that may consist of asbestos) on the basis of size, shape, and morphology. All particles with an aspect ratio (length over width) greater than three were classified as suspect asbestos fibers. The second step involves the production of SAED patterns of the individual fibers. On the basis of the patterns, the particles were put into one of these four categories:

1. Noncrystalline particles (having fibrous appearance).
2. Nonamphibole and nonserpentine fibers; these display a diffraction pattern that is distinctly different from amphiboles and serpentine. Categories 1 and 2 therefore contain nonasbestos particles.
3. Serpentine chrysotile and amphibole fibers; these are definitely identified on the basis of their SAED pattern.
4. Ambiguous fibers; these do not display a clearly identifiable SAED pattern, which may be due to insufficient size of the fibers or a preceding thermal, chemical, or mechanical treatment of asbestos that may alter the pattern.

The latter category was then analyzed using EDS. On the basis of evaluation of the intensity ratios Mg/Si, Al/Si, Ca/Si, Mn/Si, and Fe/Si and comparison of these ratios with mineral standards, an identification of asbestos fibers within the number of fibers classified as ambiguous was possible. The amount of asbestos fibers among the total particulate material was determined by relating the total size of the fibers identified as asbestos to the overall size of the whole particulate.

While Beaman and File[82] mainly demonstrated the methodology of accurate identification and determination of asbestos fibers and applied the combined technique for the analysis of a filtered particulate from water samples, other scientists have studied the contamination of air with asbestos fibers. It has been known for a long time that inhalation of high concentrations of asbestos fibers is harmful for human beings.[88-90] Lung tissues exposed to asbestos contain a large quantity of submicroscopic asbestos fibrils, frequently on the order of 300 Å in diameter and 1000 Å in length.[91] Holt and Young[92] report that a large percentage of people (on the order of 50%) who had no known connection with asbestos industry still showed a significant number of fibrils in their lung tissues. Therefore, the asbestos content of cities and rural areas (with no asbestos mining or industry nearby) has been of considerable concern. These authors did a qualitative study of several cities (e.g., London, Düsseldorf, Johannesburg, etc.) using sampling with Millipore® filters and identification of chrysotile and amphibole fibers by the TE microscopical appearance and SAED patterns. They report that the asbestos content of the air of the cities investigated is small as compared to other contamination; however, it still seems to be sufficient to explain the appearance of asbestos fibers in the lung tissues of city dwellers.

Quantitative studies of the ambient asbestos concentration in air have been carried out by Selikoff and Nicholson[93] and Nicholson and Langer.[94] They found that New York City concentrations ranged from 11 to 60 ng/m³. The problems associated with quantitative determination of asbestos can be estimated by the fact that duplicate samples differ by a factor of 2 to 3. These authors also investigated the air of 49 U.S. cities and reported that asbestos was present in all 200 samples taken.

Spurny et al.[95] determined the content of asbestos fibers of "clean air" sampled in remote rural regions. They used Nuclepore filter sampling and identification by TEM and EPMA. The authors stress the necessity of an identification of the fibers, since only about 10% of the fibrous material were found to be asbestos. The quantitative evaluation of the micrographs determined the the asbestos concentration at 10^2 to 10^3 particles per cubic meter.

Alste et al.[96] investigated the origin of asbestos in cities. They studied the morphology and chemical identify (evaluating SAED patterns) of airborne asbestos collected near a freeway in Melbourne, Australia and compared these with fibers obtained from worn brake linings. The authors found that asbestos concentration near a freeway ($\sim 5.10^5$ particles per cubic meter) were considerably higher than in other parts of the city, and that the fibers collected from air have a similar appearance and the same diffraction pattern as the fibers found in worn brake lining dust. Therefore, it was concluded that a significant amount of airborne asbestos of cities originates from the wear off of brake linings.

Investigation of biological tissues containing asbestos fibers is of central importance in asbestos studies. When fibers are deposited in the lung, they often become coated with a material composed largely of ferroprotein. Resulting structures are known as asbestos bodies or pseudoasbestos bodies.[92] Botham and Holt[97] state there is evidence that once a fiber is coated, it is no longer pathogenic; however, uncoated fibers produce the pathological effects. The rate of the coating depends on the nature of the fiber, e.g., chrysotile or glass fibers are coated more quickly than amphiboles; thus, the latter are more dangerous. Asbestos fibers in the lung are coated in the cytoplasm of macrophages and giant cells; however, if several fibers are retained by the same cell, usually only one is coated. The authors also emphasize that the number of macrophages in the lung is limited. This means that low concentrations of isolated fibers seem to have little pathological significance, but high concentrations might present a potential risk. Thus, it can be deduced that the investigation of lung tissues must include identification and study of the fibers, as well as determinations of the quantity coated with ferroprotein and of the total concentration of fibers in the tissue.

Langer and Pooley[98] established a standard method for the identification of asbestos in tissue. They employ a carbon extraction technique using ashed, thick (4 to 10 μm) sections of tissue. The sample is examined by TEM with SAED and EPMA, evaluating the morphological, structural, and chemical characteristics of the fibers. The method is especially useful for examining the asbestos content in tissue; however, does not yield information about the microanatomical relation between tissue and fiber. Suzuki et al.[91] use an STEM equipped with an EDS to study ultrathin sections of fiber containing tissues. Due to the high resolution imaging provided in an STEM and the capability of quantitative analysis of submicron areas, the structure of the tissue can be studied as well as the identity of the fiber determined *in situ*. The authors investigated sections of hamster lungs exposed to chrysotile, amosite, and crocidolite. They could clearly distinguish between chrysotile fibers partially invested by ferroprotein (Figure 30), asbestos bodies consisting of an amosite fibril, and numerous iron micells or crocidolite particles in the cytoplasm of an alveolar microphage. X-Ray spectra were evaluated by comparison with asbestos standards. The authors emphasized that asbestos fibers as small as 300 Å in diameter can be identified.

Electron microscopy has also been used extensively for the characterization of urban aerosols, e.g., Henry and Blosser[50] presented electron micrographs of airborne material collected in six major cities of the U.S. Heard and Wiffen[99] used electron microscopy for the identification and semiquantitative determination of sulfate particles in the submicron component of ambient aerosols.

A further area of research where electron microscopy plays a dominant role is in studies of particle growth mechanisms in the submicron range. In this case, the high resolution imaging and the identification capacity of electron microscopy is needed. Extensive model studies on particle formation by evaporation and condensation have been carried out by Kimoto and Nishida,[87] Nishida and Kimoto,[100] Yatsuya et al.,[101] Kasukabe et al.,[102] and Wada,[103] among others.

In addition to the identification of individual submicron particles, electron diffraction patterns can be used in some instances to differentiate between two chemically identical substances of different crystal structure. Radczewski[104] showed that it is possible to determine the crystal structure of individual K_2SiF_6 particles. This substance appears in cubic form as a mineral and in a hexagonal modification of anthropogenic origin. Elec-

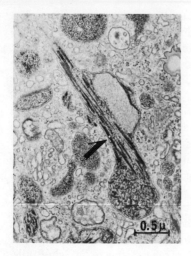

A.

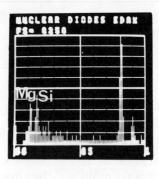

B.

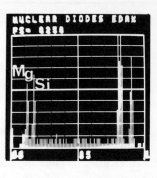

C.

FIGURE 30.   Phagosome containing chrysotile fibers and iron micelles. (A) TEM micrograph, (B) X-ray spectrum from the spot indicated, (C) X-ray spectrum of control crysotile. (From Suzuki, Y. and Ai Aita, S., *JEOL Jpn. Electron Opt. Lab. News,* 12e, 2, 3, 1974. With permission.)

tron diffraction, in this case, provides the unique possibility of determining direct information about the origin of such a fluoride.

The obvious limitations of identification of particles with SAED patterns lie in the rather tedious, and often dubious, evaluation of the diffractograms, and in the fact that the identification capability depends heavily on the identity of the substance and particle size. Larger particles may be opaque for the electron beam. For small particles, the diffracted intensity may be insufficient for visual pattern recognition, and a broadening of the diffraction lines or points is observed. The lower size limit for obtaining diffraction patterns is in the range of several hundred Angstroms.

The identification problem encountered in the evaluation of SAEDs or electron micrographs has led to combination of electron microscopic investigations and microchemical tests for specific substances. Bigg et al.[105] applied a thin film of $BaCl_2$ (300-Å thickness) to electron microscopic substrates, used this substrate for the collection of aerosols, and detected the particles consisting of sulfuric acid and ammonium sulfate. These authors also developed chemical procedures for the identification of nitrate, persulfates, and halides using the electron microscope. The special importance of performing microchemical reactions and studying the reaction products under the electron microscope is related to the fact that liquid particles (e.g., acid droplets) can be studied, and that certain anions (like persulfate and nitrate) can be identified, which cannot be done without applying chemistry.

4. Chemical Identification of Submicron Particles with Electron Energy-Loss Spectra

The chemical identification of submicron particles with electron energy-loss spectra is based on the relationship that the transmitted electrons suffer a loss of discrete amounts of energy due to ionizations of the elements in the particle and excitation of oscillations of the electron collective in the valence band of the solid (plasmon excitation). Therefore, the energy spectrum of the transmitted electrons (as shown in Figure 31) has discrete peaks within the spectrum whose energy is typical for the elements excited; e.g., the energy-loss spectrum of the aluminum coated MgO particles displays peaks at

- 15eV (1st plasmon peak of Al)
- 22eV (1st plasmon peak of MgO)
- 45eV (2nd plasmon peak of MgO)
- 50eV ($L_{2,3}$ ionization peak of Mg)
- 70eV ($L_{2,3}$ ionization peak of Al)

The spectrum can be differentiated using lock-in amplifiers, facilitating evaluation and increasing sensitivity of the results.

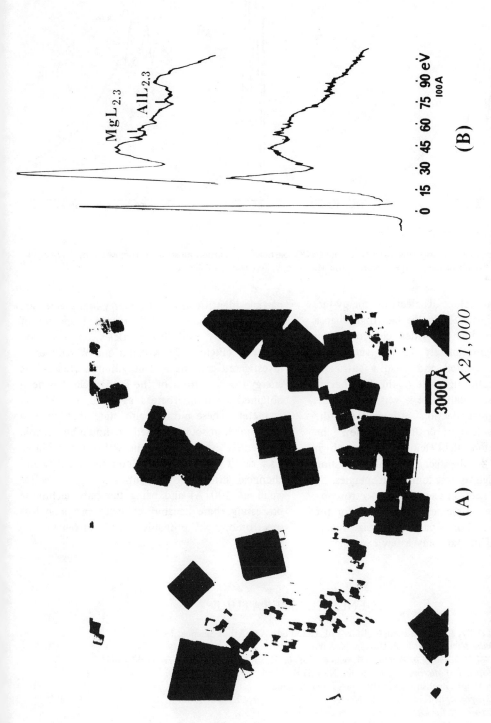

FIGURE 31. (A) TEM micrograph and (B) electron energy-loss spectra of MgO particles on evaporated Al film. The spectrum has been obtained from a specimen area of 100 Å in diameter. (From Kokubo, Y. and Wanatabe, M., *Element Analysis of Microareas by Electron Energy Analyzer, Japan Electron Optics Laboratory (JEOL), Akishima, Japan. With permission.)

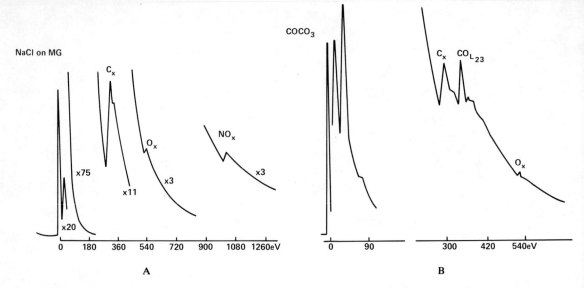

FIGURE 32. Electron energy loss spectra of NaCl and CaCO₃ particles on a carbon substrate. (From Kokubo, Y., Iwatsuki, M., and Koike, H., Sixth European Congr. on Electron Microscopy, Jerusalem, 1976.)

The analytical value of electron energy-loss spectroscopy is illustrated in the fact that when a STEM is used (which provides electron beam diameters of approximately 10 Å), very small particles can be excited and yield a sufficient intensity. The scattering angle of the TEs, having undergone inelastic collisions, is very small, thus providing a high collection efficiency of the spectrometer, which is positioned below the sample. Furthermore, light elements display a high signal yield due to the fact that the ionization cross section is high for low ionization energies.

It seems that electron energy-loss spectroscopy will be a suitable supplementary method for the identification of airborne particles in size ranges below 1000 Å. Thus far, only a few examples of

particle identification with energy-loss spectra are known. Kokubo et al.[107] recorded spectra of NaCl, CaCO₃ and BN particles on a carbon substrate (Figure 32). Kokubo and Wanatabe[106] investigated graphite and beryllium particles. The energy-loss spectrum of the Be particles has been obtained from microareas as small as 30 Å in diameter. These authors also state that when a field emission source is used, it should be possible to identify even single atoms with this technique. As a result of its ability to determine identity and chemical information about small particles (as small as 100 Å) and being the only technique possessing these capabilities, electron energy-loss spectroscopy will probably be of great value in the future.

GENERAL LITERATURE

McCrone, W. C. and Delly, J. G., *The Particle Atlas,* Vol. 1, 2nd ed., Ann Arbor Science, 1973.

Beiser, A., *Guide to the Microscope,* E. P. Dutton, New York, 1959.

Burrells, W., *Industrial Microscopy in Practice,* Morgan & Morgan, Hastings on Hudson, New York, 1961.

Clark, G. L., *Encyclopedia of Microscopy,* Reinhold, New York, 1961.

Hartley, W. G., *Microscopy,* English University Press, London, 1962.

Jones, R. M., *Basic Microscopic Techniques,* University of Chicago Press, Chicago, 1966.

Schaeffer, H. F., *Microscopy for Chemists,* Van Nostrand, New York, 1953.

Birks, L. S., *Electron Probe Microanalysis,* 2nd ed., Interscience, New York, 1969.

Beaman, D. R. and Isasi, J. A., *Electron Beam Microanalysis,* ASTM Spec. Tech. Publ. 506, American Society for the Testing of Materials, Philadelphia, 1972.

Andersen, C. A., *Microprobe Analysis,* Interscience, New York City, 1973.

Heinrich, K. F. J., Quantitative Electron Probe Microanalysis, *Nat. Bur. Stand. Spec. Publ.* No. 298, 1968.

Malissa, H., *Elektronenstrahlmikroanalyse,* Springer-Verlag, Vienna, 1966.
Goldstein, J. I. and Yakowitz, H., *Practical Scanning Electron Microscopy, Electron and Ion Microprobe Analysis,* Plenum, New York, 1975.
Reimer, L. and Pfefferkorn, G., *Rasterelektronenmikroskopie,* Springer-Verlag, Berlin, 1973.
Robinson, C. F., Ion microprobe instrumentation, in *Microprobe Analysis,* Andersen, C. A., Ed., Interscience, New York, 1973.
Andersen, C. A., Analytic methods and applications of the ion microprobe mass analyzer, in *Microprobe Analysis,* Andersen, C. A., Ed., Interscience, New York, 1973.
Morrison, G. H. and Slodzian, G., Ion microscopy, *Anal. Chem.,* 47, 932A, 1975.
Hayay, M. A., *Basic Electron Microscopy Technique,* Reinhold, New York, 1972.
Andrews, K. W., Dyson, D. J., and Keown, S. R., *Interpretation of Electron Diffraction Patterns,* Plenum, New York, 1967.
Hornbogen, E., *Durchstrahlungselektronenmikroskopie fester Stoffe,* Verlag Chemie, Weinheim, Germany, 1971.
Hall, C. E., *Introduction to Electron Microscopy,* McGraw-Hill, New York, 1953.
Vainshtein, B. K., *Structure Analysis by Electron Diffraction,* Pergamon Press, New York, 1967.
Kay, D., *Techniques for Electron Microscopy,* 2nd ed., F. A. Davis, Philadelphia, 1965.

REFERENCES

1. Dulka, J. J. and Risby, T. H., Ultratrace metals in some environmental and biological systems, *Anal. Chem.,* 48, 640A, 1976.
2. Knop, W., Heller, A., and Lahmann, E., *Technik der Luftreinhaltung,* 2nd ed., Krausskopf-Verlag, Mainz, Germany, 1972.
3. U.S. Department of Health, Education, and Welfare, Air Quality Criteria for Particulate Matter, National Air Pollution Control Boards, Publ. No. AP-49, Washington, D.C., 1969.
4. Hardy, K. A., Akselsson, R., Nelson, J. W., and Winchester, J. W., Elemental constituents of Miami aerosol as function of particle size, *Environ. Sci. Technol.,* 10, 176, 1976.
5. Lee, R. E., The size of suspended particulate matter in air, *Science,* 178, 567, 1972.
6. Natusch, D. F. S., Wallace, J. R., and Evans, C. A., Toxic trace elements: preferential concentration in respirable particles, *Science,* 183, 202, 1974.
7. Heisler, S. L., Friedlaender, S. K., and Husar, R. B., The relationship of smog aerosol size and chemical element distributions to source characteristics, *Atmos. Environ.,* 7, 633, 1973.
8. Gartrell, G. and Friedlaender, S. K., Relating particulate pollution to sources: the 1972 California aerosol characterization study, *Atmos. Environ.,* 9, 279, 1975.
9. Hidy, G. M., Appel, B. R., Charlson, R. J., Clark, W. E., Friedlaender, S. K., Hutchinson, D. H., Smith, T. B., Suder, J., Wesolowski, J. J., and Whitby, K. T., Summary of the California aerosol characterization experiment, *J. Air Pollut. Control Assoc.,* 25, 1106, 1975.
10. Husar, R. B. and Whitby, K. T., Growth mechanisms and size spectra of photochemical aerosols, *Environ. Sci. Technol.,* 7, 241, 1973.
11. McCrone, W. C. and Delly, J. G., *The Particle Atlas,* Vol. 1, 2nd ed., Ann Arbor Science, Ann Arbor, Mich., 1973.
12. Bailey, G. W. and Ellis, J. R., *Microscope,* 14, 306, 1965.
13. McHugh, J. A. and Stevens, J. F., Elemental analysis of single micrometer-size airborne particulates by ion microprobe mass spectrometry, *Anal. Chem.,* 44, 2187, 1972.
14. Nuclepore® Membranes and Quantitative Analysis, Brochure of Nuclepore Corporation, Pleasanton, Cal., 1976.
15. Spurny, K. R., Lodge, J. P., Jr., Frank, E. R., and Sheesley, D. C., Aerosol filtration by means of nuclepore filters — aerosol sampling and measurement, *Environ. Sci. Technol.,* 3, 453, 1969.
16. Spurny, K. R., Ackermann, E. R., Lodge, J. P., and Tesarova, I., Modern Physical Methods of Chemical Analysis of Aerosols — New Applications of Analytical Nuclepore Filters in Polluted Atmosphere, *Proc. 3rd Int. Clean Air Congress,* VDI-Verlag, Düsseldorf, 1973.
17. Frank, E. R., Spurny, K. R., Sheesley, D. C., and Lodge, J. P., The use of nuclepore filters in light and electron microscopy of aerosols, *J. Microsc.,* 9, 735, 1970.
18. Denee, P. B. and Stein, R. L., An evaluation of dust sampling membrane filters for use in the scanning electron microscope, *Powder Technol.,* 5, 201, 1971/72.
19. Heard, M. J. and Wiffen, R. D., An automatic device for continuous sampling of the atmospheric aerosol for electron microscopy, *Atmos. Environ.,* 6, 343, 1972.
20. Renoux, A., Butor, J. F., and Madelaine, G., Use of electron microscopy for determination of the granulometric distribution of atmospheric aerosols with radii between $3.5.10^{-3}$ μm and 1 μm, *Chemosphere,* 3(3), 119, 1974.

21. **Schütz, L.,** Analysis of Atmospheric Aerosol Particles Using a Scanning Electron Microscope, *Berichte zur Elektronenmikroskopischen Direktabbildung von Oberflächen (BEDO),* Vol. 9, Verlag Remy, Münster, Germany, in press.

22. **McCrone, W. C. and Delly, J. G.,** *The Particle Atlas,* Vol. 2, 2nd ed., Ann Arbor Science, Mich., 1973.

23. **Brown, K. M., McCrone, W. C., Kulm, R., and Forlini, A. L.,** Dispersion staining. I. Theory, method and apparatus. II. The systematic application to the identification of transparent substances, *Microsc. Cryst. Front,* 13, 311, 1963; 14, 39, 1963.

24. **Crossmon, G. C.,** Dispersion staining microscopy as applied to industrial hygiene, *Am. Ind. Hyg. Assoc. Q.,* 18, 341, 1957.

25. **Grabar, D. G. and Principe, A. H.,** Identification of glass fragments by measurement of refractive index and dispersion, *J. Forensic Sci.,* 8, 54, 1963.

26. **Grabar, D. G.,** Application of dispersion staining to microscope identification of settled dust, *J. Air Pollut. Control Assoc.,* 12, 560, 1962.

27. **Thaer, A.,** Ein Beitrag zur lichtmikroskopischen Mineralbestimmung in Feinstäuben, insbesondere des Steinkohlenbergbaus, *Staub,* 38, 30, 1954.

28. **Giever, P. M.,** Analysis of number and size of particulate pollutants, in *Air Pollution,* Vol. 2, Academic Press, New York, 1968, 249.

29. Mikro-Videomat 2, Informationsblatt Mikro No. 12, Carl Zeiss, D-7082 Oberkochen, Germany, 1976.

30. Leitz-Textur Analyse System, E. Leitz GmbH, D-6330 Wetzlar, Germany, 1976.

31. **Riediger, G.,** Teilchenzählung und Teilchengrössenanalyse mit dem quantitativen Fernseh-Mikroskop Quantimet 720, *Staub,* 32, 3, 96, 1972.

32. **Levy, J. D.,** An image analysis system, *Int. Lab.,* p. 81, May/June 1976.

33. **Malissa, H., Kaltenbrunner, J., and Grasserbauer, M.,** Ein Beitrag zur Gefügeanalyse mit der Mikrosonde, *Mikrochim. Acta, Suppl.,* 5, 453, 1974.

34. **Alex, W.,** Prinzipien und Systematic der Zählverfahren in der Teilchengrössenanalyse, *Aufbereit. Tech.,* 13(2, 3, 10, 11), 1972.

35. **Exner, H. E.,** Analysis of grain and particle size distributions in metallic materials, *Int. Metall. Rev.,* 17, 111, 1972.

36. **Exner, H. E.,** European Instruments for Quantitative Image Analysis, in *Stereology and Quantitative Metallography,* ASTM Special Tech. Publ. 504, American Society for the Testing of Materials, Philadelphia, 1972.

37. **Underwood, E. E.,** *Quantitative Stereology,* Addison-Wesley, Reading, Mass., 1970.

38. **Malissa, H.,** Stereometrische Analyse mit Hilfe der Elektronenstrahlmikroanalyse, *Z. Anal. Chem.,* 273, 449, 1975.

39. **Fischmeister, H. F.,** Applications of quantitative microscopy in material engineering, *J. Microsc.,* 95, 119, 1972.

40. **Gahm, J.,** Die mikroskopische Bildanalyse in der Mineralogie, *Fortschr. Mineral.,* 53, 1, 79, 1975.

41. **Bartel, W.,** *Korngrössenmesstechnik,* Springer-Verlag, Berlin, 1970.

42. **Herdan, G. and Smith, M. L.,** *Small Particle Statistics,* Butterworths, London, 1960.

43. **Brenneis, H. J.,** Überqualitative Mikroelektrolysen mittels kleiner Elektroden, *Mikrochemie,* 9, 385, 1931.

44. **Seeley, B. K.,** Detection of micron and submicron chloride particles, *Anal. Chem.,* 24, 576, 1952.

45. **Malissa, H. and Benedett-Pichler, A. A.,** *Anorganische Qualitative Mikroanalyse,* Springer-Verlag, Vienna, 1958.

46. **Malissa, H.,** Über die Empfindlichkeit mikroanalytischer Reaktionen, *Mikrochim. Acta,* 35, 266, 1950; 38, 33, 1951.

47. **McCrone, W. C. and Delly, J. G.,** *The Particle Atlas,* Vol. 4, 2nd ed., Ann Arbor Science, Ann Arbor, Mich., 1973.

48. **Malissa, H. and Grasserbauer, M.,** Die Bedeutung physikalischer Mikromethoden zur Untersuchung von Stäuben, *Mikrochim. Acta,* 2, 325, 1975.

49. **McCrone, W. C. and Delly, J. G.,** *The Particle Atlas,* Vol. 3, 2nd ed., Ann Arbor Science, Ann Arbor, Mich., 1973.

50. **Henry, W. M. and Blosser, E. R.,** A Study of the Nature of the Chemical Characteristics of Particulates Collected from Ambient Air, Contract CPA 22-69-153; EPA Report PB 220 401/4, Durham, N.C., 1970.

51. **Cheng, R. J., Mohnen, V. A., Shen, T. T., Current, M., and Hudson, J. B.,** Characterization of particulates from power plants, *J. Air Pollut. Control Assoc.,* 26, 787, 1976.

52. **Pietzner, H. and Schiffers, A.,** Mineralogical and chemical investigations of firing residues and depositions from the furnace of pulverized coal fired power plants, *Sonderheft der VGB,* Technische Vereinigung der Grosskraftwerksbetreiber, Essen, Germany, 1972.

53. **Malissa, H.,** Integrated dust analysis by physical methods, *Angew. Chem. Int. Ed. Engl.,* 15, 123, 1976.

54. **Schulz, E. J., Engdahl, R. B., and Frankenberg, T. T.,** Submicron particles from a pulverized coal fired boiler, *Atmos. Environ.,* 9, 111, 1975.

55. **Stöber, W., Flachsbart, H., and Hochrainer, D.,** Der aerodynamische Durchmesser von Latexaggregaten und Asbestfasern, *Staub,* 30(7), 277, 1970.

56. **Buckle, E. R. and Pointon, K. C.,** Growth and sedimentation in cadmium aerosols, *Atmos. Environ.,* 8, 1335, 1974.

57. **Ricks, G. R. and Williams, J. H.,** Effects of atmospheric pollution on decidous woodland. II. Effects of particulate matter upon stomatal diffusion resistance in leaves of *Quercus Petraea* (Mattuschka) Leibl., *Environ. Pollut.,* 6, 87, 1974.

58. **Maggiore, C. J. and Rubin, I. B.,** Optimization of an SEM X-ray spectrometer system for the identification and characterization of ultramicroscopic particles, *Scanning Electron Microscopy* (Part 1), Illinois Institute of Technology, Research Institute, Chicago, Ill., 1973.

59. **Wittry, D. B.,** Resolution of electron probe microanalyzers, *J. Appl. Phys.,* 30, 953, 1959.
60. **Russ, J. C.,** *Energy Dispersion X-Ray Analysis,* Spec. Tech. Publ. 485, American Society for the Testing of Materials, Philadelphia, 1971.
61. **Bayard, M.,** Applications of the electron microprobe to the analysis of free particulates, in *Microprobe Analysis,* Andersen, C. A., Ed., Interscience, New York, 1973.
62. **Dorn, C. R., Pierce, J. O., Phillips, P. E., and Chase, G. R.,** Airborne Pb, Cd, Zn and Cu concentration by particle size near a Pb Smelter, *Atmos. Environ.,* 10, 443, 1976.
63. **Armstrong, J. T. and Buseck, P. R.,** Quantitative chemical analysis of individual microparticles using the electron microprobe: theoretical, *Anal. Chem.,* 47, 2178, 1975.
64. **Fiori, C. E., Heinrich, K. F. J., Myklebust, R. L., and Darr, M. M.,** Observations on the quantitative electron probe microanalysis of particles, *Nat. Bur. Stand. (U.S.) Spec. Publ.,* 422, 1283, 1976.
65. **Heidel, R. H. and Desborough, G. A.,** Limitations on analysis of small particles with an electron probe: pollution studies, *Environ. Pollut.,* 8, 185, 1975.
66. **Landstrom, D. K. and Kohler, D.,** Electron Microprobe Analysis of Atmosphere Aerorols, EPA Report PB 189 282/BE, Durham, N.C., 1969.
67. **Advanced Metals Research Corp.,** Burlington, Mass.: Electron Microprobe X-ray Analysis of Atmospheric Aerorol Particles, Contract PHD-CPA-22-69-26; EPA Report PB-189 283, Durham, N.C., 1969.
68. **Holiday, J. E.,** Soft X-ray valence state effects in conductors, *Adv. X-ray Anal.,* 13, 136, 1970.
69. **Holiday, J. E.,** Investigation of the carbon K and metal emission bands and bonding for stoichiometric and nonstoichiometric carbides, *J. Appl. Phys.,* 38, 4720, 1970.
70. **Holiday, J. E.,** The use of soft X-ray spectroscopy as a tool for studying the surface region of metals and alloys, *Adv. X-ray Anal.,* 16, 53, 1973.
71. **Nagel, D. J.,** Interpretation of valence band X-ray spectra, *Adv. X-Ray Anal.,* 13, 182, 1970.
72. **Nagel, D. J. and Baun, W. L.,** Bonding effects in X-ray spectra, in *X-Ray Spectroscopy,* Azároff, L. V., Ed., McGraw-Hill, New York, 1974.
73. **White, E. W.,** Applications of soft X-ray spectroscopy to chemical bonding studies with the electron microprobe, in *Microprobe Analysis,* Andersen, C. A., Ed., Interscience, New York, 1973.
74. **Fischer, D. W.,** Chemical bonding and valence state — nonmetals, *Adv. X-Ray Anal.,* 13, 159, 1970.
75. **Grasserbauer, M.,** Valenzbandspektroskopie mit der Mikrosonde, *Z. Anal. Chem.,* 273, 401, 1975.
76. **Grasserbauer, M.,** Die Bedeutung der Valenzbandspektren in der Elektronenstrahl — Mikroanalyse. IV. Qualitative Verbindungsidentifizierung mit Valenzbandspektren, *Mikrochim. Acta,* 2, 55, 1975.
77. **Drack, H. and Grasserbauer, M.,** Evaluation of X-ray valence band spectra by mathematical treatment, *Mikrochim. Acta, Suppl.,* 7, 289, 1977.
78. **Davison, R. L., Natusch, D. F. S., Wallace, J. R., and Evans, C. A.,** Trace elements in fly ash — dependence of concentration of particle size, *Environ. Sci. Technol.,* 8, 1107, 1974.
79. **Kaakinen, J. W., Jorden, R. M., Lawasani, M. H., and West, R. E.,** Trace element behaviour in coal fired power plants, *Environ. Sci. Technol.,* 9, 862, 1975.
80. **Linton, R. W., Williams, P., Evans, C. A., and Natusch, D. F. S.,** Determination of the surface predominance of toxic elements in airborne particles by ion microprobe mass spectrometry and Auger electron spectrometry, *Anal., Chem.,* 49, 1514, 1977.
81. **Andersen, C. A.,** The Present State of Ion Probe Microanalysis, 7th Colloquium on Metallurgical Analysis, Vienna, 1974.
82. **Beaman, D. R. and File, D. M.,** Quantitative determination of asbestos fiber concentrations, *Anal. Chem.,* 48, 101, 1976.
83. **Shiojiri, M., Kaito, C., Yotsumoto, H., and Aita, S.,** The scanning electron microscopic observation of nonmetallic smoke particles, *JEOL Jpn. Electron Opt. Lab. News,* 14e(2), 2, 1976.
84. *Analysis of Wall Repair Pastes,* Publ. 2050-27 1000 EdA 0976, AEI Scientific Apparatus, Elmsford, New York, 1976.
85. **Suzuki, M., Aita, S., and Hayashi, H.,** An attempted use of the analytical electron microscope for semiquantitative analysis of clay minerals, *JEOL Jpn. Electron Opt. Lab. News,* 13e, 1, 1975.
86. **Radczewski, O. E.,** Elektronenoptische Untersuchung feinkörniger Minerale, *Staub,* 22(8), 313, 1962.
87. **Kimoto, K. and Nishida, I.,** An electron microscope and electron diffraction study of fine smoke particles prepared by evaporation in argon gas at low pressures (II), *Jpn. J. Appl. Phys.,* 6, 1047, 1967.
88. **Merewether, E. R. A.,** The occurance of pulmonary fibrosis and other pulmonary infections in asbestos workers, *J. Ind. Hyg.,* 12, 229, 1930.
89. **Harris, R. L.,** *A Model for the Deposition of Microscopic Fibers in the Human Respiratory System,* Report, School of Public Health, University of North Carolina, Chapel Hill, 1972.
90. **Robock, K. and Klosterkotter, W.,** Investigations in the cytotoxicity of asbestos dusts, *Staub,* 33, 445, 1973.
91. **Suzuki, Y., Aita, S., Hoshino, T., and Iwata, H.,** Identification of submicroscopic asbestos fibrils in tissue by analytical electron microscopy, *JEOL Jpn. Electron Opt. Lab. News,* 12e(2), 2, 1974.
92. **Holt, P. F. and Young, D. K.,** Asbestos fibers in the atmosphere of towns, *Atmos. Environ.,* 7, 481, 1973.
93. **Selikoff, I. J., Nicholson, W. J., and Langer, A. M.,** Asbestos air pollution, *Arch. Environ. Health,* 25, 1, 1972.
94. **Nicholson, W. J., Langer, A. M., and Selikoff, I. J.,** Asbestos in the atmosphere of towns, *Atmos. Environ.,* 7, 666, 1973.

95. **Spurny, K. R., Stöber, W., Opiela, H., and Weiss, G.,** Microscope et analyse des aerosols d'amiante en air atmosphérique, *Atmos. Pollut.,* 1976.
96. **Alste, J., Watson, D., and Bagg, J.,** Airborne asbestos in the vicinity of a freeway, *Atmos. Environ.,* 10, 583, 1976.
97. **Botham, S. K. and Holt, P. F.,** The mechanism of formation of asbestos bodies, *J. Pathol. Bacteriol.,* 96, 443, 1968.
98. **Langer, A. M. and Pooley, F. D.,** Identification of single asbestos fibers in human tissues, Biological Effects of Asbestos, Proc. Int. Agency Res. Cancer, Lyon, France, 1972.
99. **Heard, M. J. and Wiffen, R. D.,** Electron microscopy of natural aerosols and the identification of particulate ammonium sulfate, *Atmos. Environ.,* 3, 337, 1969.
100. **Nishida, I. and Komoto, K.,** Electron microscope and electron diffraction study of fine particles prepared by evaporation in argon at low pressure. IV. Fine particles of tellurium, *Jpn. J. Appl. Phys.,* 14, 1425, 1975.
101. **Yatsuya, S., Kasukabe, S., and Uyeda, R.,** Formation of ultrafine metal particles by gas evaporation technique. I. Aluminium in helium, *Jpn. J. Appl. Phys.,* 12, 1675, 1973.
102. **Kasukabe, S., Yatsuya, S., and Uyeda, R.,** Ultrafine metal particles formed by gas evaporation technique. II. Crystal habits of magnesium, manganese, beryllium and tellurium, *Jpn. J. Appl. Phys.,* 13, 1714, 1974.
103. **Wada, N.,** Preparation of fine metal particles by the gas evaporation method with plasma jet flame, *Jpn. J. Appl. Phys.,* 8, 551, 1969.
104. **Radczewski, D. E.,** Bestimmung feinster anorganischer Verunreinigungen der Luft, *Siemens Z.,* 48, 2, 82, 1974.
105. **Bigg, E. K., Ono, A., and Williams, J. A.,** Chemical tests for individual submicron aerosol particles, *Atmos. Environ.,* 8, 1, 1974.
106. **Kokubo, Y. and Wanatabe, M.,** *Element Analysis of Microareas by Electron Energy Analyzer,* Special report, Japan Electron Optics Laboratory, Tokyo.
107. **Kokubo, Y., Iwatsuki, M., and Koike, H.,** Light element analysis by electron energy analyzer, Sixth European Congr. Electron Microscopy, Jerusalem, 1976.

Chapter 9
X-RAY DIFFRACTION

C. O. Ruud and P. A. Russell

TABLE OF CONTENTS

I. INTRODUCTION

Many techniques of analytical chemistry readily identify and determine the quantity of elemental constituents present but do not distinguish the chemical forms of elements present as various compounds in a mixture. X-Ray diffraction, however, can provide this information for crystalline materials.

A species must be crystalline to be analyzed by X-ray diffraction. Therefore, X-ray diffraction (XRD) is usually applicable to only solid materials. Fortunately, most inorganic compounds, including virtually all minerals and many organic solids, are crystalline. Noncrystalline materials encountered as airborne particles are organic polymers and inorganic glasses; these can produce diffuse spectra which are seldom of use in XRD analysis.

Although limited to the identification of crystalline materials, the XRD method permits identification of chemical species at relatively low concentrations and is nondestructive. The technique is often useful for the identification of crystalline materials that are hygroscopic.

II. PRINCIPLES

The X-ray diffraction method[6,12,13] exploits the fact that there is a regular crystal structure in most solids and that this structure forms a three-dimensional pattern of atoms in space. The three-dimensional pattern formed is dictated by direction and geometrical close-packing considerations inherent in bonding. The three-dimensional array of atoms can be conceived of as having planes of atoms stacked on top of each other.

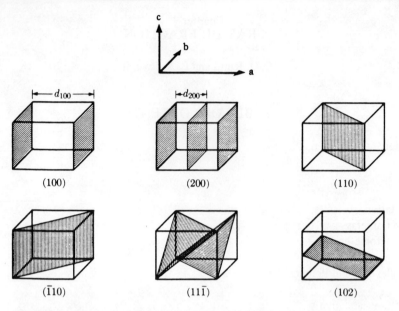

FIGURE 1. Planes, interplanar spacing (*d*), and Miller indices (h,k,1). (From Cullity, B.D., *Elements of X-Ray Diffraction,* 1st ed., Addisson-Wesley, Reading, Mass., 1956, 39. With permission.)

Each set of planes has a constant interplanar spacing (d) as seen in one view of an arrangement of atoms shown in Figure 1. The atoms can be imagined as being at the corners of the solid shapes. This particular arrangement can be viewed from many other attitudes which would show a different set of planes, each with different d values. The Miller indices identifying various planes are numerical descriptions of the planes that can be used in crystallographic computations. Each crystalline compound has a unique combination of atom positions and atom spacings, and therefore each has a unique combination of atoms constructing planes of various d spaces or interatomic distances, usually measured in angstroms.

These d spacings can be accurately measured using X-rays; thus, each crystalline species can be identified by the pattern of d spacing measurements produced by X-rays. When a collimated, monochromatic (nearly single wavelength) beam of X-rays strikes a crystalline surface at changing angles, the planes of atoms reflect part of the rays. The reflected rays are phase related and thus produce a pattern determined by destructive and constructive interference. Constructive interference usually takes place in only a few directions relative to the angle of the impinging X-rays and can be used to accurately measure interatomic

spacing. The relation which describes diffraction is Bragg's law:

$$n\lambda = 2d \sin \theta$$

which may be rewritten:

$$\frac{1}{d} = \frac{2}{n\lambda} \sin \theta$$

The analytical technique, then, for identifying compounds consists of obtaining an unknown d by determining an angle θ at which constructive interference occurs and a diffracted X-ray beam is detected. This is done by making the quantity $\frac{2}{n\lambda}$ a constant. The value λ is the X-ray wavelength being diffracted, usually measured in angstroms; n is usually unity. A characteristic X-ray wavelength which may be considered a single wavelength, or monochromatic, is usually selected as the diffracting electromagnetic radiation. This is done by selecting a metallic X-ray tube target which gives a strong characteristic X-ray emission within the wavelength range of interest. The most commonly used target in X-ray diffractometry is copper. Copper radiation is usually utilized because it is not heavily absorbed by air and produces a diffracted spectrum from most crystalline species in a range of diffraction angles (or Bragg angles, i.e., θ), which is convenient to use. Further

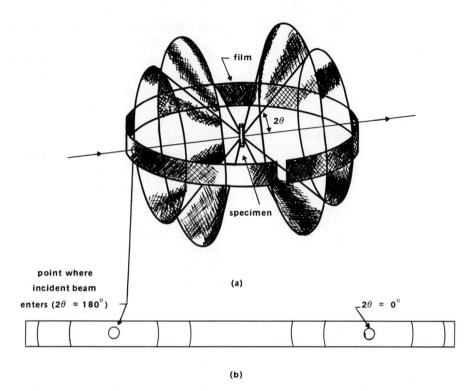

film

2θ

specimen

point where
incident beam
enters (2θ = 180°)

(a)

2θ = 0°

(b)

FIGURE 2. Illustration of a cone of diffracted radiation from a powder camera arrangement. Lower portion shows a flattened film which was used to record sections of the cones. (From Cullity, B. D., *Elements of X-ray Diffraction,* Addison-Wesley, Reading, Mass., 1956, 95. With permission.)

considerations regarding X-ray tube targets will be mentioned later.

One other very important point should be considered. The reflected X-ray intensity diffracted from a given volume of crystal is dependent upon the intensity of the incoming X-ray or incident beam and upon the various atomic densities in the planes from which X-rays are being diffracted. There are a number of other, more subtle considerations in the intensity of the diffracted X-ray beam, and these are well covered in the references.

Much of the early research using XRD was usually performed to determine crystalline structure and employed single crystal techniques. However, for the most part, the chemists, environmental engineers, mineralogists, and other applied scientists have restricted their practical interest to X-ray powder diffraction. The specimen is polycrystalline, consisting of many randomly oriented tiny crystallites. Each crystallite consists of several million atoms and its dimensions range from approximately 0.1 to 50 μm. Random orientation and uniform distribution of the crystallites is

important in powder diffractometry in that samples without these properties may produce serious deviations from predicted results.

III. INSTRUMENTATION

The object of powder X-ray diffractometry, for either quantitative determination or qualitative identification of crystalline materials in a sample, is to determine the angle θ at which diffraction, constructive interference, occurs. If there are only a few crystallites of a given species of interest in a sample, only a few orientations of those crystallites would be available to reflect X-rays in some directions, and this will usually not be sufficient to be detected by the X-ray instrumentation. Usually several thousand randomly oriented crystals of a given species are necessary so that the X-rays are diffracted in a cone as shown in Figure 2.

Figure 2 illustrates how X-rays from one set of planes are diffracted as many tiny individual spots of X-rays, each from a slightly different crystallite orientation, that combine to produce a continuous cone of X-rays. A detection device is placed in an

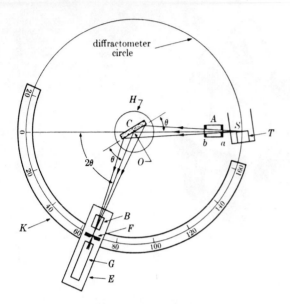

FIGURE 3.   X-Ray diffractometer (schematic). (From Cullity, B. D., *Elements of X-ray Diffraction,* 1st ed., Addison-Wesley, Reading, Mass., 1956, 179. With permission.)

orientation so as to intercept the cone of X-rays perpendicular to the direction of the beam (see Figure 2). Thus, angle θ of the diffracted beams and sometimes their intensities can be estimated. There are two types of instrumentation generally used in powder diffraction: film cameras and diffractometers.

A. Film Cameras

In a film camera, photographic film which is darkened upon X-ray exposure is orientated in such a way around the specimen that it intercepts the cones of diffracted X-rays, as shown in Figure 2. The resulting film shows segments of the cone as lines or circle segments (see Figure 2). The most common camera design used is the Debye-Scherrer camera. It requires that a sample be placed in a capillary tube or mounted on a glass fiber using petroleum jelly or some other greasy, noncrystalline material and placed in the center of the camera. The camera is then placed adjacent to an X-ray source (with a suitable beta filter to help monochromatize the incident X-rays) such that a strong, point source beam is aligned to cause X-rays to pass through the collimator of the X-ray camera. The amount of sample required for Debye-Scherrer cameras is quite small, on the order of only tens of micrograms; therefore, this technique has been most useful in situations where

only a limited amount of sample was available. In the Debye-Scherrer camera, the sample is rotated while it is being irradiated to increase the number of reflections which produces a well-defined line rather than a series of individual spots.

The relative positions of the lines on the developed photographic film are measured. From these measurements, two-theta values can be calculated which can be simply converted to d spacings. Usually the relative darkness of the lines can be used as an indication of relative X-ray intensity. A special X-ray film reader is commonly used to facilitate reading films to obtain accurate d spacings. The use of film cameras has often been applied to relatively pure samples in order to determine lattice parameters of crystalline compounds. Precise lattice parameter determinations by Debye-Scherrer cameras are used extensively throughout academia and industry.

Often individual particles of interest that are visible in an optical microscope can be extracted from the sample under investigation using a fine tungsten filament. Particles as small as 1 μm in diameter or larger can be placed at the end of a glass rod in the Debye-Scherrer camera, exposed to X-rays, and identified, if crystalline. A special type of powder camera[8] can be used to produce a line pattern without necessitating crushing. This camera rotates and permutates the single crystal sample to such a degree that a sufficient number of reflections are obtained to produce a well-defined line on the photographic film.

B. Diffractometer

The most widely used instrument for X-ray quantitative analysis or qualitative crystalline species identification is the diffractometer. It requires more sample material than a film camera. Figure 3 shows a schematic of the geometrical arrangements of this instrument. A, B, and F are slits to select photons with only certain directional vectors; C is the sample; E and H are supports; G is the X-ray detector; K is the travel of the detector graduated in degrees of two theta; O is the axis of the diffractometer goniometer; and T is the X-ray source. This source is the line source[12,13] of an X-ray diffraction tube, and the X-ray detector is usually a gas proportional or scintillation counter. The goniometer is a mechanical device which rotates the sample at its axis at θ degrees, while the detector (detector slit) is on the circumference and rotates at two-theta degrees. The use of beta filters and monochromators is not discussed herein, but their application to both diffractometers and film cameras is vital for optimization of peak height above background and sublimation or elimination of K-beta reflections. The sample for diffractometry usually is required to cover an area of approximately 1 cm^2 and be a few tenths of a millimeter thick. Therefore, a mass on the order of 0.1 g is sufficient. However, there are a number of techniques used to restrict the area of sample radiated by the X-rays; therefore, smaller sample sizes can be examined. The sample is commonly a powder consisting of particles between 0.1 and 50 μm. Larger particle sizes contribute to anomalous results due to the small number of crystal orientations contributing to X-ray diffraction. Particles less than 0.1 μm give rise to divergence of the diffracted angle vs. intensity distribution and cause peak broadening.

The operation technique for diffractometry consists of scanning, either continuous or stepwise, the goniometer over several degrees of two theta. For qualitative identification of species, several degrees two theta ranging from 100 to approximately 2 are usually scanned at a constant rate. Quantitative analysis is accomplished by slowly scanning over specific ranges of two-theta values that may contain a diffracted peak of interest and identifying peak two-theta values. It is usually necessary to integrate the counts under a peak for quantitative analysis; however, for semiquantitative analysis, the relative height of the peak is usually sufficient. Figure 4 shows a diffracted X-ray intensity vs. two-theta pattern. For very large volumes of X-ray diffraction work and certain other specific applications, the automated powder diffractometer (APD) is an extremely useful tool. A dedicated minicomputer is used to control the goniometer and analyze raw data.

The alignment of the diffractometer, goniometer, slit position, monochromator adjustment, filter selection, tube target selection, and counter tube electronic adjustments such as pulse height discrimination, time constant, scanning rates, and slit combinations are all important considerations in qualitative diffractometry and of vital importance in quantitative diffractometry. An often overlooked problem of older X-ray generators is the stability of the X-ray power supply with time. Since scanning is a time function, the incident X-ray intensity must be constant with time. Selections of the target tube for the proper

dispersion of the d-space spectrum and incoherent scattering (X-ray fluorescence) from the sample must also be considered.

IV. SAMPLE COLLECTION AND PREPARATION

Sample collection methodology will depend largely upon the sampling environment, namely, the concentration of unknown materials of interest and of materials with interfering diffraction spectra. Thus, it is necessary to collect enough material to produce concentrations sufficient for detection. By reducing the amount of possible interfering material, it is also possible to effectively increase instrument sensitivity.

Concentrations of material to be analyzed vary from extremely high levels, such as those collected in a processing mill, to the very low levels encountered in an ambient air sample. Samples are usually collected on glass fiber, thin polymer membrane, or silver filters. For ambient samples, the filter may be preceded by a cascade impactor, virtual impactor, or cyclone to eliminate the "nonrespirable" fraction of particulates. This separation may also separate compounds that produce interfering spectra.

A successful technique that provides a low background with little spectral interference utilizes silver membrane filters. The material to be examined may be deposited directly on a silver filter or redeposited from ashed polyvinyl chloride (PVC) membranes.[1,15] Detection limits for quartz using this technique are approximately 5 μgm/cm^2. The silver membrane filters are relatively expensive, and the material transfer techniques require significant handling and technical expertise. Also, if many of the particulates being deposited or redeposited are smaller than the average filter pore size, a significant amount of material may not be detected because of heavy metal shadowing by the filter itself.[10]

Recently, a technique that provides excellent sensitivity (2 μg/cm^2 for quartz) using PVC filter substrates has been documented by Henslee and Guerra.[10] This methodology permits direct examination of collected material on a much less expensive medium and eliminates possible errors caused by filter shadowing of small particles.

Sometimes materials produce overlying spectra, e.g., PbSO$_4$ and SiO$_2$ (quartz). If the materials are present as different aerodynamically sized par-

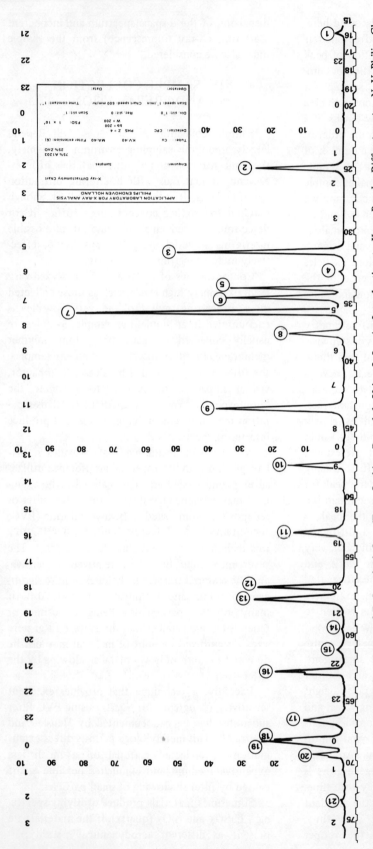

FIGURE 4. Diffraction pattern of a mixture of ZnO and Al$_2$O$_3$. (From Jenkins, R. and de Vries, J. L., *An Introduction to X-ray Powder Diffractometry*, N. V. Philips Gloeilampenfabrieken, Eindhoven, Holland, 1975, 31. With permission.)

d	3.04	2.29	2.10	3.86	$CaCO_3$	★
I/I_1	100	18	18	12	Calcium Carbonate	(Calcite)

Rad. $CuK\alpha_1$ λ 1.5405 Filter Ni Dia. Cut off I/I_1 Diffractometer I/I Ref. Swanson and Fuyat, NBS Circular 539, Vol. II, 51 (1953)					

d A	I/I_1	hkl	d A	I/I_1	hkl
3.86	12	102	1.297	2	218
3.035	100	104	1.284	1	306
2.845	3	006	1.247	1	220
2.495	14	110	1.235	2	1112
2.285	18	113	1.1795	3	2110
2.095	18	202	1.1538	3	314
1.927	5	204	1.1425	1	226
1.913	17	108	1.1244	<1	2111
1.875	17	116	1.0613	1	2014
1.626	4	211	1.0473	3	404
1.604	8	212	1.0447	4	138
1.587	2	1010	1.0352	2	0116, 1115
1.525	5	214	1.0234	<1	1213
1.518	4	208	1.0118	2	3012
1.510	3	119	0.9895	<1	231
1.473	2	215	0.9846	1	322
1.440	5	300	0.9782	1	1017
1.422	3	0012	0.9767	3	2114
1.356	1	217	0.9655	2	234
1.339	2	2010			

Sys. Hexagonal   S.G. $D_3^6 d$ - $R\bar{3}c$
a_0 4.989 b_0 c_0 17.062 A   C 3.420
α β γ Z 6 Dx 2.711
Ref. Ibid.

$\epsilon\alpha$ □ $\omega\beta$ 1.659 $\epsilon\gamma$ 1.487 Sign -
2V   D   mp   Color
Ref. Ibid.

Sample from Mallinckrodt Chem. Works.   Spect. anal.:
<0.1% Sr; <0.01% Ba; <0.001% Al, B, Cs, Cu, K, Mg, Na,
Si, Sn; <0.0001% Ag, Cr, Fe, Li, Mn.
X-Ray pattern at 26°C.
Replaces   1-0837, 2-0623, 2-0629, 3-0569, 3-0593,
3-0596, 3-0612, 4-0636, 4-0637.

FIGURE 5.   Joint Committee on Powder Diffraction Standards, file card for the mineral calcite.

ticles, they may be separated using virtual impaction or cascade impaction. It is also possible to separate them by using a density gradient technique prior to redeposition.[16]

It is also important to deposit material uniformly across the supporting substrate and eliminate preferential orientation of particles. Usually, this can be accomplished by careful collection methodology or careful mixing during redeposition. Orientation problems can be detected by measuring spectra intensities for various substrate angles of rotation and microscopy. Sample rotation (spinning) during analysis is often used to reduce quantitative errors caused by crystallite orientation or nonuniform deposition of crystallites.

V. ANALYSIS

Qualitative species identification has been applied for decades. Quantitative analysis of small amounts of fine particulate matter was initiated by Clark and Reynolds[2] and developed by a long series of investigators in various countries, including those cited and summarized by Crable and Knott,[5] Klug and Alexander,[13] Williams,[24] and Leroux et al.[14] Presently, the largest use of XRD in analyzing air pollution samples is in the analysis of dust for certain polymorphs of silica. Techniques employed to identify crystalline species and determine quantity will be discussed below.

A. Crystalline Species Identification

It has been stated that all crystalline compounds have unique X-ray diffraction spectra composed of a set of very specific constructive interference patterns of specific relative intensities. Over 30,000 of these spectra appear in the Joint Committee of Powder Diffraction Standards (JCPDS)* files. Figure 5 shows such a file card for the mineral calcite. Compounds are listed alphabetically according to chemical formula and minerals are listed alphabetically and also appear in the chemical formula list. Two search listings have also been made to aid in the identification of patterns from unknown crystalline compounds, organic and inorganic. All listings are updated each year.

For convenience, abbreviated spectral data and

*Joint Committee on Powder Diffraction Standards, 1601 Park Lane, Swarthmore, Pa. 19081.

search information are contained in a number of reference books. Each listed compound is also contained on a single index card that contains complete spectral information. These index cards list the d space values with relative diffracted X-ray intensity and Miller indices. Data also included on the cards are lattice parameters, crystal system, origin and treatment of the material, chemical composition, and certain XRD conditions used to obtain the pattern. Jenkins and de Vries[12] give a concise description of the JCPDS file and search techniques.

The search procedures found on the cards are invaluable for identification of unknown compounds; however, a blind search with no other information except that supplied by the X-ray diffraction pattern is often fruitless. This is because the exact intensity relation of a particular XRD pattern of an unknown compound usually varies from that listed in the JCPDS index cards caused by variation in sample preparation, preferred orientation, sample presentation, interfering peaks, sample history, composition, etc. Therefore, additional information such as color of the powder, origin of the material, or elemental composition is extremely helpful. For example, knowledge that the source is a combustion emission, corrosion product (metal oxide), or a local mineral can greatly simplify compound identification. Occasionally, a crystalline species cannot be identified because it is not listed in the JCPDS cards. At times, all an investigator can do is to ascertain the absence of a particular species.

B. Quantitative Analysis

To perform a quantitative or semiquantitative analysis of crystalline materials from a diffraction spectrum, it is necessary to determine the intensities of selected diffracted X-rays. Usually, the most significant peaks for which there are no interfering peaks are selected for analysis. The intensities of these peaks are determined by measuring maximum intensity above background or by integrating the total intensity above background, respectively referred to as peak height and integrated intensity. Because the latter contains more signal information, it is preferred for analyses where quantification is critical. Also, it is the necessary choice where the crystallite size distribution of the analyzed samples may vary.

It is possible to quantify Debye-Scherrer camera film spectra using microdensitometer and

photographic-microphotometric techniques.[2] These devices basically consist of a collimated light source and photomultiplier which is used to scan the film to integrate the amount of light absorbed by exposed photographic film at the darkened bands.[13] The amount of absorption is related to the actual intensity of the diffracted X-ray beam but not by a simple proportionality. The diffractometer, however, is the favored instrument for quantitative XRD analysis because of its superior precision, analysis time, overall convenience, and automation capability.

In the application of X-ray diffractometry to quantitative analysis, a number of points should be mentioned. First, the absolute intensity that a given weight or volume of a pure crystalline species will diffract from a given set of atomic planes is usually not available. Moreover, there are a number of variables that are not easily quantifiable such as diffractometer goniometer alignment, particle size, detector efficiency, X-ray source contamination, and X-ray power supply uncertainties. The relative amount of absorption of the incident and diffracted beam by the sample in its path through the sample, usually several tens of micrometers, must be considered. Preferred orientation of the crystallites along certain crystallographic axes can cause severe intensity anomalies of the diffracted X-ray beam. The last important consideration is background radiation caused largely by incoherent scattering of the incident radiation in the sample.

These considerations make it necessary to apply one of three general techniques to measure the relative weight fraction of a crystalline species present in a sample. These techniques use internal standards, dilution (known addition), or external standards and are mainly used to reduce errors caused by X-ray absorption.

The internal standard method consists of comparing the intensity of the diffracted peak or peaks in the unknown against the peak intensity of a known amount of material usually added to the sample powder. This technique assumes the samples are of infinite thickness. Any sample over approximately a millimeter thick is considered infinitely thick to the X-ray beam. The technique of mixing in an internal standard produces uncertainties because of weighing, mixing, and the resultant lower intensity of the peaks. Another useful technique is the use of thin powder samples deposited on a metal substrate, the metal then gives a diffracted peak against which other peaks produced by the sample may be compared.[24] This technique has been improved upon through the availability of metallic membrane filters, usually silver, upon which airborne particulates may be directly collected. Upon obtaining the ratio of the diffracted peak of interest to that of the standard, this quantity is compared against a calibration curve for conversion of the raw intensity data to a weight fraction.[6,12,13]

The external standard method neglects possible variations in X-ray absorption and determines quantity present by comparison against pure or mixed standards. This works best with thin-layer samples, such as fine particulate monolayers, where absorption of X-rays is minimal.

The two techniques described above require the construction of a calibration curve from standard materials for each crystalline species. The known addition method requires the intensity data from known additions to be plotted vs. the amount added. This plot is then extrapolated back to zero addition to obtain the amount of material originally present.

Misalignment of the diffractometer can cause serious errors in the intensity determination of diffracted peaks especially for peaks several two-theta degrees apart. It is prudent to choose standardization peaks that are close to the unknown peak in order to minimize the errors. Preferred orientation is a constant problem, and by and large, the analyst has no control over this factor; however, sample preparation should be consistent so that its effect is always the same[6,13] Some analysts have attempted to promote preferred orientation as a remedy to controlling anomalies it causes. Preferred orientation effects are often minimized by use of a spinning device which rotates the sample at its center about a normal axis to its surface.[6,12,13] Particle size also causes anomalies, especially in particulate samples where the sample is thin and the substrate rough.[10]

VI. APPLICATIONS OF XRD

X-Ray diffraction analysis has many advantages over other analytical methods. These include identification and quantification of polymorphs of the same compound, e.g., α-quartz and crysto-balite and compounds consisting of similar elements. It is also nondestructive; sample prepara-

tion is minimal; the spectra produced are inherently simple; and in airborne particulate analysis, the depth of sample analyzed is usually on the order of the thickness of that collected. Disadvantages include anomalies caused by particle size discrepancies, preferred orientation, and peak overlap due to interferences of other crystalline species. The minimum detectability limit is dependent upon the matrix material, the sample presentation method and the relative X-ray scattering ability of the species being analyzed.

Silica — Presently, in the U.S., the major applications of XRD quantitative and qualitative analyses to airborne particulates involve silica dust. A number of filter materials have been used as the collection media and the filtrate analyzed either directly upon the media or after transfer to silver membrane filters.[1,10,16,23] These techniques differ markedly from the classic internal standard method described by Klug and Alexander[13] and are preferred because of elimination of inaccuracies caused by weighing and mixing and the smaller amount of sample that can be analyzed. The newly developed methods produce a minimum detectability limit on the order of 5 μg/cm^2 of sample area. The automated powder diffractometer has found significant application to the analysis of dusts for silica because its consistency, speed, and automation is of great advantage to laboratories analyzing a thousand or more silica samples per month.

Asbestos — Jenkins[11] implied that X-ray diffraction patterns have been the primary means of identification and classification of asbestos minerals in the recent years. Even with the importance of electron microscopy for the identification and counting of individual asbestos mineral fibers, X-ray diffraction is still an important analytical technique. Rickards[22] has described a method for the quantitative determination of chrysotile asbestos on filter membranes in quantities as small as 10 μgs. He evaluated his method by suspending chrysotile fibers in water and depositing them on 25 mm Millipore® membrane filters, using nickel-filtered copper X-radiation. Rickards concluded that the detection limit for an external standard method was 10 μg on the filter with a standard deviation of ± 10%.

Cook[3] described the use of X-ray diffraction for quantitative analysis of amosite asbestos in water. His method used an external standard and in some circumstances could measure concentra-

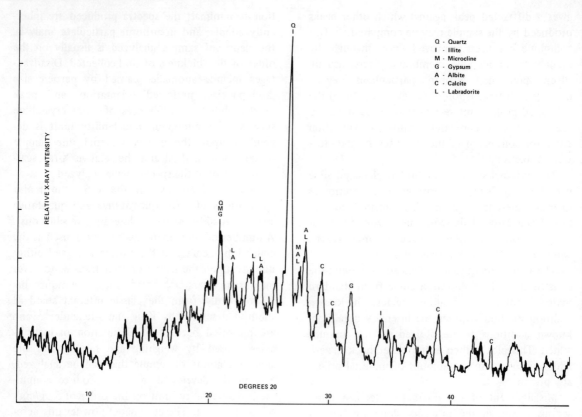

FIGURE 6.   An XRD pattern of airborne particulates collected in Denver, Colorado. Q, quartz; I, illite; M, microcline; G, gypsum, A, albite; C, calcite; L, labradorite.

tions as low as 0.5 μg/l. On comparing electron microscopy with X-ray diffraction results, he concluded that the latter was a viable method of the determination of amphibole concentration in water samples. Crable et al.[4,5] have also published a number of papers regarding the measurement of trace amounts of chrysotile, crocidolite, and amosite in industrial dusts.

XRD results provide only an estimate of mass, and often fiber number and size distribution are required parameters. The only viable method for estimating these is transmission electron microscopy coupled with selected area electron diffraction. Furthermore, many nonfibrous forms of a mineral give the same XRD patterns or very similar to those of the fibrous or asbestos form.[18]

Ambient minerals – During the winter of 1973, ambient air samples were collected in and around Denver, Colorado. In conjunction with an intense overall study, material collected on HiVol® glass-fiber filters was analyzed using X-ray diffraction. Peaks for quartz, calcite, muscovite, microcline, and cristobalite were identified. Figure 6 shows an

X-ray diffraction pattern from a particulate aerosol sample containing quartz, calcite, and clay minerals.

Coal – Mazza and Wilson[17] have used X-ray diffraction to analyze feed coals and their combustion products. Minerals examined included aluminosilicates, carbonates, sulfides, quartz, feldspars, and gypsum. Bulk samples of combustion products were collected, carefully ground, low temperature ashed, ground again, and packed into a sample holder.

Lead compounds – Olsen and Skogerboe[21] examined lead compounds produced by automotive sources in soil using X-ray powder diffraction. The lead compounds were preconcentrated using gradient density and magnetic separation techniques. Fractions showing the highest lead concentrations were examined using an X-ray diffraction film camera. Here, finely powdered material was mounted on an amorphous quartz filament. Foster and Lott[7] examined lead smelter and ambient air samples collected and analyzed on filter paper. Minerals and lead compounds associated with smelter operations were readily observed.

VII. CONCLUSION

There is a growing concern about compounds associated with air pollution because of the differential health effects of compounds which have similar elemental composition (e.g., sulfur series). However, the methods for analyzing compounds are limited; few are nondestructive. Thus, X-ray diffraction can be expected to become an increasingly utilized tool in air pollution analysis. Its most serious limitations will continue to be (1) sensitivity limitations and (2) restriction to crystalline compounds. Major advances will be in the areas of (1) instrumentation, (2) sample preparation, and (3) selected species preconcentration techniques.

REFERENCES

1. **Bumstead, H. E.,** Determination of alpha-quartz in the respirable portion of airborne particulates by x-ray diffraction, *Am. Ind. Hyg. Assoc. J.,* 34, 150, 1973.
2. **Clark, G. L. and Reynolds, D. H.,** Quantitative analysis sample of mine dusts, *Ind. Eng. Chem.,* 8, 36, 1936.
3. **Cook, P. M.,** Semiquantitative determination of asbestiform amphibole mineral concentrations in western Lake Superior water samples, *Advances in X-ray Analysis,* Vol. 18, Plenum Press, New York, 1975.
4. **Crable, J. V.,** Quantitative determination of chrysotile, amosite and crocidolite by x-ray diffraction, *Am. Ind. Hyg. Assoc. J.,* 27, 293, 1966.
5. **Crable, J. V. and Knott, M. J.,** Quantitative x-ray diffraction analysis of crocidolite and amosite in bulk or settled dust samples, *Am. Ind. Hyg. Assoc. J.,* 27, 449, 1966.
6. **Cullity, B. D.,** *Elements of X-ray Diffraction,* Addison-Wesley, Reading, Mass., 1956.
7. **Foster, R. L. and Lott, P. F.,** The x-ray identification and semi-quantification of toxic lead compounds emitted into air by smelting operations, *Nat. Bur. Stand. (U.S) Spec. Publ.,* Proc. 8th Materials Research Symp. on Methods and Standards for Environmental Measurement, National Bureau of Standards, Gaithersburg, Md., September 1976.
8. **Gandolfi, A.,** Discussion upon methods to obtain x-ray "powder patterns" from a single crystal, *Mineral. Petrogr. Acta,* 13, 67, 1967.
9. **Guy, A. G.,** *Elements of Physical Metallurgy,* Addison-Wesley, Reading, Mass., 1951.
10. **Henslee, W. W. and Guerra, R. E.,** Direct quantitative determination of silica by x-ray diffraction on PVC membrane filters, *Advances in X-ray Analysis,* Vol. 20, Plenum Press, New York, 1977.
11. **Jenkins, G. F.,** Asbestos, in *Seeley W. Mudd Series — Industrial Minerals and Rocks,* American Institute of Mining and Metallurgical Engineers, New York, 1960.
12. **Jenkins, R. and de Vries, J. L.,** *An Introduction to X-ray Powder Diffractometry,* N. V. Philips Gloeilampenfabrieken, Eindhoven, Holland, 1975.
13. **Klug, H. P. and Alexander, L. E.,** *X-Ray Diffraction Procedures,* John Wiley & Sons, New York, 1954.
14. **Leroux, J., Davey, A. B. C., and Poillard, A.,** Proposed standard methodology for the evaluation of silicoses hazards, *Am. Ind. Hyg. Assoc. J.,* 34, 409, 1973.
15. **Leroux, J. and Powers, C. A.,** Direct x-ray diffraction quantitative analysis of quartz in industrial dust films deposited on silver membrane filters, *Staub Reinhalt. Luft,* 29(5), 26, 1969.
16. **Markham, M. C. and Wosczyna, K.,** Determination of microquantities of chrysotile asbestos by dye absorption, *Environ. Sci. Technol.,* 9, 930, 1976.
17. **Mazza, M. H. and Wilson, J. S.,** X-ray diffraction examination of coal combustion products related to boiler tube fouling and slagging, *Advances in X-ray Analysis,* Vol. 20, Plenum Press, New York, 1977.
18. **Mumpton, F. A.,** Characterization of chrysotile asbestos and other members of the serpentine group minerals, *Siemens Rev.,* Vol. 41, 7th Special Issue, 75, 1974.
19. **Nenadic, C. M. and Crable, J. V.,** Applications of x-ray diffraction to analytical problems of occupational health, *Am. Ind. Hyg. Assoc. J.,* 32, 529, 1971.
20. **NIOSH** criteria document, Recommendations for an occupational exposure standard for crystalline silica, *Fed. Regist.,* 39, 250, 1974.
21. **Olsen, K. W. and Skogerboe, R. K.,** Identification of soil lead compounds from automotive sources, *Environ. Sci. Technol.,* 2, 227, 1975.
22. **Rickards, A. L.,** Estimation of trace amounts of chrysotile asbestos by x-ray diffraction, *Anal. Chem.,* 44(11), 1872, 1972.
23. **Thatcher, J.,** Estimation of trace amounts of chrysotile asbestos by x-ray diffraction, Information Report No. 1021, U.S. Department of the Interior, Mine Environmental Safety Administration, Washington, D.C., 1975.
24. **Williams, P. P.,** Direct quantitative diffractometric analysis, *Anal. Chem.,* 31, 1842, 1959.
25. **Wyckoff, R. W. G.,** *Crystal Structures,* Vol. 1, Interscience, New York, 1963.

Chapter 10
IDENTIFICATION AND DETERMINATION OF PARTICULATE COMPOUNDS: ELECTRON SPECTROSCOPY*

T. Novakov

TABLE OF CONTENTS

I. INTRODUCTION

It is desirable that wet chemical and other microanalytical procedures be complemented by nondestructive physical methods. X-ray photoelectron spectroscopy, also known as ESCA (Electron Spectroscopy for Chemical Analysis),[1] whose application to chemical characterization of pollution particles is described in this chapter, is one such method. For example, application of this method has helped to uncover the presence of significant concentrations of reduced nitrogen species other than ammonium in ambient aerosol particles.[2] This group of species contains, among others, certain amines and amides[3] that are not soluble in water or such solvents as benzene and therefore could not be detected by wet chemical methods. Most analyses of pollution aerosol particles have employed wet chemical methods. On the basis of this kind of measurement, different

workers have concluded that the principal particulate nitrogen species are ammonium and nitrate ions[4] and have suggested that the most likely combination of these is ammonium nitrate and ammonium sulfate.[5] The species uncovered with the aid of electron spectroscopy have thus escaped observation by means of wet chemistry.

In this chapter, we will describe the use of X-ray photoelectron spectroscopy for chemical characterization of ambient and source-enriched aerosol particles. These analyses involve measurement of the chemical shift, core electron level splitting, relative concentrations, and volatility (in vacuum) of different particulate species. Because the method of photoelectron spectroscopy has been described in great detail in a number of papers and monographs, only the fundamentals of the technique relevant to this topic will be reviewed here. Since most of the mass of airborne pollution particles consist of compounds of car-

*This work was done with support from the U.S. Department of Energy and the National Science Foundation.

bon, nitrogen, and sulfur, special emphasis will be placed on characterization of C, N, and S species. Attempts to chemically characterize some trace metals, such as lead and manganese, both originating in fuel additives, will also be described.

II. X-RAY PHOTOELECTRON SPECTROSCOPY

X-ray photoelectron spectroscopy is the study of the kinetic energy distribution of photoelectrons expelled from a sample irradiated with monoenergetic X-rays. The kinetic energy of a photoelectron (E_{kin}) expelled from a subshell (i) is given by $E_{kin} = h\nu - E_i$, where $h\nu$ is the X-ray photon energy and E_i is the binding energy of an electron in that subshell. If the photon energy is known, experimental determination of the photoelectron kinetic energy provides a direct measurement of the electron binding energy.

The electron binding energies are characteristic for each element. The intensity of photoelectrons, originating from a subshell of an element, is related to the concentration of atoms of that element in the active sample volume. In principle, this feature enables the method to be used for quantitative elemental analysis. The binding energies, however, are not absolutely constant but are modified by the valence electron distribution, so that the binding energy of an electron subshell in a given atom varies when this atom is in different chemical environments. These differences in electron binding energies are known as the chemical shift. The origin of the chemical shift can be understood in terms of the shielding of the core electrons by the electrons in the valence shell. A change in the charge of the valence shell results in a change of the shielding, which affects the core electron binding energies. For example, if an atom is oxidized, it donates its valence electrons and thus becomes more positively charged than the neutral configuration. Some of the shielding contribution is removed, and in general, the binding energies of the core electron subshells are increased. Conversely, the binding energies will show an opposite shift for the reduced species. Therein lies the usefulness of chemical shift determinations in the analysis of samples of unknown chemical composition. In practice, the measurements of the chemical shifts are complemented by the determination of relative photoelectron intensities,

from which the stoichiometric information can be inferred.

The relation $E_{kin} = h\nu - E_i$ is unambiguous for gaseous samples. In solid samples, however, the photoelectron has to overcome the potential energy barrier at the surface of the sample. This potential energy barrier is known as the work function of the sample, ϕ_{sample}. However, if the solid sample is in electrical contact with the electrically grounded spectrometer, the Fermi levels of the sample and of the spectrometer are equalized. On entering the spectrometer, a photoelectron is accelerated by $e[\phi_{sample} - \phi_{spect}]$, and as it reaches the detector, it acquires the kinetic energy $E_{kin} = h\nu - E_{i,f} - \phi_{spect}$. In the experiment, therefore, the kinetic energy is determined by the spectrometer work function, ϕ_{spect}, and by the binding energy referenced to the Fermi level of the spectrometer, $E_{i,f}$.

We shall briefly describe some of the theoretical results on chemical shifts that relate to the chemist's intuitive conception of bonding and molecular structure and to the subject of analytical applications of photoelectron spectroscopy.

In the early stages of photoelectron spectroscopy, it was realized that chemical shifts can be related to the oxidation state of the atom in a molecule. Subsequently, attempts were made to correlate the binding energy shifts with the estimated effective atomic charges. Electronegativity difference methods,[6,7] the extended Hückel molecular orbital method,[8] and the complete neglect differential overlap (CNDO) method[9-11] were used to calculate the effective charge. Only rough correlations were obtained, however. It was subsequently found that the poor correlations result from neglect of the potential generated by all charges in the molecule.

Chemical shifts can be adequately described by the electrostatic potential model in which the charges are idealized as point charges on atoms in a molecule. The electron binding energy shift, relative to the neutral atom, is equal to the change in the electrostatic potential resulting from all charges in the molecule, as experienced by the atomic core under consideration. Different approaches to the potential model calculations were used by Gelius et al.,[12,13] Siegbahn et al.,[7,14] Ellison and Larcom,[10] and Davis et al.[11] Detailed theoretical analyses of the potential model were given by Basch[15] and Schwartz.[16]

In practice, line broadening and the small

magnitude of the chemical shifts may make the determination of even the oxidation state difficult in some cases. Because of this difficulty in cases of transition metal compounds, the "multiplet splitting" effect can be employed to infer the oxidation state. Multiplet splitting of core electron binding energies[17] is observed in the photo-electron spectra of paramagnetic transition metal compounds. In any atomic or molecular system with unpaired valence electrons, the 3s-3d exchange interaction affects the core electrons differently, according to the orientation of their spin. This causes the 3s core level to be split into two components. For example, in an Mn^{2+} ion whose ground state configuration is $3d^5$ ^6S, the two multiplet states will be ^7S and ^5S. In an Mn^{4+} ion having the ground state configuration $3d^3$ ^4F, the two spectroscopic states will be ^5F and ^3F. In the first approximation, the magnitude of the multiplet splittings should be proportional to the number of unpaired 3d electrons. Hence the Mn(3s) splitting will be greatest for Mn^{2+} ions and least for Mn^{4+}.

Because of the low energy of photoelectrons induced by the most commonly used Mg or Al X-rays, the effective escape depth for electron emission, without suffering inelastic scattering, is small. Recent studies have given electron escape depths of 15 to 40 Å for electron kinetic energies between 1000 and 2000 eV.[18] This renders the ESCA method especially surface sensitive and thus useful in surface chemical studies. Because of the high energy resolution, electrons that have escaped the solid sample without energy loss are well separated from lower energy electrons whose energy has been degraded by the inelastic colli-sions. The chemical shift measurements are per-formed only on electrons with no energy loss.

III. ESCA AS AN ANALYTICAL TOOL

The principal sources of uncertainty and error in ESCA analyses are related to: (1) electron energy loss processes, (2) electron escape depth variations, (3) binding energy calibration proce-dures, and (4) sample exposure to vacuum and X-rays during analysis. The first two of these problems may have major repercussions for the determination of concentrations of elements and species in chemically heterogeneous samples. The third problem influences the validity of the assign-

ment of chemical states through the determination of chemical shifts. Finally, the spectrometer vacuum and heating of the sample by the X-ray source may be the cause of losses of volatile species.

In this section, experimental results and conclu-sions on some of the problems that relate to the analytical application of ESCA will be reviewed. These results deal with different specific objectives that are felt to be equally applicable to the topics of this chapter.

Wagner[19] has determined a table of relative atomic sensitivities that enables the conversion of photoelectron peak intensities into relative atomic concentrations of elements in a sample. Wagner's study, as well as the one by Swingle,[20] indicates that ESCA can be used as a semiquantitative ($\leqslant50\%$ relative error) or even as a quantitative ($\leqslant10\%$ relative error) method when comparing chemically similar samples. Unfortunately, both of the above authors found that the relative photo-electron intensities obtained with chemically dis-similar samples show wide variations. For example, Wagner's[19] data on the Na(1s)/F(1s) intensity ratios (corrected for stoichiometry) for a number of sodium- and fluorine-containing compounds were found to vary by as much as a factor of 2.

Swingle[20] has attributed the observed varia-tions in the apparent sodium and fluorine atomic sensitivities to the differences in the structure of the photoelectron energy loss spectrum. In cases of chemically similar samples, however, the same mechanism should be responsible for inelastic electron scattering.

The effect of the chemical form of an atom on the relative intensity of the photoelectron peak arising from that atom has been studied in detail by Wyatt, Carver, and Hercules.[21] They have demonstrated that different lead salts show differ-ent atomic sensitivities and have suggested that the escape depth for lead salts depends on the crystal-line frame surrounding the lead cation and its coordination number, since the observed differ-ences in sensitivities could not be accounted for by inelastic electron scattering alone. These findings imply that elemental analysis must be done with great caution unless the chemical form of the elements in the sample is well defined.

The bulk sensitivity of ESCA can be estimated at approximately 0.15% based on bulk percentage. Therefore ESCA is not a "trace element" tech-nique in the usual sense. However, because of the possibility of detecting as little as 0.1% of a monolayer[22] (about 10^{12} atoms), the absolute sensitivity of ESCA is in the picogram range. Thus, ESCA is a unique trace method if the analysis is confined to species that preferentially locate at the surface.

The topics discussed so far in this section pertain essentially to the determination of relative concentrations of elements and species. The importance and uniqueness of ESCA, however, are in its capability to measure the chemical shifts that contain implicit information about the molecular forms. The principal difficulty encountered in chemical shift measurements pertains to the calibration procedure used to account for the electrostatic charging of the sample. Hercules and Carver[23] list four basic calibration approaches that have been used. These make use of (1) contamination carbon, (2) admixed species, (3) vapor deposition of noble metals, and (4) sample constituents as internal calibration standards.

IV. APPLICATION OF ESCA TO CHEMICAL ANALYSIS OF PARTICULATES

A. Effect of Sample Composition at Relative Intensities

It is important to assess the usefulness of (1) photoelectron peak intensities for inferring the likely stoichiometry of certain compounds, for example, to distinguish between ammonium sulfate and ammonium bisulfate; and (2) the relative photoelectron peak intensities for determination of relative concentrations of elements and chemical species in particulate matter. Dod et al.[24] have investigated the possible effects of the chemical composition and the surrounding matrix on photoelectron peak intensities, using heterogeneous samples that reasonably simulate the situation found in ambient particulates.

The first set of these experiments investigated the constancy of the intensity ratios of S(2p) and N(1s) photoelectron peaks, corrected for stoichiometry, for a number of pure sulfur- and nitrogen-containing compounds. The mean value for peak intensity ratios of these samples is 1.65 ± 0.21 (1σ). The results indicate that within an uncertainty of ±15% the photoelectron peak intensities of nitrogen and sulfur reflect the relative abundances of these atoms in compounds.

The second set of experiments[24] investigated

the effects of matrix dilution on photoelectron peak intensities of nitrogen and sulfur from ammonium bisulfate. Activated carbon, lead chloride, and graphite mixtures were used as the diluent matrix. As before, such mixtures were assumed to simulate the conditions normally found in atmospheric particulates. The results of these dilution experiments demonstrated a constancy in the sulfate to ammonium peak ratio to within ±10%, i.e., well within the error limit for peak intensity determination.

Thus, based on these results, it seems justifiable to use relative photoelectron peak intensities to infer the apparent stoichiometry of sulfur- and nitrogen-containing compounds of the kind commonly associated with air pollution particulates. It will be shown later in the text that photoelectron peak intensities can be used to determine the concentration ratios of certain elements and species.

Before outlining the results for chemical states of sulfur, nitrogen, and carbon species in atmospheric particulates, we shall first justify the use of apparent carbon (1s) binding energy to correct for sample charging. As discussed later in more detail, carbon peaks from atmospheric particulates appear essentially as a single peak with a binding energy corresponding to a neutral charge compatible with condensed hydrocarbons and soot-like material. Since carbon is by far the most abundant element in particulates, practically the entire ESCA C(1s) signal is due to the sample itself, rather than to hydrocarbon contamination of the sample in the spectrometer.

In order to test the validity of using the carbonaceous content of a sample as the internal binding energy reference, the apparent binding energies of C(1s) and Pb($4f_{7/2}$) were determined from a number of ambient samples collected near a major San Francisco Bay Area commuter route. (The samples were collected in 1972 in the course of the Aerosol Characterization Experiment of the California Air Resources Board.) The results are shown in Figure 1 where these binding energies are plotted against time of day corresponding to sample collections. (The samples were collected for 2 hr on silver membrane filters.) The figure shows that the variations in the carbon and lead peak positions are similar. Assuming that the chemical composition of lead and carbon species is similar throughout the episode, we can conclude that the binding energy error caused by the

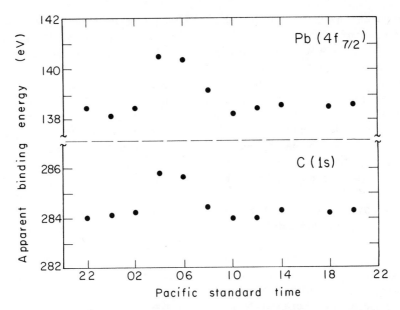

FIGURE 1. Apparent binding energies (not corrected for sample charging) of carbon (1s) and lead ($4f_{7/2}$) photoelectrons from a number of 2-hr samples collected near a major San Francisco Bay Area commuter route. The apparent binding energies are plotted against time of day corresponding to sample collection.

use of C(1s) as the reference should not exceed ±0.25 eV. It appears, therefore, that use of the C(1s) peak binding energy as an internal reference is adequate to determine the chemical states of major species associated with atmospheric particulates.

B. Chemical States of Particulate Sulfur and Nitrogen

The feasibility of using X-ray photoelectron spectroscopy for the chemical characterization of particulates was first explored by Novakov, Wagner, and Otvos.[25] These authors have determined the elemental composition of particulate samples and have found that most of the particulate nitrogen is in the reduced chemical state. The observed nitrogen (1s) photoelectron lines were of complex structure, indicating the presence of several different reduced nitrogen species. The sulfur (2p) peak appeared to have a single component, consistent with sulfate.

A more systematic ESCA study of the chemical states of sulfur and nitrogen as a function of particle size and time of day was performed by Novakov et al.[2] The particle separation was accomplished by use of a Lundgren cascade impactor. The samples for analysis were collected on Teflon® films covering the rotating impactor drums. Two size cuts, 2.0 to 0.6 μm and 5.0 to 2.0 μm, corresponding to the fourth and third impactor stages, were used in this experiment.

The sulfur (2p) spectra, indicating the presence of two components, were assigned to sulfate and to sulfite. The 6+ and the 4+ state were present in both size ranges. However, their relative abundance changed with particle size and time of day.

The ESCA spectra from this study have revealed the presence of four different chemical states of particulate nitrogen: nitrate, ammonium, and two reduced species tentatively assigned to an amino-type and a heterocyclic nitrogen compound. The nitrate was observed only in the larger particle size range but not in the smaller one.

Hulett et al.[26] used ESCA for chemical characterization of coal fly ash and smoke particles. These authors analyzed specimens of smoke particles collected on filter paper from coal burned in a home fireplace and found three distinct components in the sulfur (2p) peak: a single reduced state assigned to a sulfide and two species of higher oxidation, corresponding to sulfite and sulfate. The sulfur (2p) peak of fly ash particles was also found to consist of two components. These were assigned to sulfates and/or adsorbed

SO_3, since their binding energies were much higher than those for sulfite ions.

Araktingi et al.[27] used ESCA to analyze particulates collected in Baton Rouge, Louisiana. A number of elements were identified in this work, the most abundant of which were sulfur, nitrogen, and lead. The N(1s) peak appeared to consist of at least two components with binding energies at about 399.4 and 401.5 eV. The authors did not propose an assignment to these peaks. It would appear, however, that these are similar to the nitrogen peaks identified in samples collected in Pasadena, California by Novakov et al.[2] The S(2p), according to these authors, was entirely in the form of sulfate.

In summary, these early experiments demonstrated that there is considerable variety in the chemical states of sulfur and nitrogen associated with air pollution particulates. For example, particulate sulfur may exist in both oxidized (sulfate, adsorbed SO_3, and sulfite) and reduced (sulfide) states, while nitrogen species include nitrate, ammonium, and two other, previously unrecognized, reduced forms. This obviously suggests a more complex situation than the one inferred from wet chemical results, i.e., that the only significant sulfur and nitrogen species are sulfate, nitrate, and ammonium ions.

Craig et al.[28] have attempted to formulate an inventory of chemical states of particulate sulfur based on their examination of ESCA spectra of more than a hundred samples collected at various sites in California. The samples used in that study were collected both with and without particle size segregation. All samples were collected for 2 hr as a function of time of day. The sulfur spectra of these samples were of varying degrees of complexity, sometimes covering a wide range of binding energies. The binding energies deduced from the analysis of particulate samples were assigned to certain characteristic chemical states with the help of ESCA results obtained with a number of pure compounds and with certain surface species produced by the adsorption of SO_2 and H_2S on several solids.

The chemical states of particulate sulfur identified in the work of Craig et al.[28] are, in terms of the chemical shift, equivalent to adsorbed SO_3, SO_4^{--}, adsorbed SO_2, SO_3^{--}, $S°$, and possibly two kinds of S^{--}. The assignment of $S°$ as neutral (elemental) sulfur was made because of the similarity of its binding energy to that of elemen-

tal sulfur. It is possible, however, that sulfur in species designated by $S°$ is actually in the -2 oxidation state. Differences in binding energies between different sulfides are expected because of the differences in the corresponding bond ionicity. Naturally, not all of these species did occur at all times and locations. In each case, sulfates were found to be the dominant species, although at times concentrations of other forms of sulfur were comparable to the sulfate concentrations.

As we mentioned earlier, ESCA analyses of ambient particulates uncovered the existence of previously unsuspected reduced nitrogen species[2] whose binding energies are similar to certain amines and/or heterocyclic nitrogen compounds. For simplicity, we shall denote these species by N_x. Further studies on the chemical structure of N_x species were performed by Chang and Novakov[3] by means of temperature-dependent ESCA measurements. The experimental procedure consisted of measuring ESCA spectra of ambient samples and gradually increasing sample temperatures. The samples were collected on silver membrane filters that could withstand the temperatures used in the experiment.

The results of one such measurement for an ambient particulate sample, collected in Pomona, California, during a moderate smog episode (24 October 1972), are shown in Figure 2. The spectrum taken at a sample temperature of 25°C shows the presence of NO_3^-, NH_4^+, and N_x. At 80°C the entire nitrate peak is lost, accompanied by a corresponding loss in the ammonium peak intensity. The shaded portion of the ammonium peak in the 25°C spectrum represents the ammonium fraction volatilized between 25 and 80°C. The peak areas of the nitrate and the volatilized ammonium are approximately the same, indicating that the nitrate in this sample is mainly in the form of ammonium nitrate. The ammonium fraction still present at 80°C but absent at 150°C is associated with an ammonium compound more stable than ammonium nitrate, such as ammonium sulfate. At 150°C, the only nitrogen species remaining in the sample is N_x. At 250°C the appearance of another peak, labeled N_x', is seen. The intensity of this peak continues to increase at 350°C. The total $N_x + N_x'$ peak area at 150, 250, and 350°C remains constant, however, indicating that a part of the N_x is transformed into N_x' as a consequence of heating.

N_x' species will remain in the sample even if its

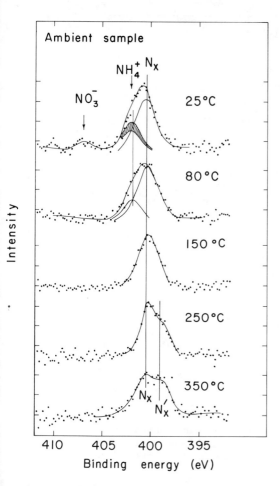

FIGURE 2. Nitrogen (1s) photoelectron spectrum of an ambient sample as measured at 25, 80, 150, 250, and 350°C. (From Chang, S. G. and Novakov, T., *Atmos. Environ.*, 9, 495, 1975. With permission.)

temperature is lowered to 25°C, provided that the sample has remained in vacuum. However, if the sample is taken out of vacuum and exposed to the humidity of the air, N_x' will be transformed into N_x. It was concluded that N_x' species are produced by dehydration of N_x:

$$N_x \underset{+H_2O}{\overset{-H_2O}{\rightleftarrows}} N_x'$$

Based on the described temperature behavior and on laboratory studies[3] of reactions that produce species identical to those observed in the ambient air particulates, N_x was assigned to a mixture of amines and amides. (N_x photoelectron peaks are broad indications of the presence of

more than one single species.) Dehydration of the amide results in the formation of a nitrile, N_x'.

Application of temperature-dependent ESCA measurements[3,29] also revealed the presence of a previously unrecognized form of ammonium characterized by its relatively high volatility in vacuum. Temperature-dependent studies have also indicated that nitrate in ambient samples may occur in a volatile form different from common nitrate salts. Tentatively, such a nitrate is assigned to nitric acid adsorbed on the filter material or on the particles.

These conclusions are illustrated with the aid of the spectra shown in Figure 3. The spectrum shown in Figure 3(a) (collected in 1973 in West Covina, California) has been obtained with the sample at −150°C. The sample was kept at a low temperature in order to prevent volatile losses in the ESCA spectrometer vacuum. It will be shown later that volatilization in vacuum was suspected as one reason for the apparent inconsistency between the ammonium and nitrate determination by ESCA and by wet chemical techniques. Individual peaks corresponding to nitrate, ammonium, and N_x (amines and amides) are clearly seen in the spectrum.

Figure 3(b) shows the same spectral region of the same sample after its temperature was raised to 25°C. This spectrum shows only a trace of the original nitrate and about a 60% decrease in the ammonium peak intensity. Considering the nitrate and ammonium peak intensity in the spectra from the sample kept at a lower temperature, it is estimated that, at most, about 15% of the total ammonium could be associated with nitrate as NH_4NO_3. The volatile ammonium component is therefore not ammonium nitrate. Ammonium sulfate and ammonium bisulfate were found to be stable in vacuum at 25°C during time intervals normally used to complete the analysis. Furthermore, since no detectable decrease in the sulfate peak was observed over the same temperature range, it was also concluded that ammonium sulfate (and/or bisulfate) is not being volatilized in the spectrometer vacuum.

The limited volatility of ammonium salts and the behavior of the ambient samples suggest that a major fraction of ammonium in these samples is present in a previously unrecognized form. The volatility properties of nitrate in this sample and in other samples suggest the possibility of the existence of adsorbed nitric acid in accordance with the wet chemical results of Miller and Spicer.[30]

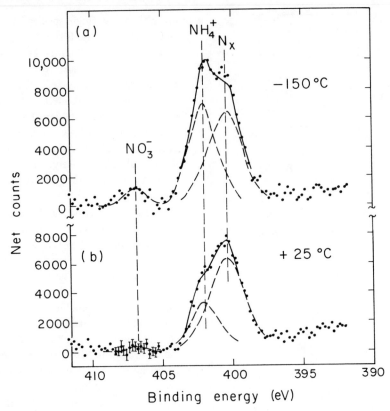

FIGURE 3.   (a) Nitrogen (1s) photoelectron spectrum of an ambient sample as
measured at –150°C. (b) The spectrum of the same sample as measured at 25°C.
(From Chang, S. G. and Novakov, T., *Atmos. Environ.,* 9, 495, 1975. With
permission.)

ESCA analysis of particulates enables the possibility not only of detection of specific ions and functional groups but also of their mutual relationship. This is achieved by the measurement of the ESCA chemical shift augmented by the determination of relative concentrations and by study of the volatility properties of certain particulate species.

The capability of ESCA for a straightforward differentiation of different forms of atmospheric sulfates was recently demonstrated by Novakov et al.[31,32] Figure 4 shows the nitrogen (1s) and sulfur (2p) regions in ESCA spectra of two ambient samples. One was collected in West Covina, California, in the summer of 1973, and the other was collected in St. Louis, Missouri, in the summer of 1975. The peak positions corresponding to NH_4^+, $-NH_2$, and SO_4^{--} are indicated. The solid vertical bar indicates the ammonium peak intensity expected under the assumption that the entire sulfate is in the form of ammonium sulfate.

Obviously, the observed ammonium content in the West Covina sample is insufficient to account for the sulfate by itself. This is in sharp contrast with the St. Louis sample where the observed ammonium intensity closely agrees with that expected for ammonium sulfate.

These results demonstrate that ammonium sulfate in the aerosols can easily be distinguished from other forms of sulfate, such as the one found in the West Covina case. However, wet chemical analyses[33] performed on West Covina samples collected simultaneously with the ESCA samples resulted in ammonium concentrations substantially higher than those suggested by the ESCA measurements. As mentioned earlier, this apparent discrepancy between the two methods was subsequently explained by the volatility of some ammonium species in the ESCA spectrometer vacuum. That these volatile losses are not caused by the volatilization of ammonium sulfate is evidenced by the St. Louis case, where no volatile

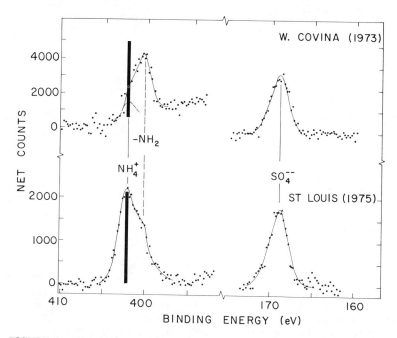

FIGURE 4. Nitrogen (1s) and sulfur (2p) regions in X-ray photoelectron spectra of two ambient samples, one from West Covina, California and one from St. Louis, Missouri. The peak positions corresponding to NH_4^+, $-NH_2$ (N_x), and SO_4^{--} are indicated. The solid vertical bar represents the ammonium intensity expected under the assumption that the entire sulfate is in the form of ammonium sulfate. The difference in the relative ammonium content of the two samples is obvious. The sulfate and ammonium intensities in the St. Louis sample are compatible with ammonium sulfate. The ammonium content in the West Covina sample is insufficient to be compatible with ammonium sulfate. Both samples were exposed to the spectrometer vacuum for about 1 hr. (From Novakov, T., Chang, S. G., Dod, R. L., and Rosen, H., APCA Paper 76-20.4, presented at the Air Pollution Control Assoc. Annual Meeting, Portland, Oregon, June 1976; Lawrence Berkeley Laboratory Report LBL-5215, June 1976.)

losses were observed. Similarly, ammonium nitrate (negligible in these samples) and ammonium bisulfate were found to be stable in vacuum during the time periods usually required to complete the analysis. Therefore, species other than these must be responsible for the apparent loss of ammonium in vacuum.

That the volatile ammonium is not necessarily associated with sulfate or nitrate ions is illustrated by results[31],[32] represented in Figure 5. Here the changes in the nitrogen (1s) spectrum of a sample collected in a highway tunnel (Caldecott Tunnel near Oakland, California) are shown as a function of the sample vacuum exposure time. Obviously, the ammonium peak intensity decreases with the vacuum exposure time of the sample. The amino-type nitrogen species intensity remains constant, however. The amount of nitrate in this sample was negligible compared with ammonium. The maximum ammonium peak expected, assuming that the

entire sulfate is ammonium sulfate, is indicated by the solid vertical bar in Figure 5. It is obvious, therefore, that the counterions for this ammonium are neither nitrate nor sulfate.

Figure 6 summarizes the findings about ammonium volatility in the three samples discussed above. The shaded bars at the far left of the figure indicate the expected ammonium intensity based on the assumption that all of the sulfate in the sample is in the form of ammonium sulfate. It is evident from the figure that only the St. Louis sample contains ammonium sulfate, while the West Covina and tunnel samples contain a different kind of ammonium that volatilizes in the spectrometer vacuum.

Novakov et al.[31],[32] have applied this procedure routinely to analyze a number of ambient samples. The results of such measurements for six St. Louis samples are shown in Figure 7 where the ratio of the observed ammonium peak intensity to

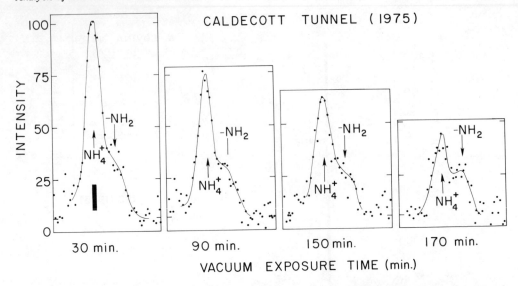

FIGURE 5. The variation in the observed ammonium peak intensity with vacuum exposure for a sample collected in a highway tunnel (Caldecott Tunnel near Oakland, California). The decrease in the peak intensity is caused by the volatilization of the ammonium species present in the sample. The solid vertical bar represents the ammonium intensity expected assuming that the sulfate in this sample is in the form of ammonium sulfate. The amount of nitrate in this sample is also small compared to ammonium. The ammonium in this sample is considerably in excess of that expected for ammonium sulfate or ammonium nitrate. (From Novakov, T., Chang, S. G., Dod, R. L., and Rosen, H., APCA Paper 76-20.4, presented at the Air Pollution Control Assoc. Annual Meeting, Portland, Oregon, June 1976; Lawrence Berkeley Laboratory Report LBL-5215, June 1976.)

the peak intensity expected of the ammonium in the form of ammonium sulfate is plotted as a function of the sample vacuum exposure time. From inspection of this figure, it is evident that in addition to the cases of practically stoichiometric ammonium sulfate (samples 913 and 914), there are cases where the observed ammonium is found in excess of ammonium sulfate. The excess ammonium consists of volatile ammonium species that decay until ammonium sulfate is the only ammonium species left (sample 917).

The anions corresponding to the volatile ammonium species cannot be identified with certainty at this time. One possibility is that these species are produced by the adsorption of ammonia on fine soot particles to form carboxyl- and hydroxyl-ammonium complexes that have been shown to have volatility properties similar to those of ambient particulates.[3] Another possibility is that these species could be due to ammonium halides which are also volatile in vacuum.

In conclusion, ESCA analysis of ambient samples allows for a straightforward differentiation of various forms of atmospheric sulfate- and ammonium-containing species. The following distinctly different cases have been identified:

1. Ammonium sulfate accounts for the entire ammonium and sulfate content of the sample.

2. Ammonium appears in concentrations above those expected for ammonium sulfate (and nitrate). The excess ammonium is volatile in vacuum.

3. Ammonium appears mostly in a volatile form independent of sulfate and nitrate.

C. Chemical Characterization of Particulate Carbon

ESCA has also been used in attempts to chemically characterize particulate carbon.[3] In most instances, the carbon (1s) peak of ambient particulates appears essentially as a single peak with a binding energy compatible with either elemental carbon, or condensed hydrocarbons or both. As seen in Figure 8, where the carbon (1s) spectrum of an ambient air particulate sample is shown, chemically shifted carbon peaks, due to oxygen bonding, are of low intensity compared with the intense neutral chemical state peak. From the standpoint of air pollution, it is important to distinguish the volatile hydrocarbon-type (mostly secondary species) carbon from the mostly non-volatile soot-like (primary species) carbon.

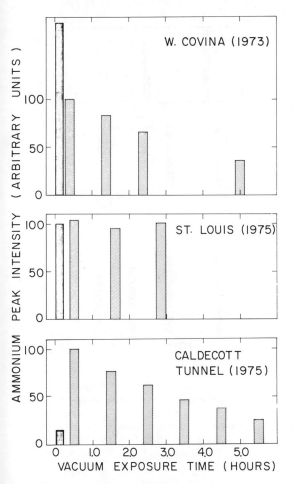

FIGURE 6. Volatility properties of West Covina, St. Louis, and automotive ammonium aerosol. The shaded bars on the far left of the figure indicate the expected ammonium intensity if the entire sulfate were ammonium sulfate. (From Novakov, T., Chang, S. G., Dod, R. L., and Rosen, H., APCA Paper 76-20.4, presented at the Air Pollution Control Assoc. Annual Meeting, Portland, Oregon, June 1976; Lawrence Berkeley Laboratory Report LBL-5215, June 1976.)

Chang and Novakov[3] have attempted to distinguish these species by comparing the carbon (1s) peak obtained with the sample at 25°C to the carbon (1s) peak with the sample at 350°C. The difference between the low-temperature and the high-temperature runs should give the fraction of volatile carbon. Figure 8 shows the result of one such experiment for a sample collected in West Covina in 1975. This experiment suggests that most of the ambient particulate carbon is nonvolatile in vacuum at 350°C. Assuming that the secondary hydrocarbons will have substantial vapor pressure at 350°C, the authors have suggested that a large fraction of the total particu-

late carbon is of a primary (soot-like) nature. The conclusion about a high nonvolatile carbon content could be erroneous if a large fraction of particulate carbon volatilizes even at 25°C in vacuum. This seems unlikely, however, because reasonably close agreement has been found between the total carbon concentration as measured by ESCA and by a combustion technique.[29]

D. Chemical States of Trace Metals in Particulates

The application of chemical shift measurements to the chemical characterization of metals is difficult because of the small differences in binding energies between different metal compounds. Chemical characterization of particulate lead species was attempted by Araktingi et al.[27] These authors attempted to determine the relative abundance of lead oxide and lead halide in samples collected in Baton Rouge. Because of the small chemical shift between oxide and halide, the two suspected components could not be resolved, however.

Harker et al.[34] used ESCA to determine the chemical state of manganese in particulate emissions from a jet turbine combustion burning 2-methyl cyclopentadiene, manganese tricarbonyl (MMT) as jet fuel additive. The oxidation state assignment was made by examining the multiplet splitting of the (3s) core level in the exhaust sample and some manganese compounds. In addition to oxides, MnF_2 was also studied, since it is the most ionic compound of divalent manganese, and therefore its Mn^{2+} ion should exhibit the largest possible (3s) splitting.

In Figure 9, comparative (3s) spectra for MnF_2, exhaust particulates, and MnO_2 are shown. Based on the magnitude of the splitting, it is concluded that the oxidation state of the manganese in the combustor exhaust is +2 as MnO. Other +2 manganese compounds are eliminated by the fact that oxygen is the only negatively charged species present in sufficient concentrations to balance the manganese.

E. Intercomparison of ESCA with Other Analytical Methods

An intermethod comparison of ESCA with analytical methods of proven accuracy and precision was undertaken by Appel et al.[29] in order to validate the quantitative aspects of ESCA analysis. This work focused on validation of sulfate, nitrate, ammonium, and carbon data as

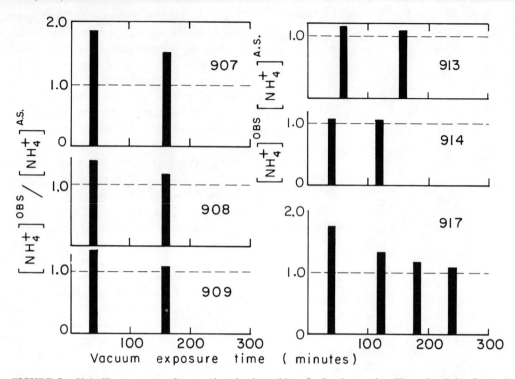

FIGURE 7.   Volatility property of ammonium in six ambient St. Louis samples. The ratio of the observed ammonium peak to the one expected under the assumption that the entire sulfate in these samples is ammonium sulfate vs. vacuum exposure time is shown. Note the cases of apparently stoichiometric ammonium sulfate (samples 913 and 914) and the cases where the volatile ammonium component is found in excess of that required for ammonium sulfate. (From Novakov, T., Chang, S. G., Dod, R. L., and Rosen, H., APCA Paper 76-20.4, presented at the Air Pollution Control Assoc. Annual Meeting, Portland, Oregon, June 1976; Lawrence Berkeley Laboratory Report LBL-5215, June 1976.)

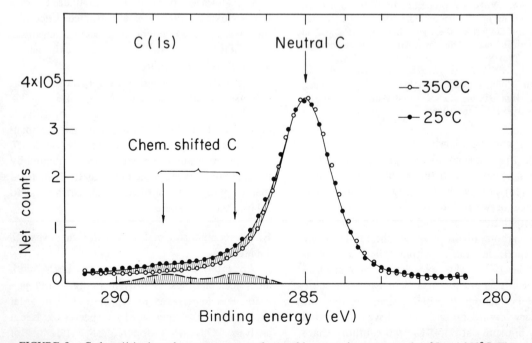

FIGURE 8.   Carbon (1s) photoelectron spectrum of an ambient sample as measured at 25 and 350°C. The shaded area represents the difference between low- and high-temperature spectra. The apparent volatile losses are mainly confined to the chemically shifted component of the carbon peak. (From Chang, S. G. and Novakov, T., *Atmos. Environ.*, 9, 495, 1975. With permission.)

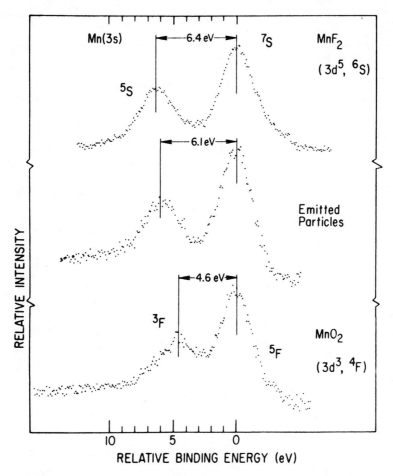

FIGURE 9. Manganese (3s) spectra of MnF_2, exhaust particles, and MnO_2 showing the multiplet splitting of the 3s core level. (From Harker, A. B., Pagni, P. J., Novakov, T., and Hughes, L., *Chemosphere*, 6, 339, 1975. With permission.)

obtained by ESCA in a program to characterize ambient California particulates.

Samples for this study[29] were collected on 47-mm Gelman® GA-1 cellulose-acetate filters for sulfate analysis and on 47-mm, 1.2 μm pore size silver membrane filters for carbon and nitrogen species analysis. In addition to 24-hr samples on high-volume (Whatman® 41) filters, 2-hr samples were collected.

ESCA analyses were conducted only on the membrane filters while, depending on sensitivity, wet chemical analyses were conducted on either the 2-hr or 24-hr high-volume filters. Direct comparisons involved analyses of sections from the same filters, while indirect comparison required comparison of calculated 24-hr average values from ESCA analysis of 12 2-hr filters with wet chemical analysis of the high-volume filter. Three wet chemical methods[29] were used:

1. Stanford Research Institute (SRI) used a microchemical method, which measures total water-soluble sulfur, to analyze 2-hr filters.

2. Barium chloride turbidimetric analysis was used to analyze water-soluble sulfate in high-volume samples.

3. Air and Industrial Hygiene Laboratory (AIHL), California State Department of Health, uses a method for 2-hr samples that employs an excess of a barium-dye complex in acetonitrile-water solution. A decrease in absorbance due to the formation of barium sulfate is measured.

ESCA analysis included the determination of relative concentrations of sulfate and lead from the measured peak areas corrected for elemental sensitivity. Relative concentrations were converted to micrograms per cubic meter of sulfate by normalization to lead concentrations (in micro-

TABLE 1

Comparison of Sulfate Data ($\mu g/m^3$) ESCA (Cellulose-ester filters, $\bar{\Sigma}$ 2-hr Low Volume) vs. Wet Chemistry (Whatman®41, 24-hr High Volume)

Site	Date	High vol wet chemistry	$\bar{\Sigma}$ 2-hr low vol ESCA	High vol $\overline{\bar{\Sigma}$ 2-hr low vol}
San Jose	8-17-72	1.6 ± 0.4	1.1 ± 0.2	1.5 ± 0.5
San Jose	8-21-72	1.0 ± 0.2	1.4 ± 0.3	0.7 ± 0.2
Fresno	8-31-72	4.2 ± 1.0	4.0 ± 1.1	1.1 ± 0.4
Riverside	9-19-72	5.9 ± 1.5	6.4 ± 4.6	0.9 ± 0.7

Note: Ratio of means = 1.0; Spearmans ρ = 0.80; linear regression slope = 1.0; intercept $\equiv$ 0.

From Novakov, T., *Proc. Second Joint Conf. on Sensing Environ. Pollutants,* Instrument Society of America, Pittsburgh, Pa., 1973, 197. With permission.

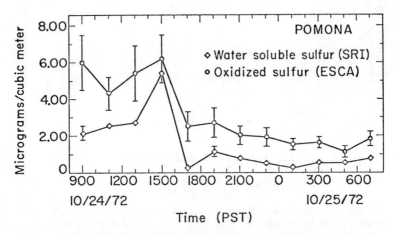

FIGURE 10.   Diurnal variations of sulfate as measured by ESCA and by a wet chemical method (SRI). The data are for a 24-hr sampling period during a smog episode in Pomona, California. (From Appel, B. R., Wesolowski, J. J., Hoffer, E., Twiss, S., Wall, S., Chang, S. G., and Novakov, T., *Int. J. Environ. Anal. Chem.,* 4, 169, Gordon and Breach, 1976. With permission.)

grams per cubic meter), which were determined by XRF analysis.[35]

Table 1 lists the results of the comparisons of ESCA sulfate determinations with those by the SRI and AIHL methods and by the $BaCl_2$ turbidimetric method. The ratio of means between wet chemical methods and ESCA varies from 0.5 to 1.0. It was concluded that ESCA provides sulfate analyses that are correct within a factor of 2. A qualitative indication of the precision of ESCA results is obtained by comparing the diurnal patterns for sulfate determined by ESCA and by alternate procedures.

Figure 10 shows sulfate diurnal patterns as measured by ESCA and by the SRI method. The data are for a 24-hr sampling period during a moderate smog episode in Pomona, California. ESCA and SRI procedures yield strikingly similar diurnal patterns, suggesting a sufficient precision for the ESCA method.

More recently, Harker[36] compared the results of ESCA sulfate analysis to the results obtained by XRF and a wet chemical method. In the Los Angeles area, a number of 4-hr samples were collected on Fluoropore® filters. The results of Harker's study are shown in Figures 11 and 12. These results prove that the ESCA technique can determine a good portion of the bulk composition of atmospheric particulates, in spite of its surface sensitivity.

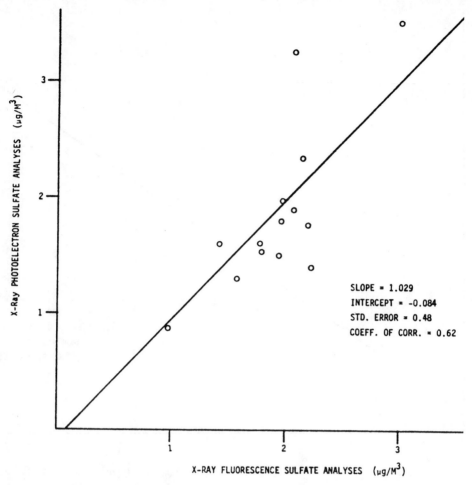

FIGURE 11. Comparison of quantitative particulate sulfate analyses by ESCA and X-ray fluorescence spectroscopy. (From Harker, A. B., Internal Technical Report, Science Center, Rockwell International, Thousand Oaks, Cal., 1976.)

Carbon analysis by ESCA has also been validated, and the results are described in the above-mentioned paper by Appel.[29] Twenty-nine samples collected on silver membrane filters were analyzed for total carbon, both by ESCA and by a combustion technique. A mean ratio of 0.9 ± 0.1 was found between the combustion method and ESCA, suggesting that average carbon analyses are reasonably accurate.

The results of ESCA analyses for nitrate were also compared with the results of wet chemical procedures conducted on the same filters and on 24-hr filters.[29] For nitrate, using both comparative wet chemical techniques, the ESCA results were lower by a factor of about 5. This result is consistent with the volatilization of adsorbed nitric acid, as discussed above.

Similarly, ESCA analyses systematically under-

estimate the ammonium concentrations. The reasons for this discrepancy are related to the volatility of certain ammonium species in vacuum.

V. CONCLUSION

X-Ray photoelectron spectroscopy is a viable analytical tool for the chemical characterization of atmospheric particulates. Analysis by this method is nondestructive and requires no sample preparation. The only requirement for the sampling substrate material (filters or other collection media) is that it cannot contain the same elements as those found in the particulates.

Ample experimental evidence suggests the justification of using relative photoelectron peak intensities to infer the apparent stoichiometry of

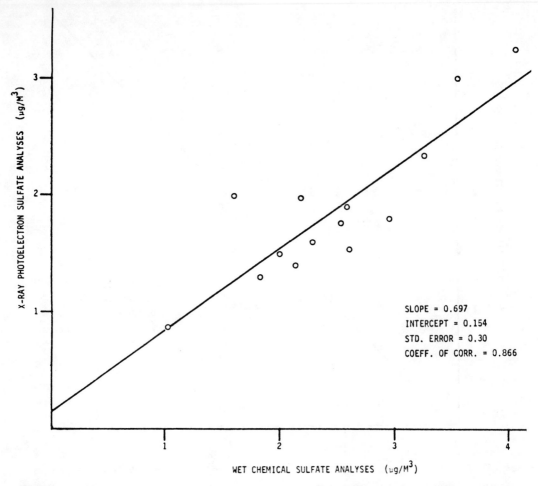

FIGURE 12.   Comparison of quantitative particulate sulfate analyses by ESCA and a colorimetric wet chemical procedure. (From Harker, A. B., Internal Technical Report, Science Center, Rockwell International, Thousand Oaks, Cal., 1976.)

certain molecular species commonly associated with particulate matter.

Determination of chemical shifts, together with information concerning stoichiometry and volatility, allows for a complete characterization of the principal sulfur- and nitrogen-containing compounds. Multiplet splitting, to determine the oxidation state of certain transition metals, can be used in cases where the chemical shift by itself does not appear to be sensitive enough.

Use of X-ray photoelectron spectroscopy has led to a better understanding of the chemical composition of atmospheric particulates; this is illustrated by the following observations and results:

1. Particulate sulfur may exist in a variety of chemical states, under certain atmospheric conditions.

2. Substantial concentrations of previously unsuspected reduced particulate nitrogen species of the amine, amide, and nitrile type have been identified.

3. Ammonium species associated with anions other than sulfate and nitrate have been identified.

4. Much of the particulate nitrate may be present in the form of adsorbed nitric acid.

5. The relative surface concentrations of sulfate, lead, and particulate carbon are similar to their bulk concentration.

REFERENCES

1. Novakov, T., Chemical characterization of atmospheric pollution particulates by photoelectron spectroscopy, in *Proc. Second Joint Conf. on Sensing Environmental Pollutants,* Instrument Society of America, Pittsburgh, Pa., 1973, 197.

2. Novakov, T., Otvos, J. W., Alcocer, A. E., and Mueller, P. K., Chemical composition of photochemical smog aerosol by particle size and time of day; chemical states of sulfur and nitrogen by photoelectron spectroscopy, *J. Colloid Interface Sci.,* 39, 225, 1972.

3. Chang, S. G. and Novakov, T., Formation of pollution particulate nitrogen compounds by NO-soot and NH_3-soot gas-particle surface reactions, *Atmos. Environ.,* 9, 495, 1975.

4. Junge, C. E., *Atmospheric Chemistry and Radioactivity,* Academic Press, New York, 1973.

5. Hidy, G. M. and Burton, C. S., Atmospheric aerosol formation by chemical reactions, *Int. J. Chem. Kinet., Symposium,* 1, 509, 1975.

6. Thomas, T. D., X-ray photoelectron spectroscopy of halomethanes, *J. Am. Chem. Soc.,* 92, 4184, 1970.

7. Siegbahn, K., Nordling, C., Johansson, G., Hedman, J., Heden, P. F., Hamrin, K., Gelius, U., Bergmark, T., Werme, L. O., Manne, R., and Baer, Y., *ESCA Applied to Free Molecules,* North-Holland, Amsterdam, 1969.

8. Schwartz, M. E., Correlation of core electron binding energies with the average potential at a nucleus: carbon 1s and extended Hückel theory valence molecular orbital potentials, *Chem. Phys. Lett.,* 7, 78, 1971.

9. Pople, J. A., Santry, D. P., and Segal, G. P., Approximate self-consistent molecular orbital theory. I. Invariant procedures, *J. Chem. Phys.,* 43, S129, 1965.

10. Ellison, F. O. and Larcom, L. L., ESCA: a new semi-empirical correlation between core-electron binding energy and valence-electron density, *Chem. Phys. Lett.,* 10, 580, 1971.

11. Davis, D. W., Shirley, D. A., and Thomas, T. D., K-electron binding energy shifts in fluorinated methanes and benzenes: comparison of a CNDO potential method with experiment, *J. Chem. Phys.,* 56, 671, 1972.

12. Gelius, U., Heden, P. F., Hedman, J., Lindberg, B. J., Manne, R., Nordberg, R., Nordling, C., and Siegbahn, K., Molecular spectroscopy by means of ESCA, *Phys. Scr.,* 2, 70, 1970.

13. Gelius, U., Roos, B., and Siegbahn, P., Ab initio MO SCF calculations of ESCA shifts in sulphur-containing molecules, *Chem. Phys. Lett.,* 4, 471, 1970.

14. Siegbahn, K., Nordling, C., Fahlman, A., Nordberg, R., Hamrin, K., Hedman, J., Johansson, G., Bergmark, T., Karlsson, S. E., Lindgren, I., and Lindberg, B. J., ESCA – atomic, molecular and solid state structure by means of electron spectroscopy, *Nova Acta Regiae Soc. Sci. Ups., Ser. IV,* 20, 1967.

15. Basch, H., On the interpretation of K-shell electron binding energy chemical shifts in molecules, *Chem. Phys. Lett.,* 5, 337, 1970.

16. Schwartz, M. E., Correlation of 1s binding energy with the average quantum mechanical potential at a nucleus, *Chem. Phys. Lett.,* 6, 631, 1970.

17. Fadley, C. S., Multiplet splittings in photoelectron spectra, in *Electron Spectroscopy,* Shirley, D. A., Ed., North-Holland, Amsterdam, 1972, 781.

18. Fadley, C. S., Baird, R., Siekhaus, W., Novakov, T., and Bergstrom, S. A. L., Surface analysis and angular distribution measurements in X-ray photoelectron spectroscopy, *J. Electron Spectrosc. Relat. Phenom.,* 4, 93, 1974.

19. Wagner, C. D., Sensitivity of detection of the elements by photoelectron spectrometry, *Anal. Chem.,* 44, 1050, 1972.

20. Swingle, R. S., II, Quantitative surface analysis by X-ray photoelectron spectroscopy, *Anal. Chem.,* 47, 21, 1975.

21. Wyatt, D. M., Carver, J. C., and Hercules, D. M., Some factors affecting the application of electron spectroscopy (ESCA) to quantitative analysis of solids, *Anal. Chem.,* 47, 1297, 1975.

22. Brundle, C. R. and Roberts, M. W., Surface sensitivity of ESCA [electron spectroscopy for chemical analysis] for submonolayer quantities of mercury adsorbed on a gold substrate, *Chem. Phys. Lett.,* 18, 380, 1973.

23. Hercules, D. M. and Carver, J. C., Electron spectroscopy: X-ray and electron excitation, *Anal. Chem.,* 46, 133R, 1974.

24. Dod, R. L., Chang, S. G., and Novakov, T., unpublished results, 1976.

25. Novakov, T., Wagner, C. D., and Otvos, J. W., Analysis of Atmospheric Particulates by Means of a Photoelectron Spectrometer, paper presented at the Pacific Conference on Chemistry and Spectroscopy, San Francisco, Calif., October 1970; abstr., *Vortex,* 31, 7, 46, 1970.

26. Hulett, L. D., Carlson, T. A., Fish, B. R., and Durham, J. L., Studies of sulfur compounds adsorbed on smoke particles and other solids by photoelectron spectroscopy, in *Proc. Air Quality,* Plenum, New York, 1972, 179.

27. Araktingi, Y. E., Bhacca, N. S., Proctor, W. G., and Robinson, J. W., Analysis of airborne particulates by electron chemistry for chemical analysis (ESCA), *Spectrosc. Lett.,* 4, 365, 1971.

28. Craig, N. L., Harker, A. B., and Novakov, T., Determination of the chemical states of sulfur in ambient pollution aerosols by X-ray photoelectron spectroscopy, *Atmos. Environ.,* 8, 15, 1974.

29. Appel, B. R., Wesolowski, J. J., Hoffer, E., Twiss, S., Wall, S., Chang, S. G., and Novakov, T., An intermethod comparison of X-ray photoelectron spectroscopic analysis of atmospheric particulate matter, *Int. J. Environ. Anal. Chem.,* 4, 169, 1976.

30. Miller, D. F. and Spicer, C. W., Measurement of nitric acid in smog, *J. Air Pollut. Control Assoc.,* 25, 940, 1975.

31. **Novakov, T., Dod, R. L., and Chang, S. G.,** Study of air pollution particulates by X-ray photoelectron spectroscopy, *Fresenius Z. Anal. Chem.,* 282, 287, 1976.

32. **Novakov, T., Chang, S. G., Dod, R. L., and Rosen, H.,** Chemical Characterization of Aerosol Species Produced in Heterogeneous Gas-particle Reactions, APCA Paper 76-20.4, presented at the Air Pollution Control Assoc. Annual Meeting, Portland, Oregon, June 1976; Lawrence Berkeley Laboratory Report LBL-5215, June 1976.

33. **Spicer, C. W.,** personal communication, 1976.

34. **Harker, A. B., Pagni, P. J., Novakov, T., and Hughes, L.,** Manganese emissions from combustors, *Chemosphere,* 6, 339, 1975.

35. **Giauque, R. D.,** data presented in Characterization of Aerosols in California (ACHEX), Vol. 3, Final Report to Air Resources Board, State of California, Science Center, Rockwell International, Thousand Oaks, Cal., September 1974.

36. **Harker, A. B.,** Quantitative Comparison of the XPS Technique with XRF and Wet Chemical Sulfur Analyses, Internal Technical Report, Science Center, Rockwell International, Thousand Oaks, Cal., 1976.

Chapter 11

IDENTIFICATION AND DETERMINATION OF PARTICULATE COMPOUNDS: INFRARED SPECTROSCOPY, EXTRACTION, AND CHROMATOGRAPHY

R. Kellner

TABLE OF CONTENTS

I. INTRODUCTION

Airborne dust is, in general, a complex mixture of inorganic and organic compounds. In the case of urban aerosols, ammonium and alkali nitrates, sulfates, halides, and carbonates, silicate, quartz, asbestos, and organics can occur, in addition to metal oxides, soot, and adsorbed gases. To characterize the chemical properties of airborne particles, the use of compound-specific methods of analysis, such as infrared (IR) and Raman spectrometry or electron spectroscopy for

chemical analysis (ESCA) (see Chapter 10) is recommended, since the chemical behavior of the airborne particles is a function of compounds rather than elements (see Table 1).

Previous investigations[1] have shown that the distribution of the cited compounds is not uniformly a function of particle diameter. It was found that the secondary particles, originating from gas-solid conversion reactions,[2] contain mainly ammonium and alkali nitrates, sulfates, and hydrocarbons and show the highest concentrations in the diameter range of 0.01 to 1.0 μm. In the

TABLE 1

Comparison of IR, Raman, and Electron Spectroscopy

Method	Principle information	Minimum sample weight	Relative detection limit	Relative reproducibility	Impactor compatible	Solvent required	Limitations
IR	Molecular structure	1 μg	0.1—1%	±10%	+	KBR, CsJ	Band overlapping
Raman	Molecular structure	1 μg	1—10%	±50%	+	0	Band overlapping
ESCA	Atomic and oxidation numbers	≈10 ng	≈1%	±10%	+	0	

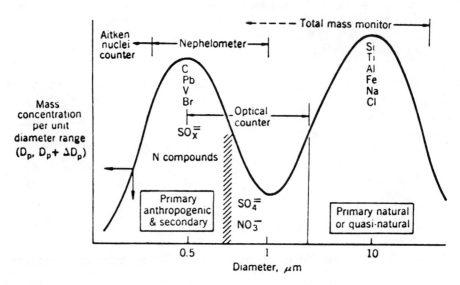

FIGURE 1.   Idealized mass/size distribution for urban aerosols. (From Hidy, G. M., *J. Air Pollut. Control Assoc.*, 25, 1114, 1975. With permission.)

1.0- to 20-μm range, mostly primary particles, produced directly by different chemical or physical processes, can be detected. These particles vary in their chemical composition according to the emission source (see Figure 1).

In principle, IR spectrometry can be used for the analysis of polyatomic molecules or ions whose dipole moment changes during vibration. Nitrates, sulfates, carbonates, silicates, ammonium salts, oxides, and hydrocarbons show characteristic IR-active vibrations. As shown in Section II, IR spectrometry provides information on the structure of compounds via the symmetry-governed number and the bond strength- and mass-determined position of IR-active absorption bands. This fact distinguishes between compounds having the same elemental composition but different structure. Moreover, one main advantage of IR spectrometry applied to airborne particle analysis is its ability to differentiate between Na_2SO_4,

$(NH_4)_2SO_4$, and the respective acid salts $NaHSO_4$ and $(NH_4)HSO_4$. This structural information is often disturbed by band overlapping in complex mixtures. Therefore, the use of IR spectrometry for dust analysis is recommended only in combination with grain size differentiating methods, such as cascade impactors. When employing this method, secondary particles can usually be effectively separated from primary particles and nitrate, sulfate, ammonium, and the sum of the hydrocarbons (after extraction) readily analyzed. For further characterization of the organic extract, which can consist of more than 100 individual compounds, a chromatographic separation is necessary (see Section 4).

IR spectrometry can also be used for a screening of the IR-active matrix of primary airborne particles. However, careful consideration must be given to the influence of band overlapping in a complex mixture and of grain size effects on the intensity of IR absorption bands (see Section II.C.).

II. IR SPECTROMETRY AS A METHOD OF PARTICLE ANALYSIS

The references at the end of this chapter discuss the principles of IR spectrometry and its application in analytical problems.[3-8] The most common application of IR spectrometry is in characterization of pure organic compounds or simple organic mixtures. The system to be treated here — airborne particles — is mainly of an inorganic nature. Therefore, this chapter will attempt to provide a survey of principle information to be gained by an IR spectrum of inorganic substances and of the variable parameters that influence accuracy and repeatability of IR spectrometric measurements with respect to solid particle analysis.

Transmission spectra of the samples to be analyzed are generally recorded after grinding the dust with an alkali halide (KBr, CsJ, or NaCl) and compressing it into a pellet (pellet diameter 13 to 1 mm, sample amount 1 mg to 1 μg). For direct investigation of filter-collected dust samples, the attenuated total reflectance (ATR) method or

transmission of the cellulose-nitrate filter (see Section III.B.4. quartz analysis) are occasionally used.

A. Vibrational Spectra and Structure of Gases and Solid Compounds

A nonlinear, isolated molecule or ion with N atoms is known to have 3 N – 6 normal modes of vibration (3N – 5 for a linear molecule). Not all of these vibrations cause an absorption band in the IR or Raman spectrum in higher molecular symmetry.[3] Selection rules indicate that IR activity of a vibration implies a change of the dipole moment, Raman activity a change in the polarizability of the molecule during vibration. Furthermore, it can occur that two or more active vibrations have the same frequency. These vibrations are degenerate and give rise to only a single absorption band in the spectrum.

Symmetry properties of a given molecule determine the number of absorption bands appearing in a vibrational spectrum. The application of group theory to symmetry operations allows the computation of the number of IR- or Raman-active species of normal vibrations in a molecule as a function of its point group and the number of atoms. A representation of the structure sensitivity of IR spectrometry is illustrated below for two 3-atomic gas molecules, CO_2 (linear) and H_2O (bent):

CO_2 (point group $D_{\infty h}$)

$$v_s \quad \vec{O} = C = \overset{\leftarrow}{O} \quad \text{IR-inactive}$$

$$v_{as} \quad \vec{O} = \overset{\leftarrow}{C} = \vec{O} \quad 2350 \text{ cm}^{-1}$$

$$\delta \quad \overset{\uparrow}{O} = C = \overset{\uparrow}{O} \quad 667 \text{ cm}^{-1}$$
$$\downarrow$$

$$\delta \quad \overset{+}{O} = \overset{-}{C} = \overset{+}{O} \quad \text{(two doubly degenerated vibrations)}$$

H_2O (point group C_{2v})

v_s — O, H H — 3650 cm^{-1}

v_{as} — O, H H — 3756 cm^{-1}

δ — O, H H — 1600 cm^{-1}

The number of IR-active vibrations clearly indicates the linear structure of CO_2. Only two of the four normal modes of vibration can be observed in the IR spectrum because the symmetric vibration species in a molecule of $D_{\infty h}$ symmetry is IR inactive and the two deformation motions have equivalent energy. In the bent H_2O molecule, all three normal modes are IR active and can be observed in the spectrum.

This relationship between molecular symmetry and vibrational spectrum is somewhat more complicated for spectra of molecular crystals. Due to the static field effect, an influence of the crystal lattice on the structure of the single molecule, a shift in the frequencies of the normal modes, and a splitting of degenerate vibrations may occur. Even a fundamental vibration, inactive in free molecules, can become active by lowering of the

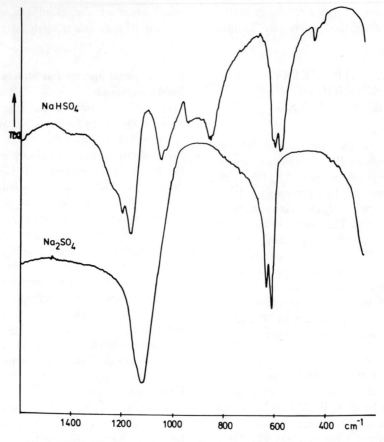

FIGURE 2.   IR spectra of Na_2SO_4 and $NaHSO_4$ in KBr.

original symmetry to the symmetry in the crystal (site symmetry). Furthermore, the dynamic field effect couples all equivalent vibrations in the molecules of the unit cell. One nondegenerate vibration can, therefore, divide into as many components as molecules existing in the unit cell. This splitting is called correlation field splitting and leads to line broadening in IR spectra characteristic of solids as compared to solutions or gases as does nonresolved rotational fine structure.

In addition to this splitting, possible changes in the shape of the internal modes (4000 to 200 cm^{-1}) and motion of the crystal lattice as a whole cause external or lattice vibrations. These usually weak absorptions can be seen in the IR spectrum below 300 cm^{-1} and generally do not couple with the structure-specific internal modes.

Despite the limitations described, the concept of IR spectrometry as a structure- or symmetry-sensitive method of analysis holds for solid-state spectroscopy as well. Secondary airborne particulates include a high content of ammonium sulfates

of varying structures.[9,10] The occurrence of acid sulfates, in addition to the regular sulfates, is described in the literature. IR spectrometry is a unique method of distinguishing between these two sulfate structures. The SO_4^{2-} ion is a 5-atomic species with tetrahedral (T_d) symmetry, whereas the HSO_4^{-} ion is of lower symmetry (C_{3v} when a molecular model with an S-O-H angle of 180° is considered or C_s for a model with an S-O-H angle different from 180°). Application of group theory results in the prediction of two IR-active vibrations for the undisturbed SO_4^{2-} ion but eight IR-active vibrations (four A_1 and E, respectively) for the HSO_4^{-} (C_{3v} model) and twelve IR-active vibrations for the HSO_4^{-} (C_s model).

Inspection of the actual spectra illustrates that the site symmetry of the SO_4^{2-} ion is lowered with respect to the tetrahedral model. This can be proved by the splitting of the 610 cm^{-1} band in the Na_2SO_4 spectrum in KBr. Despite this, the spectrum can readily be distinguished from the $NaHSO_4$ spectrum (see Figure 2) where twelve

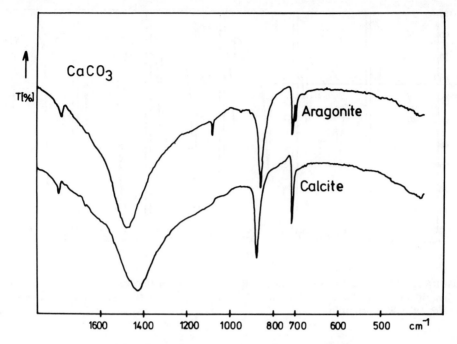

FIGURE 3.   IR spectra of Calcite and Aragonite in KBr.

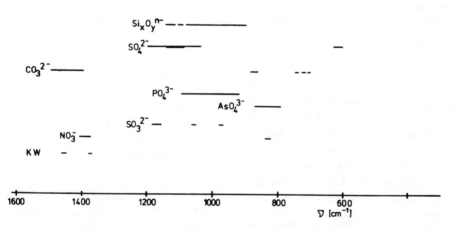

FIGURE 4.   Correlation table for qualitative IR airborne particle analysis.

strong absorption bands can be observed, which is strong evidence for C_s symmetry.

Another example of the high compound-specific power of IR spectrometry, based on symmetry properties of the molecules, is easy differentiation between calcite and aragonite, two substances having the same elemental composition (Figure 3). Calcite (D_{3h}) has only three IR-active vibrations, whereas in the aragonite spectrum, the symmetric stretching at 1080 cm^{-1} and a splitting of the degenerate deformation vibration due to lowering of the site symmetry, at around 710 cm^{-1}, are observed.

B. IR Band Position in Solid Particle Spectra

The concept of group frequencies (special molecular groups absorb in a narrow frequency region of the IR spectrum independently of the rest of the molecule) holds, in principle, for the analysis of inorganic dust components also. Thus, the frequency of an absorption band provides information on the qualitative nature of compounds in the sample. There are several limitations to the practical application of this concept, resulting from band overlapping in complex mixtures. Relevant group frequency regions for the problem under discussion are given in Figure 4.

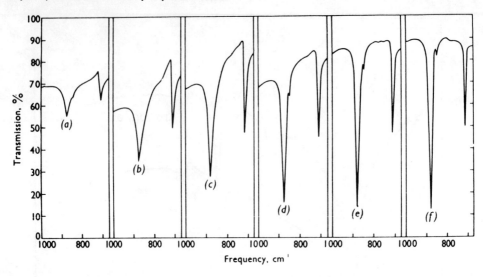

FIGURE 5.   Absorption bands of calcite in KBr. Mean values for the diameters of the particles: (a) 55 μm, (b) 40 μm, (c) 23 μm, (d) 14.7 μm, (e) 5 μm, and (f) 2.1 μm. (From Duyckaerts, G., *Analyst*, 84, 202, 1959. With permission.)

As in organic compounds, (see Colthups correlation table[4]) the group frequencies of complex molecules or ions relevant to airborne dust analysis are slightly shifted by environmental factors. These effects are occasionally helpful in further characterization of compounds. Thus, IR spectrometry provides a unique capability for differentiation of the various carbonate minerals frequently occurring in primary particles, such as calcite, dolomite, and magnesite. The ν_4-$CO_3{}^{2-}$ deformation vibration is sensitive to the nature of the cation in the substance.[11] According to the mineral standard spectra published by Moencke,[11] other carbonate minerals (i.e., $FeCO_3$ or $ZnCO_3$) that might be present in industrial dust interfere with this analysis.

A similar effect of band shifting can be observed, in certain cases, when spectra of dust components prepared in different solid solvents are compared. At the same time, a change of the band shape may occur. Keeping this in mind, the compound specific power of IR spectrometry can be fully employed in dust analysis. An illustration of this is described by Cunningham et al.,[12] who give an interpretation of IR spectra of samples collected in an impactor (see Section III.B.1.).

C. IR Band Intensity in Solid Particle Spectra

Currently, the alkali halide pellet technique[13] is generally accepted as the standard method for preparation of solid samples for qualitative IR

spectrometry. After careful optimization of the variable parameters for sample preparation, this method seems to be best suited for quantitative analysis, also. These variable parameters are

1. Grinding time (for the regulation of the grain size)
2. Grinding procedure (dry or wet)
3. Alkali halide used as solvent (influence of the refractive index)
4. Pressure applied to the disc
5. Side reactions and thermal stability

The direct transmission technique for the analysis of filter-collected airborne particles is described in the literature for quantitative work.[14] The effect of particle size and complete covering of the surface to be measured are also of main importance to the band intensity. Duyckaerts[8] reports on a rather dramatic effect that grain size of calcite particles prepared in KBr has on band intensities, as seen in Figure 5. A similar trend was demonstrated by Mangia[14] for α-quartz samples in a direct transmission experiment. In a theoretical treatment, Duyckaerts demonstrated a pronounced variation of the absorbance values as a function of grain size in the particle diameter range between 1 and 100 μm. For particles smaller than 1 μm, maximal absorption is reached, which is experimentally proved with the calcite bands at

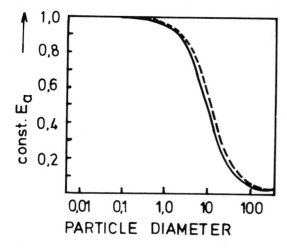

FIGURE 6. Theoretical variation of apparent absorbance E_a with particle diameter in micrometers, for spheres (dotted line) and for cubes (full time). (From Duyckaerts, G., *Analyst,* 84, 205, 1959. With permission.)

1431, 876, 710, and 320 cm^{-1}, respectively (Figure 5).

Every grinding procedure causes broad rather than uniform grain size distribution, and this is undoubtedly a source of error in quantitative measurements. Thus, omitting the grinding procedure in dust sample preparation and measuring the IR spectra of airborne particles directly after the grain size differentiating sampling procedure (as with a cascade impactor) would promote uniform distribution. Since the secondary urban particles (usually containing sulfates, nitrates, and organics) occur mainly in the diameter range < 1 μm where maximum absorbance values are reached, this direct measurement method would be highly practical. For particle diameter >1 μm, special working curves must be determined.

As pointed out by Duyckaerts,[8] linear working curves are only found when the number of particles analyzed is sufficiently high to cover the entire illuminated surface of the sample. Furthermore, it is necessary to hold constant the pressure applied to the disc during formation and to use a solid solvent whose refractive index closely matches the refractive index of the sample.

Tuddenham,[15] in a model analysis for calcite/quartz mixture analysis in KBr after wet hand grinding, reports a reproducibility of v = 2.3 to 5.8% and states that most of the minerals do not react with alkali halides. Side reaction due to thermal or mechanical instabilities are reported in nitrate analysis when KBr is used as pellet material and the grinding technique is applied. This instability seems to be the result of a redox reaction between the nitrate and the KBr. The problem can be overcome by replacing the latter with NaCl, which has a similar refractive index. There has been limited quantitative IR analysis of dust components, and the reproducibility of nitrate and sulfate analysis,[33] important in the characterization of secondary particles, is subject to improvement in the near future.

III. APPLICATION OF IR SPECTROMETRY IN THE CHEMICAL CHARACTERIZATION OF AIRBORNE PARTICULATE MATERIAL

A. Sampling Procedures

The full potential of IR spectrometry can only be utilized after preseparation of airborne dust, a multicomponent system. Chemical composition of airborne particles is not uniform as a function of grain size. This is of special importance, since grain size differentiating sampling methods can function simultaneously as separation methods (cascade impactors, centrifuges). Furthermore, the higher uniformity of grain size in each stage of the impactor (as compared to filter sampling) is essential for the application of IR spectrometry to quantitative characterization, as shown by Duyckaerts.[8] The effects of grain size on the intensity of an IR absorption band becomes negligible when the grain size drops below 1 μm.

The unique and most valuable contribution of IR spectrometry to integrated dust analysis is — in addition to a screening of the composition of the IR active matrix — the characterization of different sulfates (e.g., $CaSO_4$, $(NH_4)_2 SO_4$) and acid sulfates, nitrates, ammonium salts, and carbonates. These compounds, with the exception of the carbonates, are found mostly in secondary particles formed in the atmosphere and enriched in an aerodynamic particle diameter range smaller than 1 μm.[9,16,17,27] This characterization ability of IR spectrometry and ease of operation are reasons for employing cascade impactors more frequently as sampling devices for IR dust analysis.

Until recently, few researchers used impactors as sampling devices for IR measurements.[12,16-18] A possible explanation for this is the small sample amount provided by this technique (in urban sites,

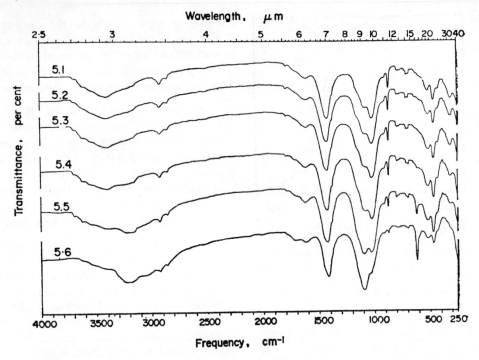

FIGURE 7. IR absorption spectra of atmospheric dust collected with an Andersen impactor sample number 5, stages 1 through 6. (From Blanco, A. J. and McIntyre R. G., *Atmos. Environ.*, 6, 561, 1972. With permission.)

several micrograms per hour in each section of the impactor). Standard preparation techniques (13-mm pellets) require a Fourier transform spectrophotometer.[12] Otherwise the use of an ultramicro pellet technique is necessary.[17,18] Most published research has been devoted to special problems such as quartz or asbestos analysis,[14,15,19-21] analysis of organic material in dust samples after extraction,[22-25] or screening of the IR-active matrix.[17,26] These difficulties can be solved for filter-collected samples, as will be illustrated in the next section of this chapter.

B. Survey of Published Applications

1. General Information

Moenke[11] has published a catalogue of mineral spectra in the IR region that can be considered, coupled with the work of Duyckaerts,[8] as the basis of the application of IR spectrometry to the characterization of inorganic particulate material. Unfortunately, the scanning region was limited (4000 to 400 cm^{-1}), and only KBr was used as pelleting material. Attempts have been made to enlarge the scanning region to the far infrared to detect characteristic vibrations in oxides and to use different solid solvents, such as CsJ and

NaCl.[27] The first application of IR spectrometry for obtaining full-scale spectra in the region of 4000 to 250 cm^{-1} from atmospheric dust samples in the 10- to 50-μg range is described by Blanco and Hoidale.[28] The authors used silver membrane filters (0.8-μm pore size) and transferred the sample to the KBr by rotating the KBr crystals with a spatula over the loaded filter to dislodge the sample. Quantitative data were not obtainable in this early study. The micropellet technique described by Andersen and Woodall[29] and a grating IR spectrophotometer were used to record this first IR spectrogram tracing of atmospheric dust.

Blanco and McIntyre[18] describe for the first time the IR spectrophotometric characterization of airborne particles sampled by a six stage Andersen impactor fractionating the particles in size ranges from 0.3 to > 10.4 μm (stages 5 and 6 < 1.5 μm, no further calibration is given). The spectra of the samples — taken in an urban environment of El Paso, Texas — showed a transition from a silicate and carbonate dominated particle fraction > 2 μm to an ammonium sulfate dominated particle fraction between 0.3 and 2 μm (Figure 7).

The components identified in the samples investigated by a direct absorption band comparison technique were quartz, kaolinite, montmorillonite, illite, calcite, dolomite, sodium nitrate and possibly potassium nitrate, ammonium sulfate, gypsum, mirabilite, thenardite, and hydrocarbons. Quantitative data were not treated in this study. Cunningham[12] published a study of the variations in chemistry of airborne particulate material with particle size and time. Samples were collected at Argonne, Illinois, a rural and suburban area without significant air pollution problems, by means of a four-stage (0.3 to 12 μm) Lundgren impactor and Fourier transform IR spectrometry. Ammonium sulfate, ammonium halide, carbonate, silicate and silica, nitrate, and hydrocarbons were determined via their characteristic group frequencies (Figure 8 and Table 2). A number of bands cannot be definitely assigned as yet, among them the "surface nitrate" band at 1384 cm^{-1}. It was found that the bands 1190, 1140, 1120, 670, 627, and 600 cm^{-1} share the same temporal variation in intensity and resemble bands due to various forms of phosphates. A most interesting point is the tentative assignment of the weak 1720 cm^{-1} band observed in fourth-stage samples with NH_4^+/SO_4^{2-} molar ratio greater than 2 to the ν_4 (bending mode) + ν_6 (lattice libration) combination band for NH_4^+ in ammonium halide. Elemental analysis for Cl and Br is indicated to support this assignment. This finding corresponds to the results of NH_4^+ analysis by ESCA.[30,31] It was found in this study that if ammonium appears in concentrations above those expected for ammonium sulfate and nitrate, this excess ammonium is volatile in vacuum, as can be expected for ammonium halides.

2. Sulfate Analysis

According to Brosset,[10] acid ammonium sulfate particles are formed by reaction of ammonia with photochemically formed sulfuric acid droplets in relatively dry air (relative humidity 30 to 40%) as found in 1973 during a so-called white episode over the North Sea, Skagerrak. When humidity rises, the acid particles are transformed into droplets that readily absorb ammonia. The following crystalline ammonium sulfate phases have been determined by chemical methods in fine particles during white episodes: NH_4HSO_4, $(NH_4)_3H(SO_4)_2$ letovicite, and $(NH_4)_2SO_4$, mascagnite. Novakov et al.[32] discuss the problem

of catalytic activity of air-suspendable carbonaceous particles for the oxidation of SO_2 into sulfate. ESCA and wet chemical methods were used to characterize the reaction products. ESCA results indicated S with an oxidation state of 6+, the solution of particle sulfate was acidic. These data alone do not characterize the actual structure of the sulfate species. According to the logarithmic diagram, the elements of the $H_2SO_4/HSO_4^-/SO_4^{2-}$ system are stable as a function of pH in aqueous solutions; in acid medium, both HSO_4^- and SO_4^{2-} are stable (Figure 9).

IR spectrometry is used for the determination of the actual structure (see Section III.B.) as demonstrated by Chang and Novakov.[16] Attenuated total reflectance (ATR) spectra were recorded for samples used in ESCA analysis, in the region between 1500 and 500 cm^{-1} and compared to reference spectra of ammonium sulfate, 1N sulfuric acid, and ammonium bisulfate in aqueous solutions. The similarity of the ATR spectra of the particles and the solution of $(NH_4)_2SO_4$ led to the conclusion that the oxidation of SO_2 on soot particles results in the formation of salt-like species having tetrahedral structure.

Malissa and Kellner[17] came to a similar conclusion during an urban aerosol characterization experiment in summer 1976 in Vienna. A four stage cascade impactor with the varying aerodynamic particle diameter ranges (first stage 0.14 to 0.4 μm, second stage 0.4 to 1.6 μm, third stage 1.6 to 6.0 μm, and fourth stage 6.0 to 25 μm) was used as a sampling device. The IR spectra were recorded with a Perkin-Elmer® 180 grating spectrophotometer after sample preparation utilizing a modified ultramicro-pellet technique (UM-pellet technique) in CsJ.[33] According to this technique, a 1-mm CsJ pellet is prepared, then covered on one side with at least 1 μg of the dust sample to be analyzed. This is pressed onto the pellet, a technique that prevents losses of sample material during grinding. Stray-light effects or unwanted irregularities of band intensity in the especially important first and second stages are not expected. In these stages, thermoanalytical methods can detect the presence of ammonium and sulfate.[34]

The IR spectra of one spot of each stage show absorption bands at about 3200, 1400, 1110, and 610 cm^{-1}, which clearly indicates the presence of $(NH_4)_2SO_4$ in these samples. The IR spectrum of bisulfate is markedly different. (Figure 10). It must be remembered, however, that silicates **must**

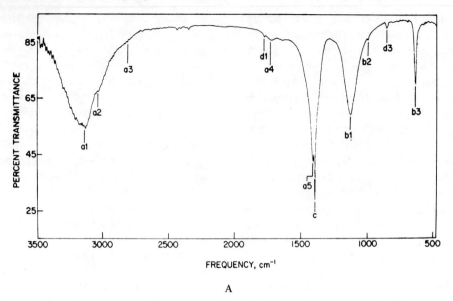

A

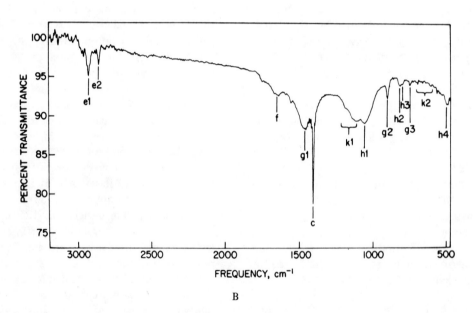

B

FIGURE 8.   Fourier Transform IR spectra of impactor collected samples of airborne particles in 13 mm KBr pellets, A. 276 μg, 0.3- to 1.2-μm diameter range, B. 54 μg, 3.0- to 12-μm diameter range. (Reprinted with permission from Cunningham, P. T., Johnson, St. A., and Yang, R. T., *Environ. Sci. Technol.,* 8, 133, 1974. Copyright by the American Chemical Society.)

be absent to allow a nondisturbed spectra evaluation in the region around 1000 cm.$^{-1}$

3. Nitrate Analysis

The results of the California aerosol characterization experiment[9] (1971 to 1974) clearly shows the great importance of nitrates to the mass concentration of airborne particles in the submicron size range besides sulfates, noncarbonate carbon, and liquid water. Nitrate and sulfate analysis were performed by ESCA. Blanco et al.,[18] Cunningham,[12] and Malissa and Kellner,[17] made use of the compound-specific power of IR spectrometry for characterization of different nitrates in KBr. Sodium, potassium, and ammonium nitrate can be differentiated by means of their IR spectra (Figure 11).

The characteristic sharp and intense NO_3 ab-

TABLE 2

Listing of Assigned Infared Bands Observed in Particulate Samples

| Frequency (cm^{-1}) | Assignment | | Designation in Figs. 8A and 8B |
	Species	Normal mode[a]	
3140	NH_4^+	ν_3	a1
3020	NH_4^+	$\nu_2 + \nu_4$	a2
2920	Hydrocarbon (C–H)	ν_3	e1
2860	Hydrocarbon (C–H)	ν_1	e2
2800	NH_4^+	$2\nu_4$	a3
1768	NO_3^- (bulk)	$2\nu_3$	d1
1720	NH_4^+ (halide)	$\nu_4 + \nu_3$	a4
1620	H_2O	ν_2	f
1435	CO_3^{2-}	ν_3	g1
1400	NH_4^+	ν_4	a5
1384	NO_3^- (surface)	ν_3	c
1360	NO_3^- (bulk)	ν_3	Overlapped
1190	PO_4^{3-}[b]	ν_3	k1
1140	PO_4^{3-}[b]	ν_3	k1
1120	PO_4^{3-}[b]	ν_3	k1
1110	SO_4^{2-}	ν_3	b1
1035	SiO_4^{4-}	ν_3	b1
980	SO_4^{2-}	ν_1	b2
880	CO_3^{2-}	ν_2	g2
840	NO_3^- (bulk)	ν_2	d3
800	SiO_4^{4-}	ν_1	h2
780	SiO_4^{4-}	ν_1	h3
728	CO_3^{2-}	ν_4	g3
670	PO_4^{3-}[b]	ν_2	k2
627	PO_4^{3-}[b]	ν_2	k2
620	SO_4^{2-}	ν_4	b3
600	PO_4^{3-}[b]	ν_2	k2
470	SiO_4^{4-}	ν_4	h4

[a]Spectroscopic designation of vibrational motion after Herzberg, G., *Molecular Spectra and Molecular Structure*, Vol. 2, Van Nostrand, New York, 1964.
[b]Assignment uncertain.
[c]This band is overlapped in Figure 8B by bands c and a5.

sorption at 1385 cm^{-1} is generally correlated to the ν_3 asymmetric stretching vibration of the NO_3^- anion. This prominent absorption is found mostly in IR spectra of airborne particles, but the weaker deformation vibration at 830 cm, $^{-1}$ indicative of the cation in the compound, is mostly overlooked (Figure 12). This conclusion was based on a model experiment published 1971 by Boehm,[35] where TiO_2 and Al_2O_3 were exposed to NO_2, the IR spectra of the oxides recorded before and after exposure were compared, and a band at 1385 cm^{-1} was found after exposure.

Cunningham[12] assigns this absorption to "surface nitrate" formed through the chemisorption of atmospheric NO_2 on the particle surface. A survey of the literature shows that the problem of nitrate formation is not clearly understood.

Chang and Novakov[36] report on an IR spectroscopic NO_3 analysis of soot particles from propane-benzene combustion in air, exhaust particulates from an internal combustion engine, airborne particles collected in a highway tunnel, and activated carbon exposed to NO in humid air. The measurements were obtained using a Perkin-

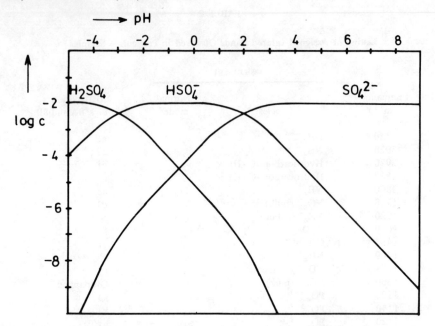

FIGURE 9.   Logarithmic diagram of the system $H_2SO_4/HSO_4^-/SO_4^{2-}$.

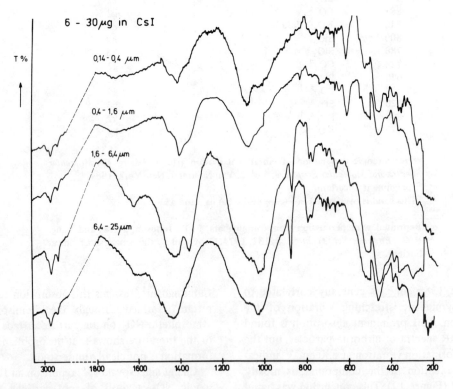

FIGURE 10.   IR spectra of airborne particles collected with a Berner impactor in Vienna, July 1976.

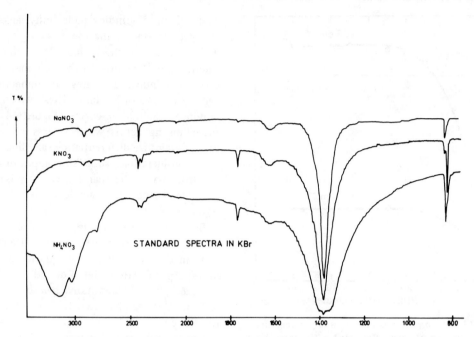

FIGURE 11.   IR spectra of NaNO$_3$, KNO$_3$ and NH$_4$NO$_3$ in KBr.

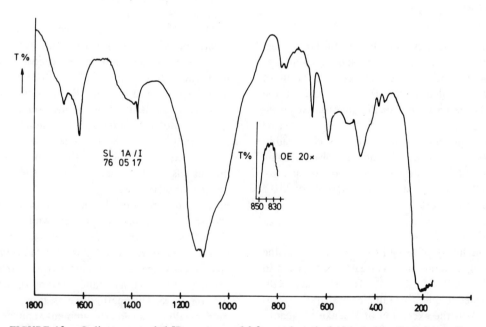

FIGURE 12.   Ordinate expanded IR spectrum of 0.9 mg of particulate material collected on a filter containing NO$_3$ in KBr.

Elmer® 621 spectrometer and the standard KBr-pelleting technique. On the basis of this analysis, the authors propose a nonphotochemical step mechanism for the formation of nitrates. They further indicate that IR spectrometry is more desirable for qualitative NO$_3^-$ analysis (despite a large amount of NH$_4^+$ in the sample) than ESCA.

Gordon and Bryan[37] report on a qualitative IR spectrophotometric analysis of ammonium nitrate in airborne particulate matter collected in Los Angeles by means of a fiber glass filter. After

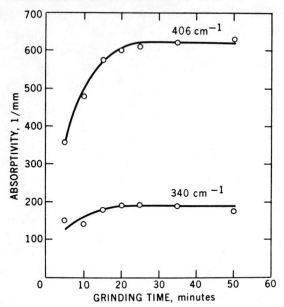

FIGURE 13.   Absorptivity as a function of grinding time for pyrite, in an agate vial. (Reprinted with permission from Estep, P. A., Kovach, J. J., and Karr, C., Jr., *Anal. Chem.*, 40, 359, 1968. Copyright by the American Chemical Society.)

extraction of benzene solubles, a still larger fraction can be extracted with methanol. Using wet chemical methods, the average concentration of NH_4NO_3 was determined as 10 to 15% of the total airborne particles. For the quantitative evaluation of IR spectra of nitrates in KBr, it must be kept in mind that redox reactions can readily occur between the sample and the solvent during grinding and even during heating in the spectrophotometer. Relative standard deviations of ±10% were recently reported.[27]

4. Quartz Analysis

The feasibility of using IR spectrometry for the analysis of minerals such as quartz was proved in the early sixties.[11,15,26,38] The standard KBr-pelleting technique for sample preparation was utilized. With the exception of Moenke,[11] quantitative analyses of quartz in mineral mixtures were developed. In agreement with the earlier findings of Duyckaerts,[8] researchers maintained that control of particle size is of vital importance. Therefore, grinding procedures must be carefully optimized for each mineral. Grinding must take place when the absorptivity curve as a function of grinding time registers a plateau (Figure 13).

To prevent structural breakdown of minerals due to constant grinding, as referred to by

Tuddenham,[15] grinding under ethyl alcohol with an agate mortar or the use of an agate ball mill with periodic cooling periods[26] is recommended. Tuddenham[15] reports further on matrix effects due to grinding mixtures of minerals having varying degrees of hardness. These effects result in nonlinear calibration curves (Figure 14). Taking these limiting factors, which are less relevant in the case of grain size differentiating sampling methods, into consideration, successful application of IR spectrometry to the quantitative analysis of quartz in mineral and dust samples is reported (mean deviation, 2%).

In addition to the application of the KBr technique to the IR quartz analysis, attempts have been made to analyze the quartz content of airborne particulate material directly on the filter material.[19,39,40] Quantitative measurements were performed at the characteristic bands for α-quartz at 800 cm^{-1}[19,39] and 690 cm^{-1}.[40] Cellulose nitrate filtering materials were used having an IR spectrum sufficiently weak in that spectral region to permit application of the IR differential technique as described by Sloane.[41]

Mangia[14] reports on a comparison of IR spectrometry and X-ray diffraction technique to the determination of α-quartz in atmospheric dust. Both the KBr and the cellulose-nitrate membrane technique were used for IR measurements. To improve the sensitivity of the method and reduce the effects of the inhomogenity, the samples were transferred from the original membrane to another, on a surface of the same shape and dimensions of the spectrophotometric entrance. Spectra of different amounts of quartz (4- to 6-μm mean diameter) on filtering membrane and the resulting calibration curve are shown in Figure 15.

The well-known effect of particle size to the shape of the quartz band at 12.5 to 12.8 μm (800 cm^{-1}) was suggested as a method to determine the mean particle diameter and the appropriate calibration curve. This method seems, however, not to be generally applicable to ambient samples with a broad grain size spectrum. In such cases, grain size differentiating methods are preferable. Despite this, Mangia[14] concludes that the IR method developed, even though not as specific as X-ray diffraction, seems to be suitable for determining even small quantities of quartz (see Table 3).

5. Asbestos Analysis

Several attempts described in the literature use IR techniques for analysis of airborne asbestos, in

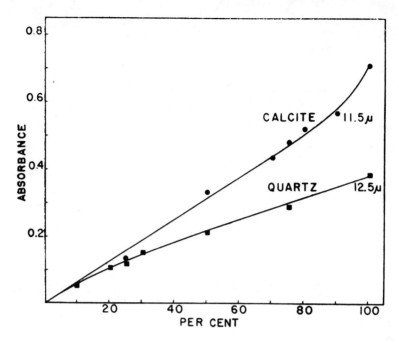

FIGURE 14. Calibration curves obtained with mixtures of quartz and calcite in KBr. (Reprinted with permission from Tuddenham, W. M. and Lyon, R. J. P., *Anal. Chem.*, 32, 1632, 1960. Copyright by the American Chemical Society.)

addition to the more common methods of electron microscopy,[42] neutron activation,[43] and X-ray diffractions.[44] A direct IR spectroscopic analysis of dust samples to evaluate the asbestos content is generally not possible, due to heavy band overlapping by other silicate minerals (see spectra collection by Moencke[11]). Bagioni[21] has published a special sample preparation procedure, based on work of Gadsen et al.,[20] to eliminate interference prior to IR dust analysis. Mineral interference is removed by centrifugation of the sample in a high-density liquid (a mixture of 1,1,2,2-tetrabromoethane and carbon tetrachloride, density adjusted to 2.45 to 2.50 g/ml at 20°C). Organic interference is removed, along with the filter material, by ashing techniques. Low temperature ashing techniques are recommended. It was shown[21] that all interference at the band at 2.72 μm can be removed. However, this method is not yet optimized for quantitative analysis.

6. Carbonate Analysis

A main advantage of IR spectrometry as a compound-specific method for screening the composition of the IR active dust matrix is its ability to differentiate between the various carbonate minerals often found in primary particles (Figure 16). This was already pointed out, for qualitative analysis, by Moencke.[11] Quantitative analysis of single carbonate components in mixtures were reported by Duyckaerts,[8] who optimized the calcite analysis at 876 and 710 cm.[-1] Tuddenham and Lyon[15] described a procedure for the analysis of binary mixtures of calcite and α-quartz (see Figure 15). Estep et al.[26] reported on a quantitative IR multicomponent analysis of commonly occurring mineral constituents of coal, including quartz, calcite, and kaolinite. Hlavay et al.[38] investigated the quantitative carbonate analysis of soils after sample preparation in KBr and by use of the absorption band at 1435 cm.[-1] These authors found a relative error of ± 0.5% for carbonate contents up to 24%. The first attempts to work out a quantitative analysis of mixtures of calcite, dolomite, and magnesite in airborne dust were recently done by Malissa and Kellner.[17]

7. Analysis of Carbon and Organic Materials

Although IR spectra of a number of carbon-containing materials (such as coals, coal pyrolysis products, chars, carbon black, and graphite) and ATR spectra of sorbates on activated carbon have been reported,[45] the transmission spectra of carbons have not been obtained because of high

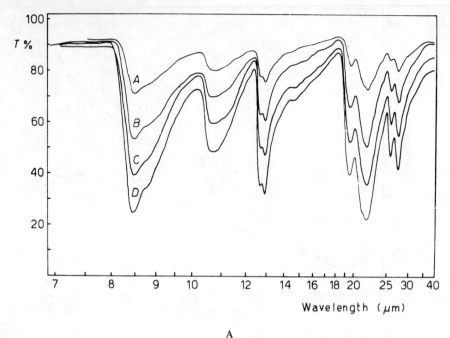

A

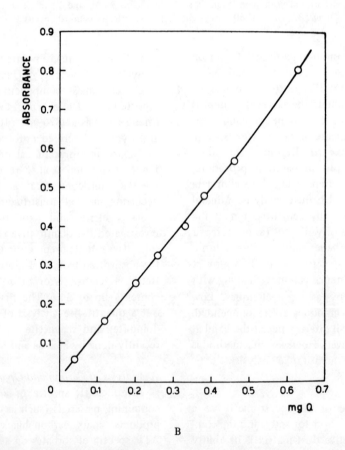

B

FIGURE 15. A. Spectra of different amounts of quartz (4- to 6-μm mean size) on cellulose-nitrate filtering membrane (A) 0.04, (B) 0.12, (C) 0.2, and (D) 0.3 mg. B. Working curve established for the 12.8-μm band (From Mangia, A., *Anal. Chem.*, 47, 928, 1975. Copyright by the American Chemical Society.)

TABLE 3

Quartz Percent and Relative Standard Deviation

	Quartz (%)	σ (%)
IR		
KBr technique	13.80	4.3
Membrane-support technique	12.92	2.3
X-Ray diffraction	12.70	1.2

Reprinted with permission from Mangia, *Anal. Chem.*, 47, 927, 1975. Copyright by the American Chemical Society.

scattering by these materials. The method developed consists of the production of a characteristic transmission spectrum through surface oxidation accompanying an extensive grinding process of the sample with KBr in air in a stainless steel vial. The grinding time was optimized at 24 hr.

A linear relationship between the absorbance of the 1580 cm^{-1} band and the mass of the carbon was found. Reproducibility of the method was reported to be excellent (better than 1% relative for fly ash samples), and interference caused by mineral constituents and organics were eliminated by an acid treatment with HCl and HF or basic

peroxide treatment, respectively. The practical value of this method seems to be rather limited, as compared with combustion techniques, due to the lengthy sample preparation step.

Application of Raman spectrometry to the problem of carbon analysis of airborne particles, as suggested by Rosen and Novakov,[46] seems to be promising with regard to structural information, although the effect of particle size on the intensity of Raman lines is not well understood. The spectra shown in Figure 17 give strong evidence for the existence of physical structures similar to activated carbon in samples studied.

Direct application of IR spectrometry to the characterization of organic constituents in airborne particulate material is generally not acceptable due to band overlapping problems. There are, however, some very interesting reports in the literature on IR spectrometric studies of organic materials extracted from airborne particles by different organic solvents, solvent mixtures, or water. Grosjean and Friedlander[23] investigated, in a study of "Gas-Particle Distribution Factors for Organic and Other Pollutants in the Los Angeles Atmosphere," the composition of the organic carbon fraction extracted by isooctane-isopropyl

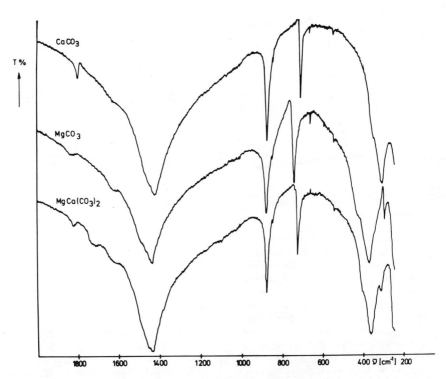

FIGURE 16.   IR spectra of carbonate minerals.

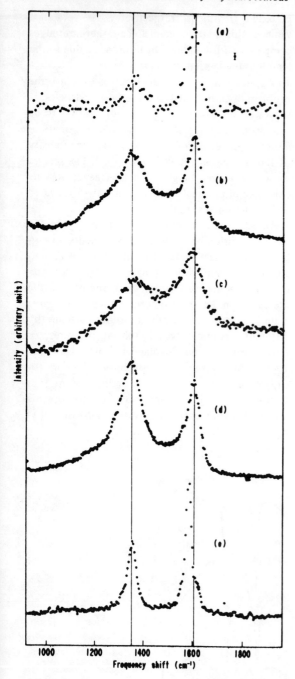

alcohol (50-50% v/v) and water after cyclohexane (extraction efficiency 94-100%). They determined, with IR spectrometry, the presence of organic nitrates, carbocyclic acids and esters, carbonyls, and polymeric peroxidic material. A preponderance of polar organics, neutral oxygenates, and acids (constituting 2/3 of the organic fraction) was derived. Strong correlation between the O_3 content of the air and both the organic carbon fraction and C=O IR absorption band were obtained as a basis of a diurnal sampling procedure (see Figure 18).

Della Fiorentina et al.[24,25] describe an IR spectrometric method for the determination of the nonvolatile organic matter associated with airborne particles. The method consists of sampling utilizing the filtration principle (2 m^3 air per minute, glass-fiber filter 400 cm^2 effective surface) and extraction of the dry filter with carbon tetrachloride in a soxhlet apparatus. This is followed by IR spectrophotometric determination of

FIGURE 17.   Raman spectra between 1200 and 1700 cm^{-1} of (a) ambient sample collected in 1975. The sample was collected on a dichotomous sampler and was in the small size range fraction; (b) automobile exhaust collected from 100 cold starts of a 1974 Pinto using lead-free gas having no catalytic converter; (c) Diesel exhaust; (d) Activated carbon; and (e) polycristalline graphite. The slit width for samples b to e was 3 Å; while for sample a, 7-Å slits were used to improve signal to noise. (From Rosen, H. and Novakov, T., Lawrence Berkeley Laboratory Annu. Rep. 5214, Berkeley, Cal., 1975-76, 21. With permission.)

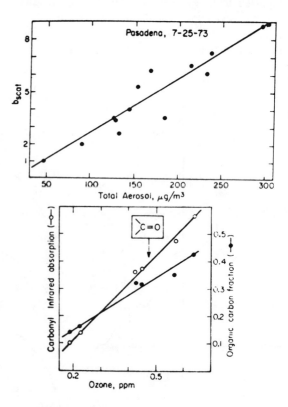

FIGURE 18.   Organic carbon fraction and carbonyl infrared absorption band intensities (v – C = 0) vs. O_3 concentration. (From Grosjean, D. and Friedlander, S. K., *J. Air Pollut. Control Assoc.*, 25, 1038, 1975. With permission.)

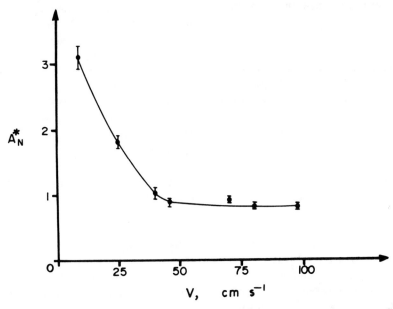

FIGURE 19.   Absorbance in the C-H stretching region, normalized to 1000 m³ filtered air of a 25-cm³ CCl₄ extract as a function of flow through speed for a sampling duration of 24 hr. (From della Fiorentina, H., De Graeve, J., and De Wiest, F., *Atmos. Environ.*, 9, 517, 1975. With permission.)

the aliphatic CH groups by measuring the sum of the absorbances at 2870, 2925, and 2960 cm⁻¹, respectively, in an NaCl cell. Reproducibility of the method for concentrations higher than 0.02 mg/ml was determined (v = ± 4%). No loss of organic material was observed during the extraction process, but it was shown that the composition of the organics adsorbed on the filter varies greatly with the flow rate and duration of sampling. Most compounds with a boiling temperature lower than 300°C are progressively eliminated from the sample trapped on the filter as a function of sampling duration and speed of the air aspirated at the filter (see Figure 19). With a sampling duration of 24 hr or more, however, only heavy aliphatic substances remain on the filter and the composition becomes stable. It was determined[24,25] that the composition of these heavy organic materials was stable within deviations of 15%. Thus, the method developed can be employed in rapid estimation of the concentration of these materials in airborne particles.

A further typical application of extraction combined with IR spectrometric characterization of the dissolved organic compounds is described by Ciaccio et al.[47] for the Hi-Vol and Andersen sampler collected airborne particulate material. The authors applied the benzene extract to a silica gel chromatographic column (length 4 cm) and eluted with the following solvents in the order given: hexane plus 5% benzene, chloroform 2-propanol, methanol, acetone, and hot acetone. IR spectra of these fractions showed the presence of aldehydes, ketones, acids, hydroxylic groups and possibly oximes, organic nitrogen substances, and aza-heterocyclic compounds. The oxygen carbonyl, nitrogen, and substances containing double bonds were found to be concentrated in the 2-propanol Anderson fraction (particle diameter 0.43 − 7.0 μm). However, this method must be considered as a screening method, and compound-specific analysis of organic materials in airborne dust requires the application of chromatographic separation techniques as described in Section IV.

IV. APPLICATION OF EXTRACTION AND CHROMATOGRAPHY TO CHEMICAL CHARACTERIZATION OF AIRBORNE PARTICULATE MATERIAL

Numerous papers have appeared dealing with characterization of organic compounds in airborne

particles. This section will describe the principle conditions of analytical methods used in this field and some typical examples mentioned in the literature.

A. Sampling Procedures

Organic compounds significantly contribute to the total particulate matter in urban aerosols. Sum concentrations up to 43% are reported;[9] however, compound-specific analysis requires the characterization of the different single components. More than 100 compounds have been determined in the organic fraction (important when considering health effects) occurring in the parts per million range, e.g., some polynuclear aromatic hydrocarbons (PAH). Therefore, characterization of organics in urban aerosols involves trace separation and identification by gas and high-pressure liquid chromatography, methods having sensitivity in the nanogram range. The sample amount needed for chromatographic separation of organics in airborne particles is in the milligram range,[48,49] allowing analysis of substances in parts per million concentrations. Usually, high-volume samplers with glass-fiber filters are used for dust collection to provide the needed sample amount.

To extract the organic material, benzene was applied as a standard solvent in a soxhlet apparatus. Extraction efficiencies of 25 different solvents and 24 binary mixtures were investigated by Grosjean.[22] He determined that extraction with benzene or other nonpolar solvents usually[50] leads to serious underestimation of aerosol organics, especially of the polar secondary (photochemical) products like carbonyl compounds, organic nitrates, or carboxylic acids, as illustrated by IR spectrometric analysis of the different extracts. The use of binary mixtures for extraction or successive extractions using a nonpolar and a polar solvent are strongly recommended. This leads to a higher organic carbon extraction efficiency (in comparison to benzene as solvent) than with both single polar and nonpolar solvents. The organic carbon content of the different extracts was determined in an organic carbon analyzer after the organic solvent was removed by vaporization. Ninety-five to one hundred percent of the aerosol organic carbon is extracted and measured with this method, which does not allow identification of single components.

B. Gas Chromatography

Gas chromatographic (GC) separation of organic extracts of airborne particles requires the application of preseparation steps, such as thin-layer chromatography[51] or liquid - liquid extraction.[48,49] Primary extraction is generally carried out by means of single solvents such as benzene, cyclohexane, or ethers. For GC analysis of polynuclear aromatic hydrocarbons,[51] the particulate matter (200 mg) cyclohexane extract is applied on a silica gel plate to remove disturbing components. Only the PAH fraction undergoes gas chromatography on a packed column in isothermal conditions (250°C). It was determined that a 5-m gas chromatographic column (stainless steel, inner diameter 2 mm, filled with 1% SE 52 Chromosorb® G 60-80 mesh) is comparable, in terms of resolution and analysis time, to a capillary column. A gas chromatogram with 30–40 peaks has been obtained. Nine PAH have been quantitatively analyzed in the range of 0.1 to 39 ng/m^3 air. A typical procedure including solvent extraction for preseparation is described by Ketseridis et al.[49] The authors report on a severe loss of sample material due to absorption at the glass-fiber filters.

The application of gas chromatography coupled with mass spectrometry for the analysis of benzene-extractable compounds in airborne particles is described in detail by Cautreels and Van Cauwenberghe.[48] This work led to the identification of more than 100 compounds in urban aerosols. The benzene-extractable compounds — 5.8% of total particles — were separated into neutral, acidic, and basic substances. The acidic fraction was converted to the methylated derivatives for GC analysis. In the neutral fraction, 22 saturated aliphatic hydrocarbons, 36 polynuclear hydrocarbons, and 13 polar oxygenated substances were identified. In the acidic fraction, 19 fatty acids and 19 aromatic carboxylic acids were identified; in the basic fraction, 15 peaks of nitrogen-containing analogues of the PAH were identified.

Single component concentrations were determined to be between 0.1 and 720 ppm (quinoline and di-2-ethylhexylphthalate, respectively). No correction for losses of substances with large vapor pressure during sampling, as described by della Fiorentina,[24,25] was made. Neutral compounds comprise 59.3% of the total weight of collected benzene solubles; the acidic substances, 22.5%; the basic compounds, 9.7%; and the water solubles 8.5%. Tables 4 through 8 survey the organic compounds determined in this study.

TABLE 4

Saturated Aliphatic Hydrocarbons in Airborne Particulates

Structure	Mol wt	Relative retention time[a]	Concentration range[b]
n-Tetradecane	$C_{14}H_{30}$ (198)	0.252	–
n-Pentadecane	$C_{15}H_{32}$ (212)	0.328	–
n-Hexadecane	$C_{16}H_{34}$ (226)	0.405	–
n-Heptadecane	$C_{17}H_{36}$ (240)	0.482	25
n-Octadecane	$C_{18}H_{38}$ (254)	0.560	28
n-Nonadecane	$C_{19}H_{40}$ (268)	0.639	26
n-Eicosane	$C_{20}H_{42}$ (282)	0.718	26
n-Heneicosane	$C_{21}H_{44}$ (296)	0.793	37
n-Docosane	$C_{22}H_{46}$ (310)	0.865	74
n-Tricosane	$C_{23}H_{48}$ (324)	0.935	162
n-Tetracosane	$C_{24}H_{50}$ (338)	1.000	265
n-Pentacosane	$C_{25}H_{52}$ (352)	1.065	351
n-Hexacosane	$C_{26}H_{54}$ (366)	1.129	<300
n-Heptacosane	$C_{27}H_{56}$ (380)	1.193	381
n-Octacosane	$C_{28}H_{58}$ (394)	1.256	265
n-Nonacosane	$C_{29}H_{60}$ (408)	1.335[c]	390
n-Triacontane	$C_{30}H_{62}$ (422)	1.424	185
n-Hentriacontane	$C_{31}H_{64}$ (436)	1.533	310
n-Dotriacontane	$C_{32}H_{66}$ (450)	1.663	115
n-Tritriacontane	$C_{33}H_{68}$ (464)	1.830	145
n-Tetratriacontane	$C_{34}H_{70}$ (478)	–	–
n-Pentatriacontane	$C_{35}H_{72}$ (492)	–	–

[a]Dexsil® 300 column, relative to $C_{24}H_{30}$.
[b]Expressed in ppm weight of dry particulate matter.
[c]First peak appearing in the isothermal part of the chromatogram.

From Cautreels, W. and Van Cauwenberghe, K., *Atmos. Environ.*, 10, 447, 1976. With permission.

TABLE 5

Polynuclear Aromatic Hydrocarbons in Airborne Particulates

Structure	Mol wt	Relative retention time[a]	Concentration range[b]
Simple polyaromatics			
Naphthalene	128	–	–
Biphenyl	154	0.320	1
Phenanthrene, anthracene	178	0.730	2
Fluoranthene	202	1.000	20
Pyrene	202	1.049	15
Benzo[a]fluorene	216	1.114	3
Benzo[c]fluorene	216	1.128	3
Benzo[c]phenanthrene	228	1.258	5
Benzo[ghi]fluoranthene	226	1.282	3
Benz[a]anthracene, chrysene	228	1.329	39
β,β'-Binaphthyl	254	1.456	3
Benzo[k]fluoranthene, benzo[b]fluoranthene	252	1.653[c]	83

TABLE 5 (continued)

Polynuclear Aromatic Hydrocarbons in Airborne Particulates

Structure	Mol wt	Relative retention time[a]	Concentration range[b]
Benzo[a]pyrene, benzo[e]pyrene	252	1.768	51
Perylene	252	1.815	5
Benzo[b]chrysene, o-phenylenepyrene	276	2.444	17
Dibenzanthracenes	278	2.400	3
Picene, benzo[c]tetraphene	278	2.487	3
Benzo[ghi]perylene, anthranthrene	276	2.655	12
Dibenzopyrene	302	–	2

Alkylated polyaromatics

Structure	Mol wt	RRT	Conc
Methylphenanthrene	192	0.838	1
Methylanthracene	192	0.862	1
Ethylphenanthrene	206	0.943	1
Ethylanthracene	206	0.963	1
Methylfluoranthene	216	1.150	2
Methylpyrene	216	1.172	2
Methylbenz[a]anthracene	242	1.369	3
Methylchrysene	242	1.424	3
Methylbenzo[k,b]fluoranthene	266	1.824	3
Methylbenzo[a,e]pyrene	266	1.880	3

Hydroderivatives

Structure	Mol wt	RRT	Conc
Dihydrobenzo[c]fluorene	218	1.059	–
Hexahydrochrysene	234	1.264	–
Dihydrobenzo[c]phenanthrene	230	1.283	–
Dihydrobenz[a]anthracene, dihydrocrysene	230	1.307	–

Nitrogen containing polyaromatics

Structure	Mol wt	RRT	Conc
Benzo[a]carbazol	217	1.324	–
Benzo[c]carbazol	217	1.387	–

[a]Dexsil® 300 column, relative to fluoranthene.
[b]Expressed in ppm weight of dry particulate matter.
[c]First peak appearing in the isothermal part of the chromatogram.

From Cautreels, W. and Van Cauwenberghe, K., *Atmos. Environ.*, 10, 447, 1976. With permission.

TABLE 6

Polar Oxygenated Compounds in Airborne Particulates

Structure	Mol wt	Relative retention time[a]	Concentration range[b]
Phenalene-9-one	180	0.678	—
Anthraquinone	208	0.862	15
7H-Benz[de]anthracene-7-one	230	1.196	5
Anth anthrone	306	1.631	—
Diethylphtalate	222	0.612	250
Diisobutylphtalate	278	0.893	—
Di-sec-butylphtalate	278	0.944	—
Di-n-butylphtalate	278	1.000	320
Benzylbutylphtalate	312	1.390	80
Di-2-ethylhexylphtalate	390	1.524	720
Three dioctylphtalate isomers	390	1.568[c]	—
		1.624	—
		1.702	—

[a]Dexsil® 300 column, relative to di-n-butylphtalate.
[b]Expressed in ppm weight of dry particulate matter.
[c]First peak appearing in the isothermal part of the chromatogram.

From Cautreels, W. and Van Cauwenberghe, K., *Atmos. Environ.*, 10, 447, 1976. With permission.

TABLE 7

Acidic Compounds in Airborne Particulates

Structure	Mol wt of methyl-derivatives	Relative retention time[a]	Concentration range[b]
Aliphatic acids			
Lauric acid	$C_{11}H_{23}COOCH_3$ (214)	0.475	91
Myristic acid	$C_{13}H_{27}COOCH_3$ (242)	0.659	103
Pentadecanoic acid	$C_{14}H_{29}COOCH_3$ (256)	0.745	61
Palmitic acid	$C_{15}H_{31}COOCH_3$ (270)	0.830	452
Margaric acid	$C_{16}H_{33}COOCH_3$ (284)	0.915	78
Oleic acid	$C_{17}H_{33}COOCH_3$ (296)	0.976	44
Stearic acid	$C_{17}H_{35}COOCH_3$ (298)	1.000	510
Nonadecanoic acid	$C_{18}H_{37}COOCH_3$ (312)	1.075	42
Arachidic acid	$C_{19}H_{39}COOCH_3$ (326)	1.152	196
Heneicosanoic acid	$C_{20}H_{41}COOCH_3$ (340)	1.224	74
Behenic acid	$C_{21}H_{43}COOCH_3$ (354)	1.294	250
Tricosanoic acid	$C_{22}H_{35}COOCH_3$ (368)	1.360	92
Lignoceric acid	$C_{23}H_{47}COOCH_3$ (382)	1.426	180
Pentacosanoic acid	$C_{24}H_{49}COOCH_3$ (396)	1.505[c]	35
Cerotic acid	$C_{25}H_{51}COOCH_3$ (410)	1.594	110
Heptacosanoic acid	$C_{26}H_{53}COOCH_3$ (424)	1.703	18
Montanic acid	$C_{27}H_{55}COOCH_3$ (438)	1.830	71
Nonacosanoic acid	$C_{28}H_{57}COOCH_3$ (452)	1.990	9
Melissic acid	$C_{29}H_{59}COOCH_3$ (466)	—	26
Aromatic acids			
Benzoic acid	136	0.186	9

TABLE 7 (continued)

Acidic Compounds in Airborne Particulates

Structure	Mol wt of methyl-derivatives	Relative retention time[a]	Concentration range[b]
Phtalic acid	194	0.457	25
Isophtalic acid	194	0.491	25
Terephtalic acid	194	0.509	150
Methylbenzoic acid	150	0.263	–
Methylphtalic acid	208	0.572	3
2-Hydroxybenzoic acid	166	0.352	6
4-Hydroxybenzoic acid	166	0.390	66
3,4-Dihydroxybenzoic acid	196	0.570	8
Naphthalene carboxylic acid	186	0.937	18
Phenanthrene carboxylic acid	236	1.132	13
Anthracene carboxylic acid	236	1.194	42
Pyrene carboxylic acid	260	1.493	3
Hydroxy-phenanthrene hydroxy-anthracene	208	1.059	2
Hydroxy-pyrene or hydroxy-fluoranthene	232	1.309	–
Tetrachlorophenol	244	0.536	20
Pentachlorophenol	278	0.727	75

[a]Dexsil® 300 column, relative to stearic acid.
[b]Expressed in ppm weight of dry particulate matter.
[c]First peak appearing in the isothermal part of the chromatogram.

From Cautreels, W. and Van Cauwenberghe, K., *Atmos. Environ.*, 10, 447, 1976. With permission.

TABLE 8

Basic Compounds in Airborne Particulates

Structure	Mol wt	Relative retention time[a]	Concentration range[b]
Quinoline	129	0.259	0.1
Isoquinoline	129 (2 isomers)	0.299	0.3
Methylquinolines	143 (6 isomers)	0.327	
Methylisoquinolines		0.363	
		0.380	
		0.404	0.9
		0.443	
		0.469	
Dimethylquinolines	157 (4 isomers)	0.476	
Dimethylisoquinolines		0.514	
		0.524	1
		0.597	
Trimethylquinolines	171 (8 isomers)	0.537	
Trimethylisoquinolines (also methylethyl-)		0.582	
		0.602	
		0.657	1
		0.700	
		0.718	
		0.768	
		0.801	

TABLE 8 (continued)

Basic Compounds in Airborne Particulates

Structure	Mol wt	Relative retention time[a]	Concentration Range[b]
Acridine		1.000	6
Benzo[f]quinoline	179 (3 isomers)	1.038	3
Benzo[x]quinolines			
Benzo[x]isoquinolines			
Phenanthridine		1.088	1
Phenoxazine	183	0.952	–
Tetramethylquinolines		0.715	
Tetramethylisoquinolines	185 (5 isomers)	0.768	0.3
(also dimethylethyl and diethyl-)		0.791	
		0.859	
		0.907	
Methylacridines		1.068	
Methylphenanthridines		1.106	
Methylbenzo[x]quinolines	193 (6 isomers)	1.123	4
Methylbenzo[x]isoquinolines		1.151	
		1.181	
		1.212	
Aza-fluoranthenes		1.342	
(ex. indeno[123-ij]isoquinoline)	203 (4 isomers)	1.365	16
Aza-pyrenes		1.438	
(ex. benzo[lmn]phenanthridine)		1.479	
Dimethylacridines		1.224	
Dimethylphenanthridines		1.239	
Dimethylbenzo[x]quinolines	207 (7 isomers)	1.267	3
Dimethylbenzo[x]isoquinolines		1.290	
		1.312	
		1.324	
		1.350	
Aza-benzo[x]fluorenes		1.418	
(ex. 11H-indeno[12-b]quinoline)		1.496	
Methyl-aza-pyrenes	217 (6 isomers)	1.529	4
Methyl-aza-fluoranthenes		1.559	
		1.574	
		1.612	
Aza-benz[a]anthracenes		1.673[c]	
(ex. benz[a]acridine	229 (4 isomers)	1.723	16
benz[c]acridine)			
Aza-chrysenes		1.791	
(ex. benzo[c]phenanthridine)		1.854	
Dibenzo[f,h]quinoline			
Dibenzo[v,h]isoquinoline			
Methylbenzacridines		1.781	
Methylbenzophenanthridines	243 (4 isomers)	1.819	3
Methyldibenzo[f,h]quinoline		1.874	
Methyldibenzo[f,h]isoquinoline		1.912	
Aza-benzopyrenes		2.123	
(ex. naphtho[amn]acridine)		2.164	
Aza-benzofluoranthenes	253 (7 isomers)	2.224	9
(ex. indeno[123-gh]phenanthridine)		2.275	
		2.348	
		2.390	
		2.491	

TABLE 8 (continued)

Basic Compounds in Airborne Particulates

Structure	Mol wt	Relative retention time[a]	Concentration range[b]
Dibenz[a,h]acridine		–	4
Dibenz[a,j]acridine	279 (2 isomers)		

[a]Dexsil® 300 column, relative to acridine.
[b]Expressed in ppm particulate matter.
[c]First peak appearing in the isothermal part of the chromatogram.

From Cautreels, W. and Van Cauwenberghe, K., *Atmos. Environ.*, 10, 447, 1976. With permission.

C. Liquid Chromatography

High-pressure liquid chromatography (HPLC) is a promising technique for separation of high molecular weight PAH. Recent development of bonded octadecylsilyl (ODS) columns of micro particle size allows[52] the near baseline separation of the carcinogenic benzo[a]pyrene (BaP) and its noncarcinogenic isomer benzo[e]pyrene. A significant increase in sensitivity over other methods was achieved by use of fluorescence spectroscopy for on-line detection. This detection method was found to be at least an order of magnitude more sensitive than absorption spectroscopy. A sensitivity of 90 pg of BaP was reached with a stainless steel Du Pont Zorbax® ODS column having a 0.25 m × 8mm I.D. (~ 3200 plates, plate height ~ 0.08 mm), in BaP elution with 7 : 3 (v/v) MeOH/H_2O at 65°C. As small a quantity as 25 pg BaP was detected by this method when deoxygenated solvents were used. The reproducibility was ± 11% relative. Particles collected on glass-fiber were extracted with benzene in a soxhlet extractor. Aliquots of the concentrated extracts were applied to the HPLC system.

Direct detection and quantitative determination of ten PAH in a heavily-travelled traffic tunnel and in a suburban air are described in Reference 52. The determined concentration range was between 3 ± 0.5 ng/m^3 (2,3,6,7-dibenzanthracene) and 120 ± 12 ng/m^3 (pyrene). Analysis of polycyclic quinones derived from PAH is easily performed. Pierce and Katz[53] describe a method of quantification of 9,10-antrachinone, benzo[a]pyrene-6,12-quinone, benzo[a]pyrene-1,6-quinone, benzo[a]pyrene-3,6-quinone and dibenzo[b,d,e,f]chrysene-7,14-chinone in airborne particles. Separation of the different compounds is performed in a silica gel column with thin-layer chromatography on polyamide or magnesium hydroxide. To improve the sensitivity of the method, the collected quinones were reduced to the parent hydrocarbons prior to fluorescence analysis. Utilizing this method, samples as small as 25 ng/g particulate were detected.

Several papers deal with network studies with the aim of monitoring the PAH content of airborne particles.[54,55] In all cases, high-volume samplers were used to collect the needed sample amount in the 50-mg range or higher. Severe difficulties occur with the application of grain size differentiating sampling methods resulting from the small sample amounts in each fraction.

REFERENCES

1.   **Whitby, K. T. and Liu, B. Y. H.,** The aerosol size distribution of Los Angeles smog, in *Aerosols and Atmospheric Chemistry,* Hidy, G. M., Ed., Academic Press, New York, 1971, 237.
2.   **Fennelly, P. F.,** Primary and secondary particulates as pollutants, *J. Air Pollut. Control Assoc.,* 25, 697, 1975.
3.   **Wilson, E. B., Decius, J. C., and Cross, P. C.,** *Molecular Vibrations, The Theory of Infrared and Raman Vibrational Spectra,* McGraw-Hill, New York, 1955.

4. Colthup, N., Daly, L., and Wiberley, St., *Introduction to Infrared and Raman Spectroscopy,* Academic Press, New York, 1964.

5. Finch, A., Gates, P. N., Radcliffe, K., Dickson, F. N., and Bentley, F. F., *Chemical Application of Far Infrared Spectroscopy,* Academic Press, London, 1970.

6. Griffiths, P. R., *Chemical Infrared Fourier Transform Spectroscopy,* Vol. 43, in Chemical Analysis Series, Elving, P. J., Winefordner, J. D., Kolthoff, I. M., Eds., John Wiley & Sons, New York, 1975.

7. Nakamoto, K., *Infrared Spectra of Inorganic and Coordination Compounds,* Interscience, New York, 1970.

8. Duyckaerts, G., The infrared analysis of solid substances, *Analyst,* 84, 201, 1959.

9. Hidy, G. M., Summary of the California Aerosol Characterization Experiment, *J. Air Pollut. Control Assoc.,* 25, 1106, 1975.

10. Brosset, C., Air-borne particles: black and white episodes, *Ambio,* 5, 157, 1976.

11. Moencke, H., *Mineralspektren,* Akademie-Verlag, Berlin, 1962.

12. Cunningham, P. T., Johnson, St. A., and Yang, R. T., Variations in chemistry of airborne particulate material with particle size and time, *Environ. Sci. Technol.,* 8, 131, 1974.

13. Schiedt, U. and Reinwein, H. Z., *Naturforscher,* 76, 270, 1952.

14. Mangia, A., Determination of α-quartz in atmospheric dust: a comparison between infrared spectrometry and X-ray diffraction techniques, *Anal. Chem.,* 47, 927, 1975.

15. Tuddenham, W. M. and Lyon, R. J. P., Infrared techniques in the identification and measurements of minerals, *Anal. Chem.,* 32, 1630, 1960.

16. Chang, S. G. and Novakov, T., Infrared and Photoelectron Spectroscopic Study of SO_2^-⇔〈↑ʒ⁺〈p− p− ↕pp⁺ ‡ʒᐃ⁺〈‖〉Ջ\'

16. Chang, S. G. and Novakov, T., Infrared and Photoelectron Spectroscopic Study of SO_2 -Oxidation on Soot Particles, Lawrence Berkeley Laboratory, Annu. Rep. 5214, Energy and Environment Division, Atmospheric Aerosol Research, Berkeley, Cal., 1975-76, 40.

17. Malissa, H. and Kellner, R., Contribution of Infrared Spectrometry to the Integrated Dust Analysis, paper presented at the Scientific Session on Environmental Analysis, Szombathely, Hungary, September 29, 1976.

18. Blanco, A. J. and McIntyre, R. G., An Infrared Spectroscopic View of Atmospheric Particulates over El Paso, Texas, *Atmos. Environ.,* 6, 557, 1972.

19. Toma, S. Z. and Goldberg, S. A., Direct infrared analysis of alpha quartz deposited on filters, *Anal. Chem.,* 44, 431, 1972.

20. Gadsen, J. A., Parker, J., and Smith, W., Determination of chrysotile in airborne asbestos by an infrared spectrometric technique, *Atmos. Environ.,* 4, 667, 1970.

21. Bagioni, R. P., Separation of chrysotile asbestos from minerals that interfere with its infrared analysis, *Environ. Sci. Technol.,* 9, 263, 1975.

22. Grosjean, D., Solvent extraction and organic carbon determination in atmospheric particulate matter: the organic extraction − organic carbon analyzer technique, *Anal. Chem.,* 47, 797, 1975.

23. Grosjean, D. and Friedlander, S. K., Gas-particle distribution factors for organic and other pollutants in the Los Angeles atmosphere, *J. Air Pollut. Control Assoc.,* 25, 1038, 1975.

24. della Fiorentina, H., De Graeve, J., and De Wiest, F., Determination par spectrometrie infrarouge de la matière organique non volatile associée aux particules en suspension dans l'air. I. Choix des conditions operatoires, *Atmos. Environ.,* 9, 513, 1975.

25. della Fiorentina, H., De Graeve, J., and De Wiest, F., Determination par spectrometrie infrarouge de la matière organique non volatile associée aux particules en suspension dans l'air. II. Facteurs influencant l'indice aliphatique, *Atmos. Environ.,* 9, 517, 1975.

26. Estep, P. A., Kovach, J. J., and Karr, C., Jr., Quantitative infrared multicomponent determination of minerals occurring in coal, *Anal. Chem.,* 40, 358, 1968.

27. Malissa, H., Kellner, R., Puxbaum, H., and Thalhammer, Ch., Integrated Dust Analysis − Infrared Techniques, paper presented at the 7th Symposium on the Analytical Chemistry of Pollutants, Lake Lanier Island, Ga., April 25 to 27, 1977.

28. Blanco, A. J. and Hoidale, G. B., Microspectrophotometric technique for obtaining the IR-spectra of microgram quantities of atmospheric dust, *Atmos. Environ.,* 2, 327, 1968.

29. Andersen, D. H. and Woodall, N. B., Infrared identification of materials in the fractional milligramm range, *Anal. Chem.,* 25, 1906, 1953.

30. Novakov, T., Dod, R. L., and Chang, S. G., Study of air pollution particulates by X-ray photoelectron spectroscopy, *Fresenius Z. Anal. Chem.,* in press, 1976.

31. Novakov, T., Chang, S. G., Dod, R. L., and Rosen, H., Chemical characterization of aerosol species produced in heterogenous gas-particle reactions, APCA 76-204, paper presented at the Air Pollution Control Association Annual Meeting, Portland, Oregon, June, 1976.

32. Novakov, T., Chang, S. G., and Harker, A. B., *Science,* 186, 259, 1974.

33. Kellner, R., Infrared spectroscopic characterization of impactor collected samples of airborne particles in the microgram range, *Mikrochim. Acta,* in press.

34. Puxbaum, H., Eine relativkonduktometrische Mikramethode zur Bestimmung von Ammonium in Stäuben, *Mikrochim. Acta,* 2, 157, 1977.

35. Boehm, H. P., *Discuss. Faraday Soc.,* 52, 264, 1971.

36. **Chang, S. G. and Novakov, T.,** Possible Mechanism for the Catalytic Formation of Nitrates in the Atmosphere, Lawrence Berkeley Laboratory Annu. Rep. 5214, Berkeley, Cal., 1975-76, 48.
37. **Gordon, R. J. and Bryan, R. J.,** Ammonium nitrate in airborne particles in Los Angeles, *Environ. Sci. Technol., 7,* 645, 1973.
38. **Hlavay, J., Elek, S., and Inczedy, J.,** Infrared spectrophotometry applied in the determination of the mineral constituents of soil, *Hung. Sci. Instrum., 38,* 69, 1976.
39. **Gillieson, A. H. and Farrell, D. M.,** *Can. Spectrosc.,* 16, 21, 1971.
40. **Dodgson, J. and Whittaker, W.,** The determination of quartz in respirable dust samples by infrared spectrophotometry,
40. **Dodgson, J. and Whittaker, W.,** The determination of quartz in respirable dust samples by infrared spectrophotometry, *Am. Occup. Hyg.,* 16, 373, 1973.
41. **Sloane, H. J.,** Infrared differential technique employing membrane filters, *Anal. Chem.,* 35, 1556, 1963.
42. **Heffelinger, K. E., Melton, C. W., and Kiefer, D. L.,** Development of a Rapid Survey of Sampling and Analysis for Asbestos in Ambient Air, Battelle Memorial Inst., Columbus, Ohio, EPA Rep. APTD-0965, February 29, 1972.
43. **Holmes, A., Morgan, A., and Sandalls, J.,** Determination of Iron, Chromium, Cobalt, Nickel and Scandium in asbestos by neutron activation analysis, *Am. Ind. Hyg. Assoc. J.,* 32, 281, 1971.
44. **Rickards, A. L.,** Estimation of trace amounts of chrysotile asbestos by X-ray diffraction, *Anal. Chem.,* 44, 1872, 1972.
45. **Smith, D. M., Griffin, J. J., and Goldberg, E. D.,** Spectrometric method for the quantitative determination of elemental carbon, *Anal. Chem.,* 47, 233, 1975.
46. **Rosen, H. and Novakov, T.,** Application of Raman-scattering to the Characterization of Atmospheric Aerosol Particles, Lawrence Berkeley Laboratory Annu. Rep. 5214, Berkeley, Cal., 1975-76.
47. **Ciaccio, L. L., Rubino, R. L., and Flores, J.,** Composition of organic constituents in breathable airborne particulate matter near a highway, *Environ. Sci. Technol.,* 8, 935, 1974.
48. **Cautreels, W. and Van Cauwenberghe, K.,** Determination of organic compounds in airborne particulate matter by gas-chromatography – mass spectrometry, *Atmos. Environ.,* 10, 447, 1976.
49. **Ketseridis, G., Hahn, J., Jaenicke, R., and Junge, Ch.,** The organic constituents of atmospheric particulate matter, *Atmos. Environ.,* 10, 603, 1976.
50. Environmental Protection Agency, Air Quality Data for Organics 1969 and 1970 from the National Air Surveillance Networks EPA Rep. APTD-1465, Research Triangle Park, N.C., June 1973.
51. **Zoccolillo, L., Liberti, A., and Brocco, D.,** Determination of polycyclic hydrocarbons in air by gas chromatography with high efficiency packed columns, *Atmos. Environ.,* 6, 715, 1972.
52. **Fox, M. A. and Staley, St. W.,** Determination of polycyclic aromatic hydrocarbons in atmospheric particulate matter by high pressure liquid chromatography coupled with fluorescence techniques, *Anal. Chem.,* 48, 992, 1976.
53. **Pierce, R. C. and Katz, M.,** Chromatographic isolation and spectral analysis of polycyclic chinones. Application to air pollution analysis, *Environ. Sci. Technol.,* 10, 45, 1976.
54. **Gordon, R. J. and Bryan, R. J.,** Patterns in airborne polynuclear hydrocarbon concentrations at four Los Angeles sites, *Environ. Sci. Technol.,* 7, 1050, 1973.
55. **Gordon, R. J.,** Distribution of airborne polycyclic aromatic hydrocarbons throughout Los Angeles, *Environ. Sci. Technol.,* 7, 370, 1973.

Chapter 12
THERMAL ANALYSIS

T. Meisel

TABLE OF CONTENTS

I. INTRODUCTION

Thermal analysis is a material testing method that measures changes occurring in a sample when it is heated or cooled. Many transformations and processes can be traced with this type of analysis. These changes can be divided into two groups:

1. Changes in physical phenomena, such as polymorphous transformations, their enantiotropic or monotropic character, their eventual relaxation processes, phase transitions (melting, solidification, evaporation, condensation, sublimation), adsorption or desorption, crystallization processes, etc.

2. Chemically oriented transformations such as chemisorption, desolvatation (dehydration), pyrolysis, oxidation-reduction processes occurring initially in the heterogeneous phase, solid-to-solid phase reactions, chemical processes in the molten phase, etc.

Generally, all of the above-mentioned transformations begin at a measurable rate and a given temperature; the process occurs relatively rapidly in a narrow temperature range. Two types of data can be obtained through analysis: the characteristic temperature range in which transformations occur and a signal proportional to the amount of the material transformed. In addition, many other properties of the material can be studied with thermoanalytical methods. Of particular interest are those changes which are a constant and predictable function of temperature, e.g., specific heat, volume, electric resistance, and various optical properties (reflection, refractive index, etc.). When measuring these properties, it must be remembered that changes occur suddenly in the course of chemical or physical transformations, in most cases.

Utilizing thermoanalytical methods, changes in enthalpy, weight, dimensions, and mechanical properties of the sample can be determined. It is also possible to monitor changes occurring during X-ray diffraction, such as alterations in the luminescence, optical, or sound-emission properties of materials. It is also possible to perform qualitative and quantitative analyses on gaseous products evolved during decomposition reactions.

A schematic diagram of a thermoanalytical processing unit is shown in Figure 1. In such a system, (1) designates the sample to be analyzed. In earlier studies, samples weighing approximately 0.1 g were used. Currently, sample amounts are limited to milligram quantities. In high-sensitivity equipment, samples of approximately 0.01 to 0.1 mg can also be measured. In general, the heat source (2) is an electric furnace. However, an IR radiator, a high-frequency heater, or a simple liquid thermostat may also be employed. Modern equipment is also combined with a regulable cooling circuit. The rate of heating and cooling is varied linearly with a programmer (3) within a wide range — usually between 0.1 to $100°C$/min. However, an isothermal or other heating method can also be used. The atmosphere control system (4) consists of a gas source, cleaner, and flow-rate measuring unit. Many instruments make measurements in a vacuum or under high pressure possible. The transducer (5) is a highly technical instrument capable of measuring given properties of the

sample as well as the following changes that occur during the course of heat treatment. The application of selective instruments providing minimal delay, appropriate sensitivity, and yielding analogous electric signals is the most favorable. Thermoanalytical methods can be classified according to the type of the measuring sensors used.

In most cases, the analogous signal obtained must be amplified, and the noise must be filtered by appropriate electrical circuits (6). Temperature measurement (7) is commonly obtained with thermocouples. However, resistance thermometers and optical thermometers can also be used for measurement. Temperature measurement is a rather complex problem in thermal analysis. The results are directly dependent upon certain variables:

1. Is the temperature measured inside the sample or in other portions of the sampling apparatus?

2. Is the recording obtained as a function of temperature or time if a linear temperature program is employed?

Modern equipment transforms analogous signals into digital signals. Dedicated computer couplings (8) with special functions are also used. Thus, thermokinetic and quantitative analytical data are obtained directly, and temperature correction is performed within the apparatus, etc. The determined relationship is monitored by either an X-Y recorder (9) or, if several properties are simultaneously measured, by means of a multichannel recorder. A plotter, magnetic tape, or a printer can also be used. In Table 1, the most important characteristics of different thermoanalytical methods are summarized.

In the literature, very few data exist for the thermal analysis of airborne particles. There may be several reasons for this, among them the fact that thermal analysis is not a widely used technique, and many laboratories do not appropriate funds for thermoanalytical equipment. This is basically because thermoanalytical equipment must be calibrated both qualitatively and quantitatively and does not yield direct results. Standardization of thermoanalytical methods and data presentation is still under development.[1] The application of different samplers — such as filter, absorbers, etc. — also limits the applicability of this method.

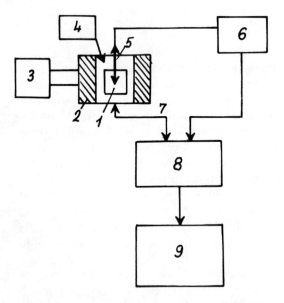

FIGURE 1.   General scheme of a thermoanalytical instrument (1) sample, (2) heat source, (3) temperature controller, (4) atmosphere controller, (5) transducers, (6) amplifier and filter, (7) temperature measuring elements, (8) data acquisition system, (9) recorder.

TABLE 1

Methods of Thermal Analysis

Technical nomenclature	Measured property in the function of temperature	Applied apparatus
Thermogravimetry (TG)	Mass	Thermobalance
Derivative thermo-gravimetry (DTG)	Rate of mass change	Thermobalance, supplied with deri-vating facility
Differential thermal analysis (DTA)	Temperature difference between the sample and a reference material	DTA apparatus
Differential scan-ning calorimetry (DSC) and diffe-rential dynamic calorimetry (DDC)	Heat flux difference between the sample and a reference material	Differential scanning calorimeter and conduction dynamic differential calorimeter
Gas evolution ana-lysis (gas evolu-tion detection and/or determination)	Volume, pressure, gas chromatographic reten-tions, thermal conduc-tivity of evolved gas, ion current during its ioni-zation, etc.	Thermal gas analyzer
Thermooptics	Refractive index, light emission, absorption, reflectance	Thermooptical device (e.g., hot stage microscope)
Thermomechanical analysis	Volume (length), tensi-le strength, penetration, viscosity, torsional mo-dule, etc.	Thermomechanical analyzer
Thermoelectrical analysis (AC and DC)	Conductance, capacitan-ce (in the function of frequency also)	Thermoelectrical device
Thermomagnetic measurement	Ferropara-, paradia-magnetic transitions (Curie or Neil point)	Thermomagnetic balance
Thermo-X-ray	X-ray diffraction pro-perties (e.g., change in d. spacings)	Thermo-X-ray diffractometer
Thermosonometry	Sound intensity, fre-quency change	Thermosonometer
Thermoemanation	Radioactive radiation	Thermoemanation analyzer

The most advantageous technique involves sampling with an impactor; very small sample amounts can be collected in a short time period. However, longer sampling period reduces the analytical data obtained.

Thermoanalytical methods have demonstrated potential for use in air pollution analysis. How-ever, based upon the results obtained to date, the special requirements of such an application are speculative.

The following are suitable methods of thermo-analysis and may be used in combined or simultan-eous application:

1. Differential thermal analysis (DTA) — differential dynamic calorimeter (DSC)
2. Thermogravimetry (TA) — derivative thermogravimetry (DTG)
3. Thermogas analysis
4. Thermo-X-ray analysis

II. DIFFERENTIAL
THERMAL ANALYSIS

Differential Thermal Analysis (DTA) is suitable for monitoring enthalpy changes of a sample by the temperature difference measurement between a sample and a reference material as a function of temperature or time. A counterconnected thermocouple, embedded in the sample and the reference material or placed in their vicinity, measures the temperature difference with a detector (Figure 2A).

Ideally, the calibrated detector shows a signal only if an exothermic or endothermic process is taking place within the sample. Otherwise, the so-called baseline is recorded. Figure 2B shows a typical DTA graph. The temperature values T_i and T_f are characteristic for the onset and completion of the process, respectively, and thus for the substance investigated. The following correlation:

$$m \, \Delta H = K \int_{T_i}^{T_L} \Delta T \, dt = K \cdot S$$

is valid, where the peak area (S) is proportional (K) to the molecular amount (m) of the sample

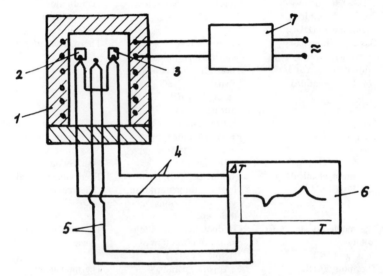

FIGURE 2.A.   General arrangement of a DTA apparatus: (1) furnace, (2) sample, (3) reference material, (4) DTA sign, (5) temperature measurement, (6) record, (7) temperature programmer. B. Typical DTA Graph: T_r = temperature of the reference material; T_s = temperature of the sample.

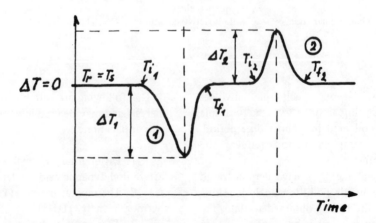

FIGURE 2.B.

and the molar transformation of heat (ΔH) in the given process.

However, exact evaluation of the curves is difficult because the recorded values are influenced by many experimental conditions such as:

1. Atmosphere of the furnace chamber (pressure, quality, streaming conditions)
2. Shape and material of the sample containers
3. Sensitivity, geometry, and heat conductivity properties of the thermocouples

Additional influencing factors relate to the substantial properties of the sample such as:

1. Grain size, distribution, and compactness of the solid sample
2. Heat conductivity (dilution)
3. Specific heat
4. Quantity
5. Prehistory of the sample

These factors influence the calorimetric sensitivity of the measurement, i.e., the magnitude of the signal in respect to the heat unit, millicalories per microvolt seconds (mcal/μVsec), and the resolution of the peaks. The above factors may also change the characteristic temperatures.

A significant obstacle of classical DTA equipment was eliminated by introduction of the differential dynamic calorimeter (DDC) and differential scanning calorimeter (DSC). With the operation principles and technical solutions provided by these instruments, transformation temperatures and reaction heat can be determined with great accuracy (ca. $\pm$ 1.0°C and 1 to 2%, respectively).

Exact measurement of heat fluxes and proportional temperature differences can be insured by strict adherence to the construction principles in DDC instruments. A measuring cell of this kind is shown in Figure 3.

DSC utilizes a power compensation to insure the equality of the temperatures of the sample and the reference material throughout the procedure. This is accomplished by means of a specially constructed and controlled individual heating circuit whose power consumption is recorded (Figure 4).

In addition to the relatively exact measurement of transformation temperatures and reaction heats, accurate measurements aid in determination of the

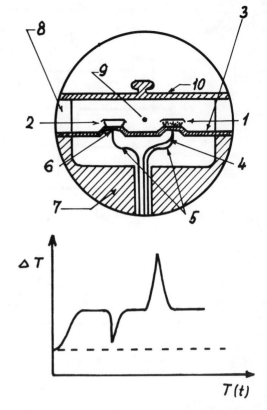

FIGURE 3. A typical Du Pont-type DSC cell arrangement (1) sample pan, (2) reference pan, (3) thermoelectric disk, (4) alumel wire, (5) chromel wire, (6) thermocouple junction, (7) heating block, (8) silver ring, (9) gas inlet, (10) lid.

specific heat of samples. On the basis of the Van't Hoff correlation, purity determinations can be conducted, and reaction kinetic data can also be obtained.

Due to constructional principles, measurements are obtained at relatively low temperatures with the following two fundamental methods of quantitative DTA. The DDC technique is suitable for measurements in the temperature range of 500 to 600°C, while the DSC method is applicable to examinations performed at 700 to 800°C. Since dust samples are often composed of inorganic matter, and their expected transformations usually occur at higher temperatures, the application of the high-temperature classical DTA method cannot be neglected in this particular field.

The capabilities of DTA provide many possibilities for its consequent application to analysis of airborne particles. Quality of the sample is determined by obtaining the so-called thermospectrum. Based upon the expected composition of dusts, for example, the quartz content can be identified and

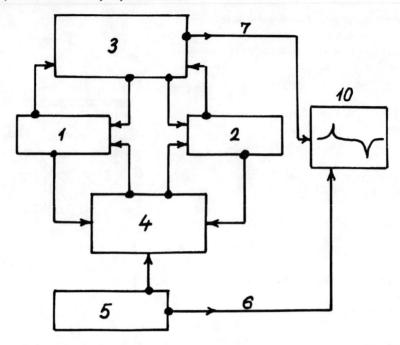

FIGURE 4.A.   Block scheme of a Perkin-Elmer type DSC system (1) sample, (2) reference, (3) differential amplifier, (4) amplifier, (5) programmer, (6) temperature sign, (7) power sign. B. Measuring head, (8) Pt temperature sensors, (9) individual heaters, (10) record.

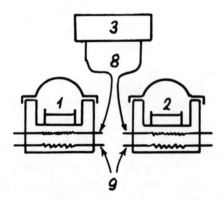

FIGURE 4.B.

determined with the DTA technique on the basis of its well-defined inversion temperature. Furthermore, airborne particles containing carbonate decompose at different temperatures to form oxides and carbon dioxides, depending on their cations. Various inorganic compounds (sulfates, silicates, nitrates, sulfides, etc.) can also be analyzed on the basis of their thermospectrum.

The identification of humidity, and the eventual water of crystallinity is also possible. Generally, the formation of these components appears during endothermic processes, in the temperature range of 100 to 250°C. The organic content sorbed on the inorganic carrier substance can also be recognized. In this case, the course of the obtained DTA curves may be significantly different, depending upon the atmosphere applied (i.e., whether the measurement is carried out in the presence of an oxidizing agent or of an inert gas). However, more significant information can be obtained if examinations of this type are completed using simultaneous measuring techniques. The desorption conditions of different gases sorbed on the solid dust carrier can also be investigated. In DTA examination of dusts, their large specific surfaces may play a significant role, especially in processes connected with mass transport.

III. THERMOGRAVIMETRY (TG)

Thermogravimetry, as a thermoanalytical method, monitors the weight changes of samples either as a function of temperature (dynamic TG) or as a function of time (static TG), when the measurement is carried out under isothermal conditions. A combination of the two functions is also possible.

The characteristic data of continuously recorded thermogravimetric curves are the T_i and T_f temperature values, which are reproducible under strictly identical experimental conditions. In processes occurring in a well-defined manner, the extent of the weight change corresponds to the stoichiometric formula. Reactions taking place in a parallel or closely overlapping, consecutive manner, may render the evaluation of TG curves difficult or even impossible. The resolution of TG curves representing partly overlapping processes can be significantly increased if the first or higher derivatives of the curve are simultaneously established, either in a mathematical or in an instru-

mental manner. The first derivative DTG (derivative thermogravimetric curve) of the TG curve is usually simultaneously recorded with the latter. TG curves describe processes connected with energy and mass transport, and as a result, the shape of the curve (i.e., the characteristic temperatures) greatly depends upon the experimental conditions. The gaseous decomposition products of the chemical transformation or the momentary concentration of the reaction partners influence the shape of the curve significantly. Concentration of the gaseous decomposition products is fundamentally determined by diffusion conditions. Accuracy of the weight measurement is basically determined, independently from the experimental conditions by the sample amount and the sensitivity of the measuring apparatus with the restriction that the exact result can only be obtained if it is corrected with the buoyancy measurement. Furthermore, the disturbing convective gas stream must be eliminated or kept at a constant value.

With modern thermobalances (Figure 5), microsized samples (1 to 5 mg) can be measured in controlled atmospheres, if necessary, under high vacuum. The heating program can be varied within a wide temperature range, and the signal of the weight changes can be obtained both in analogous and digital forms.

Important thermogravimetric investigations relevant to dust analysis include the following:

Determination of Water Content
- Water content bound by adsorption shows a weight loss generally in the temperature range between 50 to 120°C, depending on its sorption abilities.
- Release of water of crystallinity, occurring in a relatively narrow temperature range in one or more steps, can quantitatively be recognized by the height of the steps corresponding to stoichiometry. However, the decomposition temperature is significantly influenced by the partial pressure of vapor in the reaction space and generally occurs between 80 to 250°C.
- Dehydration of hydroxides of multivalent metals, i.e., the loss of so-called constitutional water, usually occurs at higher temperatures in the range of approximately 200 to 500°C.

Relatively volatile organic and inorganic materials

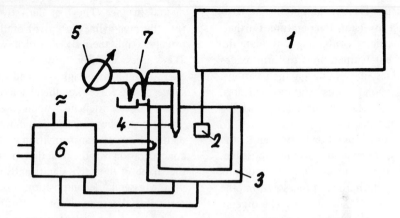

FIGURE 5.A.   General arrangement of a thermobalance (1) balance system, (2) sample, (3) furnace, (4) thermoelements, (5) galvanometer, (6) temperature programmer, (7) cold junction. B. TG and DTG traces of calcium carbonate.

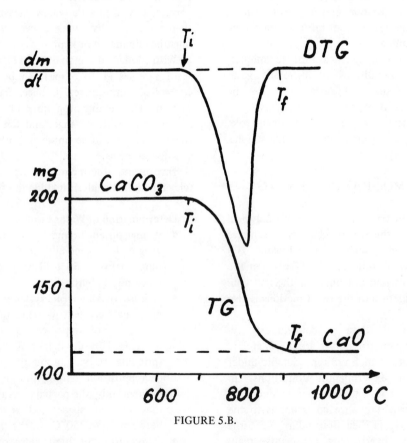

FIGURE 5.B.

bound to the surface of airborne particles by adsorption can be examined with combined thermogas analysis to determine their quality and relative quantity. Inorganic salts, which decompose when heated (such as carbonates, nitrates, sulfates, ammonium salts, etc.), can be determined on the basis of their decomposition temperature as well as by measuring the amount of the evolving decomposition products. Thermogravimetry is also suitable for model experiments within catalytic studies of reaction, initiated by heat and active surface, between solid sample and a gaseous component.

IV. THERMOGAS ANALYSIS

Thermogas analysis is suitable for qualitative evolved gas detection (EGD) and quantitative evolved gas analysis (EGA) to determine gaseous decomposition products evolved in thermal decomposition reactions. Formation and quantity of the gaseous decomposition products can be recorded either as the function of temperature or time by means of modern thermogas analytical instruments. There are several methods of performing thermogas analysis:

1. All products evolved during decomposition reactions or a portion of them are separated and analyzed with a suitable technique.
2. Changes in pressure or volume of the system, occurring during the course of the decomposition, are measured as a function of temperature or time. If more components are formed, only the total change is recorded, and in this case, no information is obtained regarding the formation of the individual components.
3. Gas analysis is simultaneously initiated with other thermoanalytical methods, such as DTA, TG, etc.
4. A selective gas analyzer (sensitive to a given component or a given type of component) is used, and the concentration of the gaseous product is continuously measured. Emission measuring devices are suitable for this purpose, e.g., the gas analyzers for sulfur dioxide, carbon dioxide, ammonia, etc. Their applicability is increased if they are fitted with a programmed temperature evaporator or pyrolysator.[2,3]
5. Gas chromatography (GC) can be applied in two different ways for thermogas analytical purposes: If a pyrolysis chamber of programmed heating is placed before the column and the decomposition products are introduced onto the column after the decomposition reaction is complete, the retention data and concentrations of the products are easily reproducible and characteristic for the sample, provided that the conditions of pyrolysis and feed are well controlled. A simultaneous DTA, TG, and gas analytical method has also been developed.[4,5] A rapidly operating gas chromatograph and programmed sampling must be

used; in addition, a computer-aided evaluation of the chromatograms is advisable.

A current method of thermogas analysis involves particle examination with a mass spectrometer (MS). A detailed description is given in Chapter 6. Therefore, only the thermoanalytical applications of MS will be discussed in this chapter.

Feeding thermally stable solid samples into the spectrometer can often be performed only through pyrolysis. In such cases, reproducible results can only be obtained if the pyrolysis is executed under carefully controlled conditions at a constant temperature.

Gohlke and Langer[6] first reported on the technique, which they called mass spectrometric thermal analysis (MTA). Essentially, this method involves analysis of the decomposition product obtained at a given temperature by means of a mass spectrometer. If a spectrometer with a relatively short scanning time is used and the samples, fed into it at different temperatures, are analyzed, repeated analyses with increasing temperature through a sufficient number of cycles will establish a temperature profile of the products.

The coupling of DTA, TG, and MS was also undertaken.[7] DTA signals of the decomposition reactions were compared with results obtained by MS. This simultaneous technique involves error due to the relative delay in the operation of the two different measurements. Furthermore, difficulty arises with the application of high vacuum, which is a primary condition of MS operation. A TG-MS system was also constructed.[8-11] A schematic modern TA-MS system, developed by Gibson,[12] is presented in Figure 6. Utilizing the

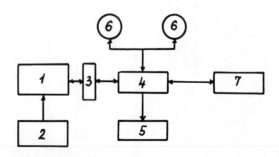

FIGURE 6.   Block scheme of a simultaneous TA-MS system (1) mass spectrometer, (2) thermal analyser TG, DTA, (3) interface, (4) computer, (5) plotter, (6) tapes, (7) teletype. (From Gibson, E. K., Jr., *Thermochim. Acta*, 5(3), 243, 1973. With permission.)

combined technique as well as other methods, Green River shale and lunar soil samples were analyzed.

The gas release pattern of the shale and soil fragments (Figure 7) is obtained by integrating the spectrums — measured at different temperatures — according to their mass numbers, and the temperature profile of the spectrum intensity of a fragment with a given mass number (Figure 8) is similarly obtained with the aid of a computer.

The sensitivity of the two methods differs significantly. The TG curve does not reveal any weight changes when the characteristic release of products has already been detected by the MS method. This rather complicated method of evaluation uses a computer that is supplied with a sufficient number of memory units and also controls the measurement in an active manner. However, this computer-aided evaluation is only possible if the rates of heating and decomposition are in accordance with the scan capabilities of the mass spectrometer.

If many products having different fragmentation patterns form in the course of the decomposition reaction, the evaluation becomes even more complicated, and it is advisable to couple the thermoanalytical equipment with a GC-MS system. For example, in cases when the carrier airborne particles attract different organic substances of nearly identical sorption abilities by adsorption on their surface, the application of a GC-MS system is advisable.

V. THERMO-X-RAY ANALYSIS

X-Ray analysis is a well-known technique for the determination of the crystalline structure of samples. Chapter 9 gives a detailed description of this method and its application. The information obtained with X-Ray analysis has been significantly increased with the introduction of X-ray cameras that allow temperatures to be programmed within wide ranges.[13] With the application of cameras capable of variable temperatures, it is possible to isothermally examine solid samples at a selected temperature, as well as to monitor the changes in *d.* spacing characteristic of the crystalline structure in the function of temperature.

With X-ray analysis, the following changes and phenomena can be measured: thermal expansion, phase transitions of uni- or multicomponent systems, recrystallization and recrystallization rate, grain size of the sample, and changes therein.[14,15] Valuable information can be obtained concerning the structure of products formed in the various periods of heat-initiated solid phase reactions using this method.

In early investigations of such transformation, the samples were measured after cooling to room temperature following heat treatment. However, this method does not lead to correct conclusions

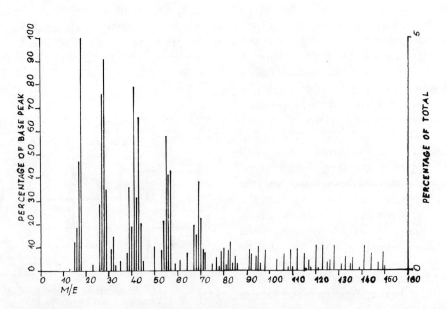

FIGURE 7.   A typical MS spectrum of organic compounds. (From Gibson, E. K., Jr., *Thermochim. Acta*, 5(3) 243, 1973. With permission.)

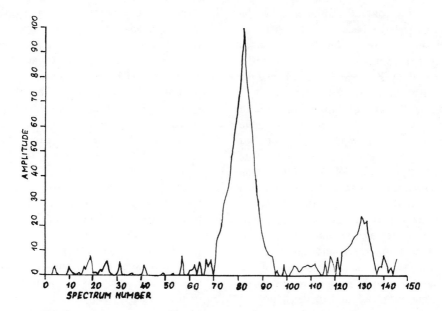

FIGURE 8. The formation pattern of a fragment with a given mass number in the function of temperature (temperature is corrected to the spectrum number). (From Gibson, E. K., Jr., *Thermochim. Acta*, 5(3), 243, 1973. With permission.)

in some cases, since the composition and structure of the investigated sample no longer correspond to those existing at the temperatures reached during heat treatment. The existence of the often occurring metastable state cannot be identified at all.

Wiedemann[16] developed an instrument for simultaneous TG and X-ray examinations. The operation principle is rather complicated, especially the TG portion of analysis. X-Ray diffractometers supplied with a heatable camera are suitable for analysis of rarely occurring airborne particles, when special requirements must be met.

VI. APPLICATIONS OF THERMOANALYTICAL METHODS IN DUST ANALYSIS

A short report on DTA examination of dust produced by industrial activity is discussed in Reference 17; included is a summary of metallic, mineral, and organic dust analysis. The following is a listing of the most important findings:

1. DTA is a suitable technique for analysis of emitted dust, and conclusions drawn from the results obtained are applicable to elaboration of technological processes in combatting air pollution.

2. Model experiments can be carried out

with the DTA method to examine pyroforous behavior and conditions of self-ignition.

3. Sampling must be performed with extreme care to ensure its representative character.

4. Dusts with large specific surfaces produce different DTA curves than those having bulky states of identical composition.

A significant portion of airborne particles consists of silicates, sulfates, carbonates, nitrates, phosphates, ammonium salts, etc. or compounds of organic origin. Occasionally, one or the other of these components is dominant; however, the sample often consists of a complicated mixture. In the vicinity of certain emitting sources (such as iron plants, ore mines, etc.), samples of unusual composition, differing from those mentioned above, are found.

Quartz is one of the most frequently occurring pollutants. DTA and X-ray analysis are suitable for both qualitative and quantitative determination. Silicium dioxide has three crystal modifications: trydymite, christobalite, and quartz. Transitions between these forms are known. However, since they are relatively slow processes, they cannot be followed by thermoanalytical methods, in most cases. A more measurable characteristic of these crystal modifications is their inversion (known as alfa-beta transition) that yields a recognizable DTA peak at about 570 to 575°C.[18]

The free quartz content of samples can be measured with relative accuracy in different matrices (in silicates also). However, experience indicates that in analyzing the quartz content of a given sample by DTA, X-ray, and chemical analysis the obtained results show significant deviation.

According to Rowse and Japson,[19] errors in quartz determination by DTA result from the different packings of the samples, the peak height construction, and measurement, coupled with the uncertainties in the reproducibility of the apparatus applied. The larger quartz content determined by chemical analysis is probably due to a portion of the accompanying feldspar, mica, and tourmaline that is measured as quartz because of incomplete dissolution. This uncertainty is increased by the fact that a small portion of the quartz dissolves or, due to its minute grain size, contaminates the mother liquid in the course of filtration. Rowse and Japson determined that the analysis can be conducted with greater precision using the DTA method than with X-ray diffraction or chemical wet analysis.

Menis and Garn[20] applied the micro-DTA technique for determination of quartz and chrysotile asbestos in airborne particles. They used 30 to 50-μg sample amounts for their measurements and obtained results with a deviation of about 15%. The analysis of asbestos is based on the fact that its three most frequently occurring modifications (chrysotile, crocidolite, and amosite) are dehydrated at significantly different temperature values. The heating rate and composition of the applied atmosphere exert a significant influence upon the course of DTA curves. Therefore, measurements should always be conducted under standard conditions, and the antecedents (origin) of quartz used for calibration should be the same, if possible, as that of the quartz present in the sample.

For analysis of quartz, one must consider that, at higher temperatures, quartz reacts with different carbonates and decomposes according to the following reaction:

$$SiO_2 + Me_2CO_3 = Me_2SiO_3 + CO_2$$

In general, one must take into consideration the eventual solid phase reactions between oxides and alkaline carbonates.

Metal carbonates are also common components of airborne particles. Decomposition of metal carbonates into metal oxides and carbon dioxide requires a relatively large amount of heat energy.[21,22] This process can be detected and measured by thermogravimetry and gas analysis, in addition to DTA measurements.

It should be remembered that the positions of both the TG and DTA signals are greatly influenced by the atmosphere, especially by the actual partial pressure of carbon dioxide, i.e., the circumstances of mass and heat transports.[21,22]

When measuring carbonates of oxidizable metal cations,[23] an inert gas atmosphere should be applied; otherwise, during the endothermic decomposition reaction, the simultaneously occurring exothermic oxidation process would be superposed, and the obtained DTA curve representing both processes could not be evaluated. These simultaneous reactions also disturb the evaluation of the TG curve. Temperature values characteristic of the decomposition of carbonates are significantly affected by the presence of small quantities of salts or by decomposition reactions of organic compounds.

Determination of carbonates by the simultaneous DTA-TG method in complex matrices involves a series of uncertainties. However, by careful selection of experimental conditions, some can be eliminated. A classical example is the decomposition of dolomite in carbon dioxide atmosphere, where two separate and characteristic decomposition processes are clearly distinguished. However, in other atmospheres, the two processes overlap to a greater extent (Figure 9). Thermoanalytical examination of other possibly occurring compounds is only briefly discussed here. For more in-depth information, see the References. Humidity and the water of crystallinity content can be determined by DTA, TG, MS, and gas analytical methods. At higher temperatures (between 400 to 1200°C) and in an inert gas atmosphere, most compounds with complex anions suffer endothermic decomposition reactions that are associated with weight loss. For example, nitrates and sulfates decompose forming nitrogen oxides and sulfur dioxides, respectively. For the decomposition of most silicates, intramolecular dehydration processes are characteristic. Ammonium salts are susceptible to sublimation. Characteristic decomposition temperatures of compounds are greatly influenced by the quality of the metal ions, the matrix effect, the possible solid- or molten-phase reactions occurring in the matrix, the grain size

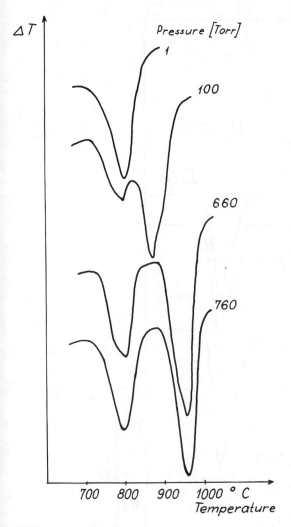

FIGURE 9. DTA curves of dolomite, $CaMg(CO_3)_2$, at different CO_2 partial pressures.

(figure labels: Pressure [Torr]; 1, 100, 660, 760; Temperature axis 700 800 900 1000 °C; ΔT)

carbon dioxide formation could also be recorded with this device.[27] A full derivatogram of a dust sample from Vienna can be seen in Figure 10. The results of the measurements are summarized in Table 2.

When evaluating the curves in Figure 10, it can be seen that the weight loss occurring at temperatures reaching 150°C is possibly due to the loss of both the humidity and the easily evaporating, sorbed organic material content. Evaporation and ignition of the organic material content occurred between 150 and 600°C. The third process associated with weight change was the decomposition of calcite and dolomite.

This type of study gives a general idea of the changes occurring in samples undergoing heat treatment. The results are useful for purposes of comparison between samples of different origin. The applicability of this method is limited, as sample amounts larger than 100 mg must be used. The introduction of microtechniques (TG, DSC, and DTA) significantly reduced this problem. The correct operation of the applied devices insured the possibility for the comparison of results, in spite of the fact that they were obtained in the course of separate measurements. With the help of these instruments, information similar in quality and quantity to those mentioned earlier can be obtained with sample amounts of a few milligrams.

Application of specific gas analyzers greatly contributed to the clarification of the nature of decomposition reactions. Employing the carbon dioxide and sulfur dioxide analyzers (DTCD, DTSD) developed by Malissa and co-workers,[28] the derived concentration profile of the formation of the two gases is recorded as the function of temperature (Figure 11).

On the basis of sulfur dioxide determinations, it can be stated that the first step corresponds to sulfur dioxide, which is probably the decomposition product of ammonium sulfate. The two peaks, obtained between 330 and 400°C, coincide with the simultaneous departure of sulfur and carbon dioxide. However, no information is given as to whether the sulfur dioxide evolution is due to decomposition of organic or inorganic substances.

The loss of sulfur dioxide above 800°C signals the decomposition of calcium sulfate. This supposition was proved by electron microprobe analysis.[29] According to the author, calcium sulfate is formed in the reaction between calcium

and its distribution, the amount of the sample, and the quality and pressure of the atmosphere, etc. These factors significantly influence the decomposition processes, sometimes in a disturbing but frequently advantageous way, yielding further valuable information on the sample material.

Finally, with the presentation of the results of Gál and co-workers,[24] the applicability of thermoanalytical methods for this special field is demonstrated. Sediment dust samples from Austrian cities (Vienna and Graz) were chosen as samples for the investigation of the combined use of various thermoanalytical techniques.[25] A Derivatograph®[26] suitable for simultaneous TG, DTG, DTA, and thermogas-titrimetric measurements was used. The temperature profile of

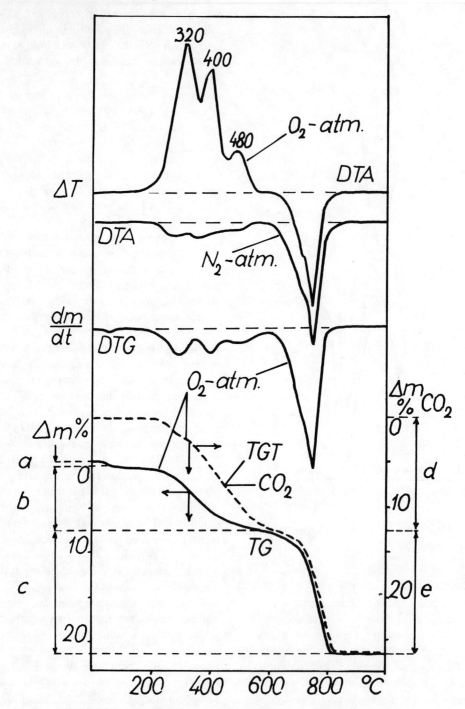

FIGURE 10.   The derivatograms of airborne particles collected in Vienna.

TABLE 2

Microthermoanalytical Data

| | Weight loss (%) | | | | Gas analysis (%) | | | | |
| | | | | | 150–570°C | | 570–800°C | | |
	25–150°C	150–570°C	570–800°C	Total	C	S	C	CO$_2$	800°C S
W1	0.6	7.4	18.7	26.8	4.1	0.11	5.8	20.1	0.22
W2	0.5	5.2	14.9	20.6	2.9	0.10	4.2	15.2	0.20
W3	0.4	5.8	14.1	20.2	3.2	0.10	4.0	14.5	0.15
G1	0.6	4.5	15.4	20.5	2.2	0.03	4.4	16.1	0.10
G2	0.6	3.9	13.0	17.5	1.7	0.04	3.8	14.0	0.06
G3	0.3	3.3	10.7	14.3	2.1	0.02	3.3	12.1	0.04

Note: W, samples from Vienna; G, samples from Graz.

From Pell, E., Puxbaum, H., and Gál, S., *Z. Anal. Chem.,* 282, 115, 1976. With permission.

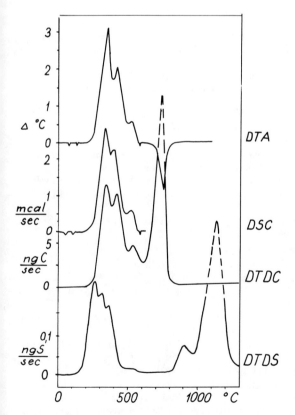

FIGURE 11. Comparison of DTA, DSC, DTDC, and DTDS traces of airborne particles collected in Vienna.

carbonate and the sorbed sulfur dioxide on the surface of the dust in the presence of humidity and air.

VII. SUMMARY

The application of thermoanalytical methods in environmental analysis is not yet common. However, as complementary techniques, they will gradually gain ground in this field. It should be pointed out, however, that valuable information can only be obtained if various thermoanalytical methods and other, principally differing, analytical methods are applied in conjunction with each other.

Acknowledgment
The author wishes to express his thanks to Dr. Puxbaum for reviewing this manuscript and for his valuable remarks.

GENERAL REFERENCES

Duval, C., Thermal methods in analytical chemistry, in *Comprehensive Analytical Chemistry,* Svehla, G., Ed., Vol. 7, Elsevier, Amsterdam, 1976.

Garn, P. D., *Thermoanalytical Methods of Investigations,* Academic Press, New York, 1965.

Lodding, W., *Gas Effluent Analysis,* Marcell Decker, New York, 1967.

McKenzie, R. C., *Differential Thermal Analysis,* Vol. 1, Academic Press, London, 1970.

McKenzie, R. C., *Differential Thermal Analysis,* Vol. 2, Academic Press, London, 1972.

Meisel, T. and Gál, S., *Methode der Thermische Analyse,* Akademische Verlagsgesellschaft, Frankfurt on the Main, in press.

Schultze, D., *Differential Thermoanalyse,* 2nd ed., VEB Deutscher Verlag der Wissenschaften, Berlin, 1972.

Smykatz-Kloss, W., *Differential Thermal Analysis,* Springer-Verlag, Berlin, 1974.

Vallet, P., *Thermogravimetrie,* Gather-Villors, Paris, 1972.

Wendtland, W. W., *Thermal Methods of Analysis,* 2nd ed., Interscience, New York, 1972.

REFERENCES

1. McAdie, H. G., Requirements and realization of thermal analysis standards, in *Thermal Analysis, Proc. ICTA Conf.,* Vol. 1, Wiedemann, H. G., Ed., Birkhäuser Verlag, Basel, 1972, 591.
2. Roberts, P. T. and Friedlander, S. K., Conversion of SO_2 to ambient particulate sulfates in the Los Angeles area, Conf. on Health Consequences of Environ. Control — Impact of Mobile Emissions Control, Durham, N.C., April 1974.
3. Leahy, D., Siegel, R., Klotz, P., and Newman, L., The separation and characterisation of sulfate aerosol, *Atmos. Environ.,* 9, 219, 1975.
4. Garn, P. D., Some problems in the analysis of gaseous decomposition products, *Talanta,* 11, 1417, 1964.
5. Wiedemann, H. G., Simultaneous TGA-DTA measurements in connection with gas-analytical investigations, in *Thermal Analysis Proc. 2nd ICTA Conf.,* Vol. 1, Schwenker, R. F. and Garn, P. D., Eds., Academic Press, New York, 1970, 229.
6. Gohlke, R. S. and Langer, H. G., Thermal analysis by mass spectrometry, *Anal. Chem.,* 37, 25A, 1965.
7. Langer, H. G., Gohlke, R. S., and Smith, D. S., Mass spectrometric differential thermal analysis, *Anal. Chem.,* 37, 433, 1965.
8. Zitomer, F., Thermogravimetric-mass spectrometric analysis, *Anal. Chem.,* 40, 1091, 1967.
9. Thermal Techniques Series, Technical Bulletin T 107, Simultaneous TGA-DTA-Mass Spectrometry, Mettler Instrument Corporation, Princeton, New Jersey.
10. Barnes, P. A., The development and application of simultaneous thermal analysis — mass spectrometry, in *Proc. of the First European Symposium of Thermal Analysis,* Dollimore, D., Ed., Heiden and Son, London, 1976, 31.
11. DiLorenzo, A., Masi, S., and Pennacchi, A., A thermogravimetric-mass spectrometric system for the analysis of organic air pollutants, in *Proc. of the First European Symposium of Thermal Analysis,* Dollimore, D., Ed., Heiden and Son, London, 1976, 37.
12. Gibson, E. K., Jr., Thermal analysis — mass spectrometer-computer system and its application to the evolved gas analysis of Green River shale and lunar soil samples, *Thermochim. Acta,* 5, 243, 1973.
13. Hillwege, K. H., *Einführung in die Festkörperphysic,* Springer-Verlag, Berlin, 1968.
14. Erdös, E., Brezina, D., Scheidegger, R., Über Eigenschaften von Chromsulfiden, *Werkst. und Korros.,* 22, 148, 1971.
15. Wefers, K., Gleichzeitige Röntgen und DTA untersuchung fester Stoffe, *Ber. Dtsch. Keram. Ges.,* 42, 35, 1965.
16. Wiedemann, G. H., Simultaneous TG and X-ray analysis method and applications, in *Thermal Analysis, Proc. 3rd ICTA Conf.,* Vol. 1, Wiedemann, H. G., Ed., Birkhäuser Verlag, Basel, 1972, 171.
17. MacKenzie, R. C. and Meldau, R., General application in industry with special reference to dusts, in *Differential Thermal Analysis,* Vol. 2, MacKenzie, R. C., Ed., Academic Press, London, 1972, chap. 46.
18. Dawson, J. B. and Wilburn, F. W., Silica minerals, in *Differential Thermal Analysis,* Vol. 1, MacKenzie, R. C., Ed., Academic Press, London, 17, 477, 1969.
19. Rowse, J. B. and Japson, W. B., The determination of quartz in clay materials, *J. Therm. Anal.,* 4, 169, 1972.
20. Menis, O., Garn, P. D., and Diamondstone, B. T., The application of thermoanalytical methods to the environmental health problems, *Proc. 4th ICTA Conf.,* Vol. 3, Buzás, I., Ed., Akadémia Kiadó, Budapest, 1975, 127.
21. Smykatz-Kloss, W., Differential-Thermo-Analyse von einiger karbonat-Mineralen, *Beitr. Mineral. Petrogr.,* 9, 481, 1964.
22. Webb, T. L. and Krüger, J. E., Carbonates, in *Differential Thermal Analysis,* Vol. 1, MacKenzie, R. C., Ed., Academic Press, London, 1972, 10.
23. Rowland, R. A. and Jonas, E. C., Variation in DTA curves of siderites, *Am. Mineral.,* 34, 550, 1949.
24. Gál, S., Paulik, F., Pell, E., and Puxbaum, H., Thermoanalytical investigation on dust, *Z. Anal. Chem.,* 282, 291, 1976.
25. Pell, E., Puxbaum, H., and Gál, S., Thermoanalytische Untersuchungen von Stäuben mit Detektion physikalischer und Chemischer Parameter, in *Z. Anal. Chem.,* 282, 115, 1976.

26. Paulik, F., Paulik, J., and Erdey, L., Derivatography – a complex method in thermal analysis, *Talanta*, 13, 1405, 1966.
27. Paulik, F., Paulik, J., and Erdey, L., Kombinierte derivatographische und thermogasanalytische Untersuchungen, *Mikrochim. Acta*, 886, 1966.
28. Malissa, H., Puxbaum, H., and Pell, E., *Z. Anal. Chem.*, 282, 109, 1976.
29. Malissa, H., Betrachtungen zur integrierten Staubanalyse mit physikalischen Methoden, *Angew. Chem.*, 88, 168, 1976.

Chapter 13
CONSIDERATION OF STANDARD REFERENCE MATERIALS

H. Malissa

TABLE OF CONTENTS

I. INTRODUCTION

The necessity of accurate measurements of the component(s) of interest (COI) in environmental systems and environmental polluting sources has been stressed by enactments of many governments and official world-wide organizations such as United Nations Educational Scientific and Cultural Organization (UNESCO), World Health Organization (WHO), and International Standardization Organization (ISO), etc. Until recently, no proper answer has been given as to how this is to be done, especially where airborne particulates are concerned. The problem places a heavy burden on the analyst. One of the primary problems is determining proper standard reference materials (SRMs) for the most urgently needed and used microtrace multi-element methods (as stressed in Chapter 1), especially when applying physical methods.

In seawater, for example, 73 elements have already been analyzed, and approximately 20 of them play an important role in the aquatic environment. Approximately 10 of them are in the range of < 1 μg/l, such as mercury. The toxicity of mercury — like many other metals — is highly dependent on the chemical species in which it

occurs. Thus it is only partly sufficient to know the total elemental concentration. At the usual concentration level of toxic elements in environmental samples, the stability of standard solutions becomes very important. In a recent paper of Rook and Moody,[1] it was shown that in a concentration level of nanograms of Hg per gram (even in washed plastics) after a fortnight only about 50% of the original concentrations could be found. However, after adding a solution of gold chloride, the initial level could nearly be recovered, and the standard solution remained constant for several months. The adsorption and chemisorption of elements from very dilute solutions has been investigated for many years, and basically, at concentrations below 10^{-4} g/l, such samples cannot be stored without special precaution.

In every case, we are bound to standards not only suited for the environmental system, but also to the rules of micro- and trace analysis, as well as to physical principles and behavior. The importance and difficulty of choosing proper SRMs becomes immediately evident by these three unavoidable demands. Proper SRM's have to be representative *and* applicable to (a) the environmental system in consideration, (b) the components of interest (COI), and (c) the analytical procedure. The first two points cannot be solved by the analyst alone; however, for the third point, the analyst is solely responsible.

National and international organizations such as ISO, International Union of Pure and Applied Chemistry (IUPAC), etc., have compiled information listing approximately 100,000 SRMs of different types. The number of certified SRMs is much less. Cali's[2] report gives more accurate figures, presenting a listing of 967 SRMs that are available from the National Bureau of Standards (NBS). Within this number, 38 belong to the environmental category; however, only 14 are liquids or solids.

Out of approximately 5000 items in IUPAC compilation[3] dealing with SRMs for trace analysis, only 22 can be considered valid for environmental analysis, and 16 of them are gases. McCrone's *Particle Atlas*[4] lists five companies (including NBS) supplying the analyst with relevant SRMs.

This situation leads to the important discussion of the methods of calibration in trace and microanalysis. As usual in analytical chemistry, the ultimate goal in environmental analysis is also to find the true value, which involves precision and accuracy. Accuracy expresses the correctness of a measurement and precision the reproducibility of a measurement. Instead of discussing the problems of classification of errors and so forth, let us consider briefly the situation of calibration methods, as it is essential to eliminate possible systematic errors. The calibration, testing, and checking can be done by the following methods:

1. Use of certified natural standards
2. Use of synthetic standards
3. Use of independently analyzed samples
4. Use of standard addition and internal standards
5. Use of fundamental relationships

II. CERTIFIED NATURAL STANDARDS

The use of certified natural standards is the best method for calibration, testing, and checking because they resemble the actual samples most closely. A set of standards suited to the more or less automated analysis of a large number of samples of complex composition and for specific questions and stations is urgently needed. The lack of such material is a primary limiting factor. NBS has prepared dried, powdered orchard leaves and bovine liver with approximately 30 analyzed elements, which can be used in calibration work, as well as glass wafers doped with 61 elements in the range of 1 ppm and 0.2 ppm. They may be used for calibration of physical methods, depending on the actual compatibility with the system under investigation.

The work done by Pella and co-workers[5] should be mentioned in connection with these standards. They reground orchard leaves, deposited the material with an area density of 5 to 91 mg/cm^2 on membrane filters, and coated this sample with a thin polymer film for protection against abrasion and moisture. More than 80% have a particle size of less than 7.5 μm; therefore, such material can be used as a standard for bulk analysis with X-ray fluorescence (XRF).

As outlined in Chapter 1 the homogenity of sample and standard is an important factor to be considered. The standard used for analysis definitely must be homogenous in respect to the method applied. Although the number of certified standard samples is increasing significantly, it is impossible to provide SRMs for many of the different types of materials, analytical problems, and methods that the environmental analyst encounters. Much time is needed for the collection, preparation, and distribution of the materials, as well as the discussion and evaluation of the data.

III. SYNTHETIC STANDARDS

Synthetic standards are also used very often. Addition of pure substances to natural standards is thought to be an especially effective method for producing synthetic standards. This approach is of real value if the analytical procedure includes a solution step or if aquatic systems are investigated. The preparation of multi-element solid standards by powder dilution techniques is often subject to considerable errors, especially if the chemical bonding conditions are ignored.

XRF is one of the widely used techniques for water and air investigations, either for the analysis of coated metals, resins, ion-exchange papers, or particulates. The simple technique of preparing a suitable standard is to impregnate filter paper with a standard solution. Baum and co-workers[6] found that after drying, a supposed homogenous standard had been obtained. These authors used a special but simple device to place 42 drops on filter paper simultaneously. They also found that Millipore® filters of different types always presented a concentration gradient in the form of the so-called ring phenomenon which means that the concentration on the edge of the paper differs from that in the center. This was previously demonstrated in 1962,[7] as was the dependence of the concentration gradient on the pH and the combination of ions. In judging this phenomenon, one must be aware of the relationship between the excited area on the paper and the concentration gradient. Often this ring phenomenon does not play a major role in conventional XRF; however, it must certainly be taken into consideration when using fine focus attachments.

Pressed or molten pellets are also often used for analysis. In this case, the equivalence of the matrix in standard and sample must be considered, as well as the problem of transfer of the sample or fractions of it, and the representativity.

Columbia Scientific Industries (CSI) produces two types of synthetic X-ray standards: (a) thin specimen X-ray calibration standards and (b) standard reference strips for air particulate analysis. For the first group, single and multi-element, homogenous standard samples on Millipore filters are available, as well as particulate and spiked particulate standards. For this, the puff technique is used and the spiked 10-μm washed quartz is treated with standard solutions and baked at 500°C. For the second group, glass filter strips are used as a supporting material and treated in a similar manner.

These synthetic samples differ widely from actual samples and do not take into account all conversion reactions that may occur during the transportation processes of pollutants. Even if good synthetic standards are used, it is practically impossible to conform to situations as they exist in actual samples (shown in Figures 1 and 2).

Figure 1 illustrates dust samples composed mainly of dolomite approximately 10 μm in size with quartz grains of approximately 1 to 5 μm diameter.[8] Which standard for the SiO_2 determination should be used in this case, or if we have calcium silicates on soot, as in Figure 2? How will a titanium determination work, knowing that Ti may occur as a pure oxide (as in rutile, anatase, brookit) or bound to Fe in ilmenite and Ti-magnetite or Ca-Ti-oxide (as in perovskite)? All of these compounds can show up in nature and from

industry, from plastic, paper, ceramics, enamels, pigments, or paints. The only method that can be calibrated for chemical analysis by using synthetic standards is neutron activation analysis. With this type of analysis only the nuclear properties, and not shape and bonding conditions, are important.

IV. INDEPENDENTLY ANALYZED SAMPLES

Calibration by means of independently analyzed samples will only be successful if all details of the reference methods are known. At least three different and independent methods should be used, with high sensitivity, known specificity, and excellent precision. If there is — according to statistical evaluation — a significant difference in comparing these methods, systematic work must be done.

This is also the method of "round robin" procedures currently used throughout the world. If there is enough material left after performance of the necessary analysis, this may be used as a certified standard. It is a time- and money-consuming procedure, requiring a great deal of cooperation from each of the participating laboratories; however, such careful research results in highly valuable reference materials. Giradi[9] and Ondov et al.[10] present details of this method and its impact.

Skogerboe and Koirtyohann[11] have formulated the problem of accuracy assurance in the analysis of environmental samples. They stressed the following question:

It is perhaps ironic in this age of environmental concern that the key strategy of operation is one of recycling the materials through the analytical process. By continuing repeat analyses of the material available, it is possible to obtain evaluations of the integrity of the results. As illustrated in Figure 3, these evaluations must be based on a combination of approaches which include spike-recovery studies, comparison of independent (in house) methods participation in collaborative test programs. Each of these approaches will provide specific types of information relevant to the overall program. Thus it is appropriate to consider each separately.

V. STANDARD ADDITION AND INTERNAL STANDARDS

Aside from sampling problems, the matrix and interelement effects are the most serious ones, if environmental analysis is mainly thought of as

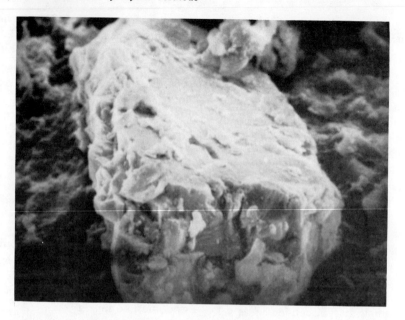

FIGURE 1.   Dust samples, containing mainly dolomite and quartz. (From Malissa, H., *Z. Anal. Chem.,* 282, 407, 1976.)

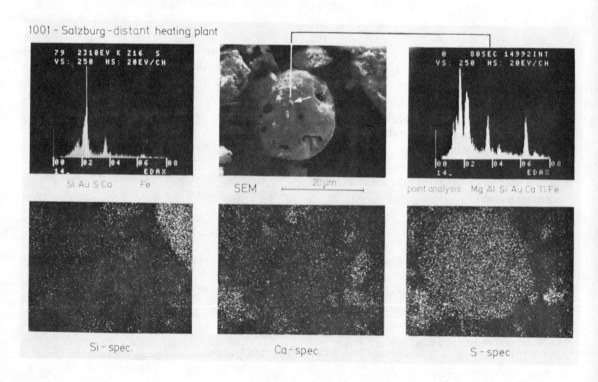

FIGURE 2.   Dust from a heating plant. (From Malissa, H., *Z. Anal. Chem.,* 282, 407, 1976.)

work in trace analysis. These effects are responsible for systematic errors. In absorption and emission spectroscopy, only the situation in ratio changing, appearance, and disappearance of line intensities must be considered. The principle of this method is the repetitive addition of known amounts of a low concentration of the COI to several samples of the unknown sample. The signal

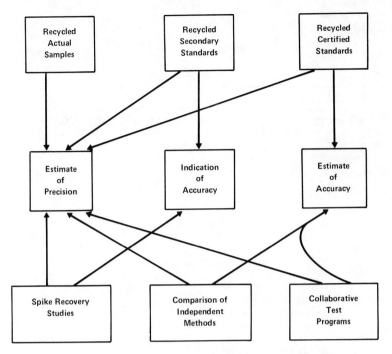

FIGURE 3. Schematic representation of the steps in an accuracy assurance program. (From Skogerboe, R. K. and Koirtyohann, R. K., Accuracy assurance in the analysis of environmental samples, *Nat. Bur. Stand. Spec. Publ.,* 422, 199, 1976.)

readings after the full analytical procedure are plotted against the added concentration, and the amount of the unknown concentration of the species can be extrapolated.

If the line is not linear, often a transformation brings an improvement. Regression and "least square lines" are also widely used in atomic absorption and spectrophotometric work; the mathematics needed can be provided as commercial software.

The internal standard method is quite different from the standard addition method. The main function of an internal standard is to check the conditions of the apparatus and measuring devices. For this procedure, elements and compounds not found in the sample are primarily used. These standards must have definite and well-known analytical properties.

VI. FUNDAMENTAL RELATIONSHIPS

The use of fundamental relationships as standards for calibration work is of primary interest because it provides the "cleanest" situation and also is needed for new techniques. But in daily analytical life they are too cumbersome to handle.

Primary standards, in the sense of analytical chemistry (such as the Coulomb or other fundamental laws, like mass action, Lambert-Beer, Clausius Clapeyron, Henry, etc.) together with practicable procedures are more commonly in use. The Commission on Electroanalytical Chemistry of the Analytical Chemical Division of IUPAC has recently reported on the status of the Faraday constant as an analytical standard.[12] Because the presently acceptable value of 96486.7 ± 0.5 Asec mol^{-1} is based especially on measurements made at NBS, and fits in very well to the theoretical value, it may serve as a primary standard to check not only measurements but also to test chemical standard substances to be used later as standards (see Section II).

A special case for the importance and limitations of the use of fundamental relationships has been demonstrated by Criss.[13] The author investigated the standard problem in XRF of filter dusts and developed a mathematical correction procedure for the absorption of the secondary X-rays within the individual dust particles. If a filter loaded with dust is analyzed using an impregnated (dried solution) filter standard, then a negative error is made due to this absorption effect.

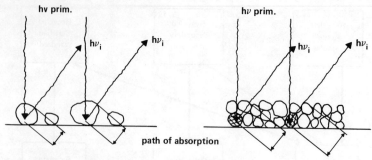

correction for absorption (Criss)

$$M_i = \frac{J_i}{J_{io}} \cdot M_{i_o} \cdot (1 + a.b)^2$$

J_i ...... X-ray intensity of element i
M_i ...... mass element i/cm^2
a ...... particle size (μm)
b ...... correction factor for composition

assumptions:
"monolayer" of dust (<100 μg/cm^2)
no absorption in standard

requirements:
particle size and composition
must be known (SEM + EDS, EPMA)

extension of formula for secondary fluorescence effects?

$$M_i = \frac{J_i}{J_{i_o}} \cdot M_{i_o} \cdot (1 + a.b)^2 (1 - a.c)^2$$

c .... fluorescence corr. factor for comp.

FIGURE 4.   XRF of dust. (From Malissa, H., *Z. Anal. Chem.*, 282, 407, 1976.)

Errors depend on particle size, composition, and excitation conditions. Typically, it is in the magnitude of 10 to 30% for medium number elements and particle sizes of several microns. For elements (e.g., Al) and larger particle sizes, the error can amount to several hundred relative percent.

In order to eliminate this error, Criss introduced a mathematical correction term with factor b. The correction, shown in Figure 4, has been developed solely by use of the fundamental physical relations governing X-ray emission and absorption. Unfortunately such theoretical correction procedures usually tend to have a limited use (as in the case of the well-known and widely applied corrections for quantitative electron and ion probe microanalysis).[14] The limitations in the case of the Criss model arise in the applicability of the correction procedure from the assumption that only a "monolayer" of dust is present on the filter. This assumption implies that absorption only takes place within the particles that contain the element to be analyzed and not by neighboring

particles. In the first case (shown on the upper left side of Figure 4), the absorption effect can be calculated, provided that the average particle size and the composition of the particles that contain the element to be analyzed are known. Although this requirement may cause severe problems, this parameter can be determined by applying scanning electron microscopy (SEM) with an energy dispersive X-ray spectrometer or an electron microprobe. In the case of multilayer filter coverage, the effects of neighboring particles on absorption are not known. The overall effect can only be treated statistically, as for XRF of powder samples. Detailed investigations are needed on a larger scale.

Another effect that may have to be considered in quantitative XRF of filter dust is the possibility of the excitation of a secondary fluorescence radiation caused by the characteristic radiation of another element in the sample, for example, the excitation of CrKα-radiation by the FeKα-radiation within a single particle. Possibly the Criss' formula could be extended by a fluorescence correction factor as indicated in Figure 4.

REFERENCES

1. **Rook, H. L. and Moody, I. R.,** Nuclear methods in environmental research, Proceedings, University of Missouri, 1974.
2. **Cali, P.,** The NBS standard reference materials program: an update, *Anal. Chem.,* 48, 802A, 1976.
3. **Koch, O. G.,** personal communication, 1977.
4. **McCrone, H. C. and Delly, I. G.,** *The Particle Atlas,* Vol. 4, 2nd ed., Ann Arbor Science, Ann Arbor, Mich., 1973, 1028.
5. **Pella, P. A., Kriehuer, E. C., and Cassalt, W. A.,** personal information.
6. **Baum, R., Gutknecht, W. F., Willis, R., and Walter, R. L.,** Preparation of standard target for X-ray analysis, *Anal. Chem.,* 47, 1727, 1975.
7. **Malissa, H.,** From spot test to spot colorimetry, in *Analytical Chemistry 1962 Int. Symp. Birmingham University (U.K.),* West, P. W., MacDonald, A. M.G., and West, T. S., Eds., Elsevier, Amsterdam, 1963, 80.
8. **Malissa, H.,** Problems of standard qualities and standards in environmental analysis, *Z. Anal. Chem.,* 282, 407, 1976.
9. **Giradi, F.,** Accuracy of Chemical Analysis of Airborne Particulates — Results of an Intercomparison Exercise, Proceedings 7th IMR Symposium 1974, *Nat. Bur. Stand. Spec. Publ.,* 422, 173, 1976.
10. **Ondov, I. M., Zoller, W. H., Olmez, I., Aras, N. K., Gordon, G. E., Rancitelli, L. A., Abel, K. H., Filby, R. H., Shah, K. R., and Ragaini, R. C.,** Four laboratory comparative instrumental nuclear analysis of the NBS coal and fly ash standards reference material, *Nat. Bur. Stand. Spec. Publ.,* 422, 211, 1976.
11. **Skogerboe, R. K. and Koirtyohann, R. K.,** Accuracy assurance in the analysis of environmental samples, *Nat. Bur. Stand. Spec. Publ.,* 422, 199, 1976.
12. IUPAC, Analytical Chemistry Division, Commission on Electroanalytical Chemistry, Status of the Faraday constant as an analytical standard, *Pure Appl. Chem.,* 45, 127, 1976.
13. **Criss, J. W.,** Particle size and composition effects in X-ray fluorescence analysis of pollution samples, *Anal. Chem.,* 48, 179, 1976.

Chapter 14

MONITORING

M. Birkle

TABLE OF CONTENTS

I. INTRODUCTION

In addition to questions concerning the composition, physical behavior, origin, residence time, etc. of particles, the necessity arises to define measurement quantities for the particle content of air and to specify appropriate limits. It should be possible to detect the measurement quantity for the particle content simply and with the smallest expenditure possible on personnel and material so that continuous monitoring of ambient air in populated areas can be performed.

The lower limit of particle size of interest in this case is determined by the transition to molecular dispersions (gases), the upper limit by the settling speed which rises rapidly with increasing particle diameter. In the range thus given of 0.001 μm to roughly 500 μm diameter,[1]

particles with a diameter d $>$ 20 μm are of subordinate importance for ambient air pollution. Investigations by Junge and Scheich[2] have shown that, in the case of an equilibrium state of pollution in air over urban areas, particle distribution is as follows: 10% of the aerosol mass with r $<$ 0.1 μm; 45% of the aerosol mass with 0.1 $\leqslant$ r $\leqslant$ 1 μm; and 45% of the aerosol mass with r $>$ 1 μm.

On the other hand, the entire surface area is at a maximum with a radius r = 0.1 μm. Particle size, which is used as a basis for all data, is a geometrical concept that, without more detailed information concerning the measurement method, has not been accurately defined. Neither radius, diameter, volume, nor surface area can be measured directly on particles of irregular shape occurring mainly in ambient air. Thus, particle size is described by means of auxiliary variables, by the so-called equivalent diameters or equivalent radii.[4] Equivalent diameter involves using as the diameter of an arbitrary particle the diameter of a spherical particle of the same material that would generate the same measurement effect, such as the same projection surface, settling speed, change in electrical resistance, or scattered light intensity. Accordingly, differentiation is made between projected equivalent diameter, dynamic equivalent diameter, optical equivalent diameter, etc. Since particles in ambient air are always polydisperse, i.e., with a more or less broad spectrum of particle size, a frequency distribution is normally presented as information concerning the particle sizes of the particle loading.

Information concerning particle content of ambient air that is not further defined can be obtained in several ways. In actual practice, however, only a few measuring methods have become widespread. Most frequently, the concentration of the particles is measured or given as mass/volume in $\mu g/m^3$. Another concentration specification, less widely used, is the number of particles per unit volume. In the U.S., in addition to concentration specification, the Coefficient of Haze (COH/1000 ft) is also measured. The COH represents, in principle, a complex variable determined from the concentration of the particles and their optical properties. The oldest technique of measuring particle content of ambient air and that which can be performed with the least technical expenditure is the dust fall measurement. The results of this measurement are expressed in mass per unit area times sampling time $(g/m^2 d)$. Accordingly, when monitoring for particles, differentiation is currently made between:

1. Measurement methods to determine the mass concentration (results expressed in $\mu g/m^3$)

2. Particle counting (results expressed in number of particles per cubic meter or in the case of particle size classification, in particles per cubic meter in the range of particle sizes)

3. Measurement methods to detect the COH (results expressed COH/1000 ft)

4. Dust fall measurements (as a rule, 30 days are used as a basis for the measurement interval; results expressed in $g/m^2 d$)

In addition to these four methods which narrowly represent the possibilities of measuring technology for detecting particle pollution of the air, note should also be made of light detecting and ranging (LIDAR) measurement within the scope of monitoring. This technique, related to radar, is rarely used to detect ambient air particle concentration. However, it offers possibilities above and beyond the customary measuring techniques to localize particle density fluctuations in the open atmosphere.

Measurement methods for determining mass concentration, for particle counting, and for determining the COH require an initial air sample obtained by means of sampling equipment.[1] Because the sample usually cannot be taken isokinetically, a change in concentration of the air sample with respect to the original condition, occurring as a result of the relative motion between particles and the carrier gas, cannot be avoided. The extent to which this effect occurs depends on the change in speed and direction of the carrier gas, as well as on the size distribution of the particles and their concentration. Thus, in practice, the sampling equipment exhibits a filtering function,[1,3] whereby efforts are made to minimize the effects of this filtering function on the desired measured quantity.[3,5] In the case of still air, geometry of the sampling aperture and air flow rate are the parameters of this filtering function. In the case of air in motion, filtering function of the sample inlet is additionally changed by the magnitude and direction of the wind velocity in the vicinity. With increasing particle size, the influence of the wind on the sampling efficiency is already noticeable at decreased wind velocities. Thus, depending upon the sampling equipment and sampling conditions (of particular importance for the larger particles), filtering functions are quite different. This must naturally have a greater effect on the measurement results with methods that determine the mass concentration than with methods that determine the number of particles. The magnitude of the resultant deviations of various sampling units thus depends strongly on the size and density distribution of particulate air pollution. Since particle concentration generally decreases with increasing particle size, differences between the results for mass concentrations thus caused are usually tolerable.

In addition to the function of taking the sample, sampling equipment also serves to feed the sample into the actual measuring unit or collection system. During transportation, a velocity profile is constructed in the sample gas. This velocity profile can either be laminar or turbulent and should be independent of ambient air motion as far as possible. If the flow is laminar, larger particles can be separated by means of settling. This effect increases with increasing residence time in the area of the laminar flow. In the case of turbulent flows, certain particle sizes also separate due to the mass forces. By means of suitable gas sampling line design (e.g., with a flow system designed as a cyclone separator), detection and measurement of only special particle size ranges (e.g., the respirable particles) can be achieved.

When using dust fall and LIDAR measurement methods, it is not necessary to take a special air sample and feed it to the measuring equipment. However, the problem of obtaining a measuring result representative of the area by selecting the set-up location or measuring area is one which all ambient air measurement systems have in common.

II. METHODS FOR MEASUREMENT OF MASS CONCENTRATION

Procedures in which the particles are collected on filters and mass of the collected material is determined are categorized as methods for determining mass concentration. The high-volume sampler and the β-radiation absorption units are most widely used devices for these methods. In the case of the high-volume sampler, relatively large quantities of air are sucked through a filter, and after appropriately specified treatment, the mass of the material sampled on the filter is determined by means of weighing. In the case of measuring units using β-radiation absorption, considerably smaller quantities of air are sucked through the filter, and the mass loading of the filter is determined by the attenuation of β-radiation.

The filters used in most current devices are planar papers.[1] In addition to fiber filters, membrane filters (e.g., cellulose nitrate) with widely varying pore widths are also used in many cases. The filters used must have sufficient collection efficiency over the entire range of flow speed of the device. As a rule, this collection efficiency must be more than 99.5% for a defined particle content of the air. When choosing the filter, in addition to collection efficiency, flow resistance of the filter and its change with increasing loading

must also be taken into account. Moreover, the filter material should not undergo any changes due to the materials customarily found in the air, and the filters must be selected with respect to any possible subsequent evaluations of the collected material.

For concentration specification, the particle mass retained on the filter must be referenced to the volume of the air which has been sucked through the filter. Strictly speaking, exact determination of the volume calls for knowledge of the parameter pressure, temperature, and humidity at the measuring location. In order to obtain a comparison of the gas volumes by these varying conditions, the determined volume is generally converted to a standard condition (standard volume) or to other appropriately specified comparison conditions.

In addition to the two measurement methods mentioned, which in principle provide gravimetrical measurements, measuring equipment that uses other measuring principles but, as a result of their calibration, display the measured value in the dimension $\mu g/m^3$ should be included in the methods for determining mass concentration within the scope of this chapter. However, only the photometric particle measurement has been important in the area of ambient air measurement technology. Thus, in addition to the high-volume sampler and the β-radiation absorption meter, photometric measurement will also be discussed.

A. High-Volume Sampler

This sampler is a particle collection unit with a relatively high airflow rate in the order of magnitude of $100 \ m^3/h$. The particles are collected on a filter and, thus, are not only accessible for weighing but, with suitable filter material, are also accessible for further analysis. For the determination of mass concentrations, (i.e., when evaluation is made by weighing) fiberglass filters are usually used.[1,6,7] With the units presently in use, air is sucked through a horizontal, weather-proofed filter.[1,6,7] As a result of the weather proofing, ambient air is sucked up from below. Therefore, particles above a certain size cannot reach the filter. As an example, Figure 1 shows the construction of a Staplex® high-volume sampler.[1] The airflow is sucked in through the annular gap between the base of the unit and the protecting cover, and the following change of direction by 180° results in a filtering effect. Coarser particles,

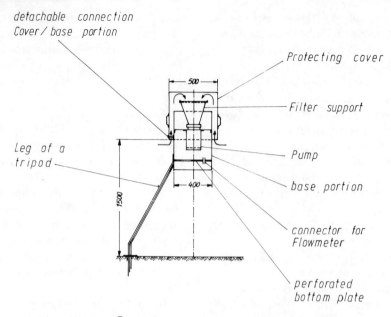

FIGURE 1.   Staplex® high-volume air sampler (schematic). (From VDI-Richtlinie 2463, VDI-Verlag, Düsseldorf, Germany, 1974. With permission.)

roughly those larger in diameter than 60 to 70 μm, are separated as a result of their mass inertia and do not reach the filter.

Determination of filter mass before and after exposure requires that the same conditions be maintained during weighing, especially with regard to moisture. Normally, the effect of moisture can be eliminated by drying the filter for 24 hr before weighing in a desiccator (e.g., with magnesium perchlorate).[1] Subsequently, it is weighed in as short a time as possible, or the dry weight is determined by means of an interpolation process. In this manner, however, only the dry substance of the particle is weighed in the first approximation. If a certain part of the moisture component of the particles should also be detected, the filter, instead of being dried in the desiccator, is brought to equilibrium before weighing under defined conditions with reference to temperature and air humidity. For evaluation of the weight of the particle mass on the filter, the treatment of the filter before weighing should always be specified. When the filter is being loaded with particles, filter resistance increases, resulting in a corresponding decrease in flow rate. A change in volume flow caused by this condition should not be more than 5%.[1] This results in an upper limit of filter loading. To achieve accurate weighing results, efforts are made to obtain as large a load as

possible, which unfortunately contributes to inaccuracies when evaluating the volume of air that has been sucked through the filter. The sampling time presently customary for ambient air measurements is between 8 and 24 hr. To maintain filter loading at a meaningful level, the high-volume sampler is sometimes operated intermittently in the case of high particle content of the air. When using a timer, the sampler can be operated for limited time periods, for example, only 40 min of every hour. If heavy filter loading is expected, it is recommended that the volume flow be checked shortly after exposure of the filter begins and after it is complete, using a rotameter flowmeter which has been calibrated previously for the system.[1]

Measurements of particle mass concentration using high-volume samplers are increasingly accepted in Europe and America as the standard reference method for determining mass concentration.[1,7,8] However, if the results of the high-volume sampler are compared with the results of other measuring methods, consideration must be given to the fact that certain particle sizes cannot be detected using the high-volume sampler. Moreover, depending upon the treatment of the filter before weighing, the moisture component of the particle is only partially weighed or sometimes excluded entirely. The particle range that can be detected is determined, with respect to the smaller particle masses, by filter properties and, with respect to the larger particle masses, by technical design of the suction equipment.

In addition to high-volume samplers, units are being used today which, although derived from the sampler in principle, have smaller airflow rates, other filters, and possibly special suction equipment.[1]

B. β-Ray Absorption Instruments

Mass concentration measurement using the high-volume sampler is a discontinuous process. For monitoring particle concentration in urban areas, however, automatic units which make continuous measurements over lengthy periods of time with low expenditure on personnel are desirable. For automated measurement, devices are being offered presently which collect particles on a filter tape which is moved forward after a specified exposure time. The increase of the weight per unit area at the place of the filter tape covered with particles is measured by means of the absorption of a β-ray.[9,12] According to Lenard's law, the attenuation of the β-ray is directly dependent on the irradiated mass per unit area, i.e., layers that have the same mass on the unit of area provide the same amount of attenuation independent of the material itself. In practice, this law holds with sufficient accuracy. Only in the case of great variations of the atomic numbers do deviations occur. The intensity of the β-radiation is expressed by the following equation:

$$I = I_o \cdot e^{-\left(\frac{\mu}{\zeta}\right)x}$$

where I_o = radiation intensity without absorber; I = radiation intensity with absorber; ζ = material density in (g/cm^3); μ = Linear coefficient of absorption (cm^{-1}); $x = d \cdot \zeta$ = mass per unit area of the absorber (g/cm^2); and d = thickness of the absorber. For small exponents, the following linear approximation is valid:

$$I = I_o \left(1 - \frac{\mu}{\zeta} \cdot x\right)$$

If the measurement of the β-radiation is performed by means of a single-pulse counting system (e.g., by means of a counter tube), the activity of the radiation source must have an upper limit so that dead time losses remain small enough to be neglected. However, there must be sufficient activity so that statistical error of the activity measurement is as small as possible. When determining the mass per unit area using β-radiation absorption, the weight per unit area of the empty filter is also measured. With ambient air measurements, irregularities in weight per unit area of the empty filter are not small enough to be neglected compared to the particle load, and therefore, the weight per unit area of the empty filter tape section must be determined and taken into consideration. In the case of low particle content of ambient air, radiation intensity will be changed only very slightly by collection of particles on the filter tape. This makes it possible to work in the linear approximation; however, it requires a very stable measuring arrangement. In addition, it must be ensured that the measurement result is not affected by disturbing effects, such as the inherent activity of the particles. This type of interference has less influence the greater the activity of the β-radiation source used for measurement. From a measuring technology perspective, a very active radiation source has definite advantages; however, it is favorable to keep the activity low enough that the units do not violate the radiation protection regulations of individual countries.

Some devices presently used for monitoring the air or offered on the market differ substantially in technical realization of the common measurement principle. As radiation sources, for example, ^{14}C sources with an activity of less than 100 μCi as well as ^{85}Kr radiation sources with 50 mCi are used. However, all devices do have one thing in common: measurement proceeds cyclically. One such measuring cycle begins when the dust-loaded filter tape section of the previous cycle is removed from the measuring arrangement. Each cycle contains a measurement of the new, unloaded filter tape section and a time span, usually adjustable, during which the sample air is allowed to flow through the filter. Normally, flow times are between 30 min and 24 hr. Selection of the most favorable flow time in each case is dependent upon the particle content of the air, the sample flow rate of the unit, and the design of the measuring equipment. Typical air flow rates[13] are between 1 m^3/hr and 2 m^3/hr with effective filter areas of approximately 1 cm^2. This high airflow rate along with the relatively small filter area increases corresponding demands of the tensile strength and collection efficiency of the filter material used. Measurement results are usually defined as the difference between weight per unit area of the empty filter and that of the filter after a certain flow time. With a flow time of 12 hr a particle content of the air of 5 $\mu g/m^3$ can still be measured.[13] Since attenuation of β-radiation is, in the first approximation, independent of the size and the chemical composition of the particle, the water component, inasmuch as it is retained in the filter, is also measured. Thus, the units are sometimes offered and operated with a heating suction pipe[12] used to heat the sample gas (for example, to 80°C) before the particles are collected on the filter. Loading of the filter with particles must either be kept so low by means of an appropriately chosen flow time that the airflow rate of the pump is not substantially changed, or the airflow rate must be held constant by means of a control system. A constant and accurate airflow rate is not only important for determination of mass concentration with β-ray absorption but also for a further evaluation of dust and particle samples collected on the filter tape, e.g., by means of X-ray fluorescence or other analytical methods.

C. Photometric Measurement

In the sector of airborne dust and aerosol measurement, optical (i.e., photometric) measurement methods offer some advantages from the technical point of view. With a suitably chosen photoelectric arrangement, it is possible to continuously record the particle content of air over lengthy periods. An optical measurement does not affect the suspension state of the particles. Airborne particles can be detected directly without problems involving possible aggregations or chemical changes of the particles on the filters. The devices used for photometric measurement are reasonably priced, simple to operate, and nearly maintenance-free. However, from the measurement effect, photometric measurement is primarily not a measurement of mass concentration.

Basically two measurement methods come into consideration for optical dust measurement:[14]

1. Light transmission method: the attenuation of a beam of light having intensity I_o is measured after passage through the measurement medium

2. Light scattering method: the intensity of the scattered light at a certain angle or angular range to the incident beam is measured

In both cases, the incident beam of intensity I_o

undergoes an attenuation corresponding to the Lambert-Beer law:

$$I = I_o \cdot e^{-Kx}$$

where I_o = intensity of the incident beam; I = intensity after passage through the sample medium; K = coefficient of extinction; and x = length of the beam path through the sample medium. The coefficient of extinction K can usually be considered as being made up of two components:

$$K = K_A + K_S$$

The absorption component K_A describes the dampening of the penetrating electromagnetic wave due to absorption. The scattering component K_S takes into consideration the attenuation of the primary beam by diffraction, refraction, and reflection.

In the case of the light transmission method (also known as the absorption measurement), attenuation of the primary beam is measured as the sum of the effects which contribute to this attenuation. Particle concentrations in ambient air and the absorption paths x, which can normally be realized, result in very small extinctions Kx so that the linear approximation $I = I_o (1 - Kx)$ can be used. In the measuring technique, this means that very small changes ΔI of the large quantity I_o must be detected. Thus, transmitted light measurements have not been widely used in the field of ambient air measurements. On the other hand, in the field of emission measurement, with its substantially higher particle concentrations, transmitted light measurement is one of the most important measurement methods whereby light is transmitted directly through the flue gas stack.

The scattered light method offers technically more favorable conditions at low particle concentrations. In this method, intensity of the scattered light in a direction other than that of the primary beam is measured. The light which reaches the photodetector is directly the measured quantity. Theoretically, scattering of light by particles is described by Mie's scattering formulas discussed in References 14–16. It is, however, not within the scope of this work to explore the details of these formulas. According to these formulas, the intensity of the scattered light is dependent upon the incident intensity, the angle at which the scattered light is measured, the quantity, size, shape, optical properties (complex refraction) of the particles, and upon the wavelength of the original incident light. Scattering angle and particle size especially can affect the scattered light intensity in certain circumstances by several orders of magnitude. For this reason the question arises whether a relationship between optical properties and total weight or total volume of the particles exists. The basic relationship of scattered light to weight or the gravimetric calibration factor of the instrument and its range of validity is of most interest. Of course, a photometer for scattered light can be calibrated gravimetrically for a specified type of particle with fixed composition and size distribution. On the other hand, it is not monetarily feasible to construct a unit that displays total weight for any arbitrary particles. In practical application of light scattering photometry for determining mass concentration, for specified instrumental parameters such as wavelength of the originally incident light, scattering angle, and type of gravimetric calibration, the effect of the variations to be expected in size distribution and composition of the particles must be estimated. However, in the field of ambient air measurement, relationships are somewhat simpler than they might first appear from the great number of influencing parameters. Under usual conditions, variously shaped particles occur with a grain size distribution which, outside the emission zones, is primarily dependent on humidity. In addition, although the particles have various chemical compositions, their optical behavior is, to a large extent, determined by an opaque solid component and accumulated water. In the first coarse approximation, it can be said that the influences of the various grain size distributions and various material compositions on the optical properties, are mainly effects of the water component. The water component can at least partially be eliminated with additional instrumental devices, e.g., by predrying the test gas with heated sample lines and detectors. Practical experience with light-scattering photometry has shown that the effect of moisture has led to remarkably higher scattered light results compared with gravimetric measurements when the relative humidity is about 65% or higher.[17] In addition, the influence of various particle parameters can be averaged in part by means of a suitably designed monitor. Units currently available on the market usually use incandescent lamps as light sources with approximately white light and do not measure at a defined

scattering angle but detect the light over an angular range. This leads to an averaging effect on wavelength and scattering angle dependencies. During the gravimetric calibration procedure, the monitor is fed with particles of a certain size and shape distribution so that even in calibration, size distribution and diversity of shapes occurring in atmospheric particles is, or can be, included. Mostly polystyrene particles with a size distribution maximum at 1 μm diameter are used for gravimetric calibration of light scattering photometers. Measuring equipment that is suitably calibrated and designed is scarcely affected by normally occurring shape fluctuations,[19] color, and index of refraction.[18] The main source of error, as compared to gravimetric measurements, is the variation of particle size distribution. Using an instrument with an incandescent lamp as a light source and an evaluated scattering angular range of 7 to 22° to the forward direction, Figure 2 qualitatively shows the dependency of the signal on the particle size. The dashed lines result as limiting curves by considering the effects of the index of refraction and other normally disturbing influences.

In order to eliminate influence of fluctuations in the light source intensity and in photodetector sensitivity, currently offered units are usually designed as dual beam monitors. Thus, the scattered light intensity is measured in relation to a reference beam which, instead of passing through the sample gas in the sensor chamber, passes through a gray filter or a light scattering standard. Air to be examined for its particle content is pumped through the sensor chamber in a flow which is as laminar as possible. The sensor chamber itself should be designed in such a manner that contamination, and thus measurement error, need not be considered or need only be considered after a long period of operation and so the chamber can be cleaned simply. The influence of contamination, in addition to the already mentioned instrumental parameters, are important criteria for assessing the technical design of various light scattering photometers.

Finally, it should be noted here that light scattering photometers directly detect optical properties of the particles. By means of the optical properties, information about mass concentration is received using a gravimetric calibration. However, in all cases in which optical properties are the primary measurement quantities, e.g., when moni-

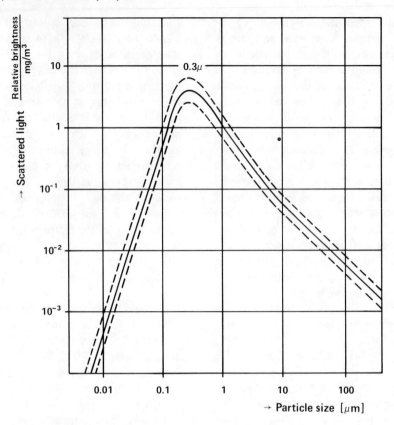

FIGURE 2.   Scattered light as a function of particle size. (From Sigrist Photometer, Zurich, Switzerland. With permission.)

toring the visibility in vehicular tunnels, the
photometric methods can be used directly.

III. PARTICLE COUNTING METHODS

The particle counting methods include methods
and instruments used to determine the number of
particles per unit volume as well as those used to
determine the particle size distribution. In the
scope of ambient air measuring technology, particle size distribution is of special interest in
particle counting. The measured results are given
as the number of particles per unit volume and per
particle size range. The number of particle size
ranges used to the approximate detection of
particle size distribution has a strong influence on
technical expenditure and cost of the instrument.
Usually, the units are equipped to count the
airborne particles in two to eight size ranges. Less
detailed information, giving only the total number
of particles per unit volume, on the other hand,
has not been widely used as a measured quantity

compared to the mass concentration measurement.
However, it has found its application in other
fields, such as surveying clean rooms. The result of
a particle count can, assuming a certain particle
shape, be converted into a mass concentration
using an average particle size and an average
specific weight. In doing so, spherical particles are
usually used as a basis.

As measurement methods for particle counting,
optical methods are principally used. Using these
methods, particles with diameters larger than 0.2
to 0.3 μm can be counted. In addition to optical
particle counters, instruments are available that
determine particle count on the basis of ion
accumulation. Using this method, detection of still
smaller aerosols is possible.

A. Optical Counting
The simplest optical particle counters are those
which measure the total number of scattering or
absorbing particles in a test volume. In principle,
these devices are constructed similarly to the

photometers mentioned in Section II.C. If, however, the particles are not only to be counted altogether but within particle size ranges, each particle must be detected individually. If, in the case of the photometric measuring equipment mentioned above, the test volume in the sensor chamber and the particle density in the sample gas is reduced until only one particle is in the sensor volume at one time, the particles can be detected individually during the flow of sample gas through the sensor volume. Similar to the photometric measurement, particle counters are also divided into units which use transmitted light and those which use scattered light.

In the case of particle counting using transmitted light,[20] the sample gas flows through a capillary tube and passes by a slit or a window correspondingly dimensioned (Figure 3). The sharply focused light beam passes through the sample gas usually perpendicular to the flow direction. The beam is detected by a photodetector. Each particle in the sample gas that flows by the window interrupts a portion of the light beam corresponding to the particle size. Thus, a specific pulsing reduction of the output signal at the photodetector is caused, with a reduction

proportional to the particle size. The pulsing reductions are counted and evaluated with a pulse-height discriminator according to their height and allocated to the various particle size ranges. Pulse width and, as a result of the frequency response characteristic of the evaluation electronics, also the pulse height depend on the flow speed or on the velocity of the particle through the light beam. Thus, correspondence of pulse height to particle size exactly applies only to a fixed sample gas flow rate. The effective sensor volume, in which the particles are detected, is defined by the cross section of the sample gas line and by the diameter of the light beam. Reduction in intensity detected at the photodetector, caused by the passing of a particle, in the first approximation, results from the ratio of particle cross-section area to test volume cross-section area.

In the case of particle counting in scattered light,[21,22] the light of an incandescent lamp is commonly used as a basis. Using slits and focusing lenses, the light beam is focused to a sharply defined spot as small as possible. The sample gas then flows through this focal spot. If a particle in the sample gas is transported through the focal spot, light scattering at the particle results and a

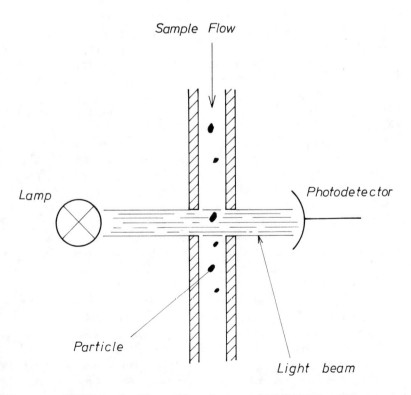

FIGURE 3.   Particle counting using transmitted light (schematic).

flash of light outside the primary beam is perceived. Depending upon the optical properties and the particle size, light is scattered to all sides with more or less intensity corresponding to the scattering theory of Mie. Conditions favorable for the apparatus are achieved if in the sensor chamber of the particle counter the scattered light is detected not only at a single angle but at a whole angular range (outside the primary beam range) using elliptical or parabolic mirrors. Usually the measurement is made in the area of the forward scattering, i.e., at angles $\alpha < 90°$ to the forward direction, because the effects of varying refraction are less noticeable here than in the range of back scattering.[21] The effective sensor volume is determined by the spatial expansion of the focal spot and by the cross-section of the sample gas flow. Flashes of light scattered from this sensor volume into the angular range detected are measured using a photodetector. The intensity of the flash of light is, in the first approximation, a measure for the particle size. Accordingly, the voltage pulses delivered by the photodetector can be evaluated using a pulse-height discriminator and then allocated to the individual particle size ranges.

Particle counters that use transmitted light, as well as those that measure with scattered light, count particles or pulses per unit time. The result is converted to particles per unit volume by means of the known sample gas flow rate. The counting process must be reinitiated after each counter period. Thus, in principle, the monitor works discontinuously and can be automated by using appropriate timers.

Errors in optical particle counting can be caused by the shape of the particles and by coincidence effects. The more the particle shape deviates from the spherical shape (i.e., the larger the difference between the major and minor diameters of a particle) the larger the effect of orientation of the particle to primary beam direction on the measured quantity. In the case of very differently shaped particles, errors thus caused are averaged only in the case of higher counts. Coincidence effects[23] as a second source of error occur if two or more particles are in the effective sensor volume at the same time. The smaller the sensor volume, the higher the particle density in the sample gas which can be detected without coincidence errors. Efforts are thus being made, above all by narrowing the beam as much as possible in transmitted light measurements or the

focal spot in scattered light measurements, to minimize the effective sensor volume. One lower limit in reducing the optically effective area is determined by diffraction at the slits and by the optical arrangement used.

Units currently available on the market vary greatly in their efficiency depending upon the technical design. One unit is available as a transmitted light particle counter[20] which can be used to detect particles with diameters $> 2\mu m$. With a sample gas flow rate of roughly 20 ml/min, particle concentrations up to 10^7 particles per liter can be counted. With measurement using scattered light, particles having diameters $>$ ca. 0.3 to 0.5 μm can be detected using the devices available[21,22] on the market. With sample gas flow rates, which vary greatly from unit to unit, particle concentrations between 10^3 and 10^5 particles per liter can be detected. Higher particle concentrations can be detected using upstream dilution systems.

B. Particle Counting by Ion Accumulation

Accumulation of ions in the air on airborne particles can, under suitable circumstances, be used for dust measurement or particle counting.[25,26,29] The effect of interest is, in principle, the same as used for the operation of ionization fire detectors. Ions produced in the air by ionizing radiation (e.g., from a radioactive substance) have a tendency to accumulate on aerosols. The term ion, as used in this chapter, means ionized air molecules on which some neutral molecules have accumulated. Mobility of these ions in an electric field is dependent upon the polarity of their charge. If constant ionization is generated in a particle-free ionization chamber by means of a long-lived radioactive substance, a constant ion current flows when a fixed chamber voltage is applied. To prevent recombination of these ions, the chamber is operated in the saturation range. If air containing particles is sucked through the ionization chamber, the ions attach themselves to the particles in the air and become large ions which, due to their reduced mobility, do not contribute to the current. This, in turn, reduces the chamber current. This reduction in current can be recorded as a measure for the particle concentration.

The relation between ionization current and particle concentration can theoretically be determined for steady state conditions.[25,26] The change of ion density with time due to creation and annihilation can be described[24-26] with:

$$\frac{dn}{dt} = q - \eta nz$$

where n = ion concentration; q = created ions per unit volume and time; η = accumulation coefficient; and z = concentration of the suspended particles. A presupposition held in this formula is that the ionization chamber is operated in the saturation range, i.e., all ions are carried away by the voltage. Recombination can be neglected and the ionization current achieves a value determined by radioactivity. Because the current i is proportional to the change in ion concentration per unit time, it follows that:

$$\frac{d^2 n}{dt^2} = -\eta z \frac{dn}{dt}$$

$$\frac{di}{dt} = -\eta zi$$

$$\ln i = -\eta zt + K$$

According to Bricard,[27] $\eta = cr_s$ whereby c is a constant and r_s is the radius of the particle. If it is assumed that i = i_o at z = o, and if the lifetime T of the ions is introduced for t, this results in:

$$\ln \frac{(i_o)}{(i)} = cr_s \cdot zT$$

$$i = i_o e^{-cr_s zT}$$

The lifetime T of the ions depends on the geometry of the ionization chamber and on the mobility of the ions whereby, for a more exact evaluation, different values must be taken for positive and negative ions.[25]

The measured quantity of this method is the product of particle concentration and radius. This means that even very small particles contribute to the measurement result as long as there is a sufficient quantity. In contrast with the optical methods, this method can also detect particles smaller than 0.1 μm. With the ion accumulation method, the range of condensation nuclei is reached. The concentration of small and minute suspended matter that varies in time and location, creates an effect in the recorded result that is superimposed in an additive manner on the results from larger suspend particles. Thus, results obtained using this method cannot be directly compared with those of other particle measurement methods without further evaluation.

Ionization of the ionization chamber (i.e., the concentration of ions created) is, when operating in the saturation range and at constant activity of the radioactive source, proportional to the density of the air. Density of the air is a function of pressure and temperature. In addition, the accumulation constant c has a temperature dependence resulting from the temperature response of the diffusion coefficient of the ions. According to Bricard,[29] c is proportional to the square root of the absolute temperature. Thus, the temperature fluctuations of the air must especially be taken into account when evaluating the results or they must be eliminated by means of thermostatic control. An additional possible source of error can be found in the conductivity of the sample air. This intrinsic conductivity of the air must be eliminated by using a so-called ion trap at the inlet. This ion trap removes all mobile ions from the sample gas before entry into the ionization chamber.

The average residence time T is inversely proportional to the chamber voltage. To increase sensitivity, it is practical to select the smallest chamber voltage possible whereby the lower limit is defined by the conditions of saturation operation. For measuring high particle concentrations, sensitivity can be reduced by increasing the chamber voltage.

In technical realization of an instrument on the basis of the principle described, a cylindrical capacitor is used as measuring and ionization chamber[28,29] through which sample air is sucked (Figure 4). Immediately after passing through the inlet opening, the air passes two metal grids at different electric potentials that serve as the ion trap. The sensor volume is also enclosed by metal grids and by the outer electrode so that the sensor chamber metal grids are at the same potential as the outer electrode. The opposite electrode of the ionization chamber is a metal rod or pipe on the central axis of the cylinder. The ionizing γ-radiation substance ($^{60}C_o$, 5 μCi) is spread on the interior surface of the outer electrode so that as homogeneous an ionization of the chamber as possible is ensured, and the formation of a space charge is avoided. The magnitude of the air flow rate is selected so that charged particles of low mobility cannot contribute to the ionization current, and the ions generated by the substance

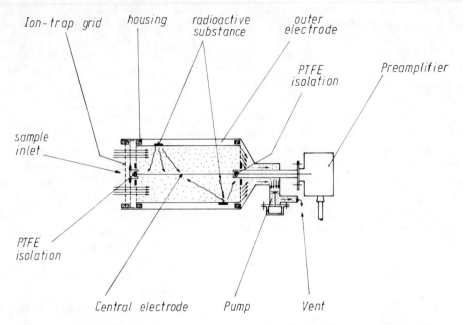

FIGURE 4.   Schematic drawing of an instrument on the basis of ion accumulation. (From Forschungsanstalt für Strahlenmesstechnik, Vaduz, Liechtenstein. With permission.)

are not sucked out. When these conditions are taken into account, suction speed of the sample gas does not affect the result. The measurement result is given in particles per unit volume whereby a fixed particle size distribution or a fixed average radius of the particles is used as a basis. Using the design mentioned here as an example with a cylindrical capacitor as sensor chamber, particles in the range of 0.02 to 5 μm at concentrations between 10^2 to 10^6 particles per cm^3 can be detected.

IV. MONITORING THE COEFFICIENT OF HAZE (COH)

The measurement of the COH has, in common with the method for determining the mass concentration using β-radiation absorption mentioned in Section 2, Part 2, air sucked through a filter, and the particle load of the filter is determined in an automated cycle using a timer and a filter tape. However, evaluation is performed in this case optically. Automated filter tape collectors with optical measurement of the particle component of the air retained on the filter have been used since 1925.[30] These are very simple units with low initial and maintenance costs, in which the values are automatically obtained at the end of each exposure time of the filter. Optical determination

of particle loading of the exposed filter, expressed as the so-called "soiling index," can be determined in two ways.[31,32] One possibility is the measurement of the COH proposed by Hemeon et al.[33] where the measurement is made by transmission, i.e., the attenuation of a light beam is determined after passage through the filter. The second possibility is presented by measurement in reflection. For this purpose, the measured quantity proposed by Gruber and Alpaugh[34] is the reflectance unit of dirt shade (RUDS).

Mainly in the U.S., the transmission measurement has become widespread for determining the COH. According to the Lambert-Beer law for transmission measurement:[33]

$$I = I_o e^{-kQ}$$

$$\log \frac{I_o}{I} = kQ$$

where I or I_o is the intensity with or without attenuating quantity of measurement respectively, k is a proportionality constant, and Q is a measure for amount of particles precipitated on the filter and therefore also a measure for the loading of the filter. The measured loading of the filter is a function of the concentration of particles in the air and is dependent upon the size of the filter area and on the volume of air sucked through the filter.

Optical density ($\log \frac{I_o}{I}$) must be normalized with reference to the filter surface and the volume of the air sample. According to ASTM D1704-61 (Standard Method of Test for Particulate Matter in the Atmosphere), the COH is thus defined as COH/1000 ft:

$$COH/1000 \text{ ft} = \frac{kQ \cdot \text{filter area} \cdot 10^5}{\text{air sample volume}}$$

where the filter area is given in ft^2 and the volume is ft^3. The measurement is taken as a difference between the loaded and the unloaded filter.

In several European countries, mainly in Great Britain, the amount of the particles precipitated onto the filter was determined by measurement of reflected light. This method was employed mainly in the past. Reflectance of a spot on the filter was specified in that the instrument was set to 100% reflectance for clean filter paper and to 0% reflectance for matte black material.

Reflectance measured in this way is converted into mass concentrations in $\mu g/m^3$ by means of a smoke calibration curve. The British Standard Smoke Calibration Curve[32] is given for values of reflectance R between 99 and 40% as:

$$C = \frac{F}{V} \cdot (91679.22 - 33320460 \text{ R} + 49.618884 \text{ R}^2$$

$$- 0.35329778 \text{ R}^3 + 0.0009863435 \text{ R}^4)$$

and for reflectances between 40 and 20% as

$$C = \frac{F}{V} \cdot (214\,245.1 - 15\,130.512 \text{ R} + 508.181 \text{ R}^2$$

$$- 8.831144 \text{ R}^3 + 0.0628057 \text{ R}^4)$$

where C = concentration in $\mu g/m^3$; V = volume of the air sucked through the filter in ft^3; and F = a constant (equals 1.00 for a 1-in. filter).

Although measurement results are given in $\mu g/m^3$ with the given conversion formulas, it should be mentioned at this point that this is not a gravimetric determination of mass concentration and cannot be one. In transmission as well as in reflection, primarily the blackening of the filter due to dust coating is measured. The Working Party on Methods of Measuring Air Pollution and Survey Techniques, Organization for Economic Cooperation and Development notes[35] that blackening of the filter spot cannot be a direct measure for mass concentration of particles in the air, and that the result of the conversion of reflectance into $\mu g/m^3$ can be viewed only as an indication for the concentration of dark-colored particles in the air. The experimental work of the group has, however, shown that this conversion yields good results for dark smoke. The same applies for the COH/1000 ft. A conversion of the concentration into COH/1000 ft or reflectance requires a consistent coloring of the particles or a consistent portion of dark-colored particles. Empirical conversion factors or calibration curves for conversion[31,32] of the quantities COH/1000 ft, reflectance, and gravimetrically determined mass concentration one into another are to be viewed as having this limitation. Generally, the color of the particles varies from source to source and from place to place.

V. MEASUREMENT OF DUSTFALL

Measurement of dustfall represents, at least in the European area, a method to detect ambient air dust loading that has been widely used for many years. The term particulate precipitation means pollution matter in the air (in a solid or liquid state)[36] that falls from the atmosphere within a certain time onto a horizontal surface near ground level. The "fall out" of these materials is due to gravity, turbulent diffusion, and washout with natural precipitation such as rain, dew, and snow. The precipitation of the particles is generally gathered with suitable collectors. After evaporating the water, dust fall is determined by weighing the dry residue. In this manner, neither water nor volatile materials contribute to the dustfall. The particle size ranges of particles detected in this way cannot be determined precisely. They depend, in a complicated manner, on external conditions, especially on meteorological conditions. Thus, for example, during a rainfall, many particles with small diameter are trapped by raindrops falling to earth and are transported to lower air layers or discharged onto the ground itself.

To date, two basically different methods have been used to detect particle-shaped precipitation: collection using collector jars and adhesive surfaces. The effective collection surface or collection inlet opening is, in both cases, placed horizontally, roughly 1.5 to 2 m above the ground. When selecting the location, care should be taken that obstacles which can influence air movement, such as trees or buildings, are at least ten times as far away from the location as the obstacle rises over the collecting surface.[36] The amount of particles precipitated onto the collection surface or via the collection inlet opening is converted into dustfall according to the following equation:

$$X = \frac{G}{F \cdot t}$$

where X is the mass of the dustfall (g/m^2 d), F is the collection surface in $m,^2$ and t is the collection time (days). As a rule, the collection time is 30 days.

A. Collector Jars

Atmospheric precipitation is, in this method, collected in sample jars with a horizontal inlet opening. Units generally consist of a collection funnel and a sampling jar connected to this funnel. In individual cases, it can also consist of the collector jar alone. In the collector jar, the solid material as well as the component bound to natural precipitation is collected. The collected materials are dried in evaporating pans and then weighed. As units of this type of collection equipment, the following should be mentioned: the English standard unit,[37] the Bergerhoff

unit,[36,38] the Hibernia unit,[36,39] and the Löbner-Liesengang unit.[36]

These units differ in material and design of the sampling surface as well as in the form of the collector jar. Precipitated particles are not again blown out of the collector jar due to turbulence of the air layer near the ground. In addition, disturbance of the results by leaves, insects, or birds should be avoided as far as possible. A very simple unit with favorable qualities is the Bergerhoff unit, introduced in Germany as the standard method and sketched in Figure 5. However, due to its large inlet opening, it is poorly protected against leaves and insects. The Löbner-Liesegang unit and the Hibernia unit are equipped with funnels (especially the flat funnel of the Löbner-Liesegang unit) that may allow already precipitated substances to blow away when wind velocity is high. In comparison to the Bergerhoff unit, both units have larger collection surfaces in the funnels. Before replacing the sampling jar, the funnels must be carefully rinsed. Effenberger has investigated the comparability of the results.[40] Accordingly, the following is valid for a coarse conversion:

$$X_B = 1.4 \, X_L$$

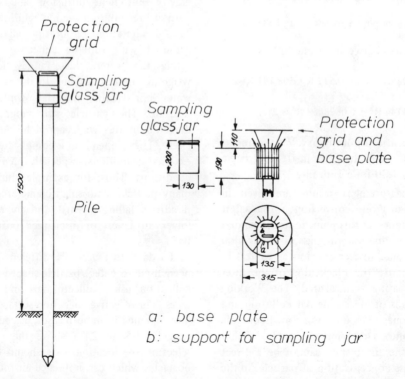

a: base plate
b: support for sampling jar

FIGURE 5.   Bergerhoff unit (schematic). (From VDI-Richtlinie 2119, VDI-Verlag, Düsseldorf, Germany. With permission.)

$$X_B = 1.09 X_H$$

where X_B = measurement result of the Bergerhoff unit; X_L = measurement result of the Löbner-Liesegang unit; and X_H = measurement result of the Hibernia unit. When more exact evaluation is required, the conversion factors must, in each case, be determined by comparative measurements with the units in question, considering the point of location and meteorological conditions.

B. Adhesive Surfaces

With the adhesive surface method, particle precipitation is collected on a horizontal, fixed surface, which is coated with an adhesive material, and gravimetrically determined. The advantage of this method is that the particles maintain the shape they had at the time of collection. This method is especially suited for detecting the propagation of specific solid particles because microscopic evaluation can be performed. Since natural precipitation (rain) of sufficient intensity causes a pronounced washout effect and because the adhesive foil units are not suitable for completely detecting the particle precipitation defined above, deviations of the results in comparison to those of the units with collector jars are to be expected. Comparisons between measurements using the adhesive foil unit according to Diem and Jurksch[41] and the units with collector jars[42,43] resulted, on the average over a lengthy period of observation of a month and more, in a particle precipitate amount greater by a factor of 1.5 to 2 for collector jars.

The adhesive foil unit,[36,41] used here as an example, consists of adhesive foil, the foil holder, and a protective grid. The foil, made of blank hard aluminum with a thickness of 0.07 mm had the dimensions of 49 mm ×83 mm. The collection surface was provided with a thin, ca. 0.2-mm thick adhesive layer of technically pure Vaselin. The basket-shaped protective grid was designed to keep birds and other animals away from the foil. The foils were dried for roughly 24 hr in a desiccator or for 2 hr in a drying cabinet at 40°C before weighing.

Collection and evaluation of dustfall is a simple method for determining the relative ambient air loading which can be performed without great expenditure. Which of the various units is chosen depends mainly upon the measurement task, for example, level measurements, specific measure-ments in the case of complaints, etc.[44] Depending upon the type and shape of the unit used, different measured results can be determined for the same amount of particle precipitation. The measured results of various devices can only be related to one another using extensive comparative measurements. In so doing, it should be remembered that the average conversion factors mentioned, which were determined from the comparison of a collective of values, can lead to large errors when converting individual values. Thus, in the case of dustfall measurements, the type of unit used should be reported and the method described precisely.

In addition to gravimetric detection of precipitated particles, proposals were also made for optical detection. The simplest possibility is to suspend the collected particles in a liquid and to measure the extinction of the liquid using a light beam.[45] Another possibility is to determine the particles precipitated onto a surface by means of the attenuation of reflected light.[46] These methods can be used if the dustfall measurement is considered to be a relative measurement for the visual effect of the dust loading instead of the dust loading itself. To date, this type of evaluation has not achieved general significance. Moreover, precipitated particles can be used for further physical or chemical examinations, regardless of whether adhesive surfaces or collector jars are used.

VI. LIDAR METHODS

Light detecting and ranging (LIDAR) measurement follows a principle similar to microwave radar using light instead of microwaves. With LIDAR, a short light impulse of high intensity, usually generated by a laser, is emitted. If this light impulse falls on a reflecting object, the distance between sender and object can be determined from the propagation time after reception of the reflected light. If the path is through the free atmosphere, a greater or lesser portion of the light will be scattered back, depending upon the composition of air the light beam passes through. The following effects can contribute to the backscatter light:[47-49] Raman scattering, Resonance-Raman scattering, Resonance or fluorescence scattering, and Mie scattering.

With the exception of Mie scattering, the effects are due to the molecular properties of the gases present in the air through which the light beam passes. Thus, they will not be discussed further in the following text. However, Mie scattering is generally due to particles (without size limitations of the particles), which is described by the scattering formulas of Mie.[16] For a particle diameter d which is small in comparison to the wavelength λ of the light, Mie's scattering formulas are converted into the scattering formula according to Rayleigh as discussed in Reference 14. In the Rayleigh range, the backscatter intensity is proportional to $1/\lambda^4$, while in the remaining range of diameters, it depends in a complicated way on the wavelength and the physical properties of the particles. In the formula of backscatter signal, the pulse of light can be treated as a stream of energy, and similar considerations can be used as in radar technology.[50-52] Thus:

$$N_e = \frac{N_s c \tau}{2} \cdot \frac{A}{4\pi R^2} \cdot \beta_{180} \cdot e^{-2\sigma R}$$

$$= \text{const} \frac{1}{R^2} \beta_{180} \cdot e^{-2\sigma R}$$

where N_e is the received power; N_s is the transmitted power; c is the velocity of light; τ is the duration of the light pulse; A is the effective area of the receiving optical system; $\frac{A}{4\pi R^2}$ is to be set equal to the solid angle from which the scattered beam is received; R is the distance to the reflecting or backscattering object. β_{180} is the volume backscatter coefficient; β_{180} at distance R is defined as fictive area per unit volume which would produce the same signal at the receiver per unit of solid angle as occurs due to natural isotropic scattering of light in the unit of volume in distance R. σR is the extinction; the factor of 2 in the exponents indicates that the beam is attenuated travelling back and forth. It is

$$\sigma R = \int_0^R K(r) \cdot dr$$

where $K(r)$ is the extinction coefficient of the atmosphere, which is dependent on location.

Outside the resonance absorption bands of the various gases, extinction is due to absorption and scattering of particles. If backscatter from fixed targets is excluded, extinction and volume backscatter coefficients are essentially determined from the same scattering processes. According to theoretical considerations,[14,16,53,54] in the Mie

range with $d \geqslant \lambda$ at a fixed wavelength and changing particle size, backscatter amplitude can fluctuate greatly, and a simple dependence of the backscatter coefficient on particle loading cannot be expected. However, for particles with considerable absorption, these fluctuations decrease rapidly with increasing absorption,[55] and the correlation between volume backscatter and particle density increases. Added to this is the effect of the average value formation in the case of statistically distributed and shaped particles.[55,56] Relative information concerning visibility, ambient air particle concentration, smoke density, etc. are thus possible in principle, but they contain some uncertainties. Disregarding the range of the first 100 or 200 m in which the Rayleigh scattering on the air molecules can still have some importance (depending upon intensity and wavelength of the emitted laser beam) the intensity of the light reflected per unit volume is, in the first approximation, dependent upon particle concentration in the regarded unit volume. When the intensity of the reflected light is recorded as a function of time, changes in particle density along the path of light become noticeable as deviations from the intensity decrease determined by

$$\beta_{180} \cdot \frac{1}{R^2} \cdot e^{-2\sigma R}$$

The intensity of the reflected light as a function of the transit time of the emitted impulse of light can be recorded with relative simplicity using a cathode ray oscilloscope. The sweep is triggered by a signal derived from the transmitted impulse of light. The output signal of the photodetector is shown on the Y axis. In order to evaluate these oscilloscope pictures, recorded within the transit time of the light, it is recommended that the oscilloscope traces be photographed.

Among currently introduced measuring equipment and methods for detection and assessment of ambient air particle loading, the LIDAR measurements have some importance because information is received which cannot be obtained or cannot be obtained in this manner with the other methods. Air space over urban areas, so vital for the protection of the population, is accessible for LIDAR measurements up to several kilometers high. Particle density fluctuation in this air space can be measured. In this manner, not only the height of clouds and fog zones can be detected, but also the height of layers with increased particle loading which occur during the formation of inversion layers.[47] A favorable aspect of LIDAR measurements is the fact that they can be repeated every few seconds, since light impulses can be transmitted every few seconds.[47,56]

Not only individual measurements with all the uncertainties connected with local atmospheric fluctuations can be made, but the behavior in time and the formation and disintegration of atmospheric layers can be recorded. However, evaluation of individual photographed oscilloscope pictures requires, for this purpose, too much time and expenditure. It is more favorable not to photograph the signal itself but to use it to modulate the brightness of a vertically running oscilloscope beam[47,57-59] and to record many of these brightly modulated beams one beside another on a single picture. In this manner, records are obtained as the one shown in Figure 6.[57] Each transmitted laser impulse of light corresponds to one of these vertical lines. The path of the impulse of light in the atmosphere is depicted by the oscilloscope beam which runs concurrently from the bottom to the top. The layers on the path of the light impulse, which reflect more light due to their particle loading, are pictured as bright areas on the oscilloscope vertical line. Brightness is a function of intensity of the backscattered light in each case. Height is obtained from the transit time of the impulse of light. The impulses of light are transmitted vertically in this case. Figure 6 shows a recording published by Uthe,[57] made on 13 August 1971 above St. Louis, Missouri, U.S. The recording shows the behavior during the day of a pronounced layer of haze. It can be seen that in the course of the morning, the haze and the cloud layer above it disintegrate. At noon a convection begins, caused by the warming of the ground. This convection carries the heavy polluted air of the city into higher layers of air and thus builds up a new layer of haze.

This type of recording is favorable for investigating haze layers and for detecting increased air pollution in inversion weather situations. It offers the possibility to study the formation, disintegration, and dynamics of pollution layers.

Producing pictures as shown in Figure 6 requires very expensive equipment compared with other particle measuring methods. This is true for the oscilloscope and the automatically timed opening of the camera as well as for the signal amplifier and the LIDAR unit itself. The layout of LIDAR units is usually based on giant-pulse lasers as light sources. In practice, ruby lasers (0.694 μm) and neodymium-yttrium-aluminum-garnet (YAG) lasers (1.069 μm) are suitable for this purpose. With these lasers, peak powers between 1 and 100 MW at pulse durations between 10 and 50 nsec can be achieved. Thus, ranges between 10 and 50 km can be reached at good visibility conditions. The YAG laser is especially recommended when constructing small, portable, air-cooled units. The suitability of various units for measurements in the field of air monitoring is strongly dependent upon the reproducibility of the output pulse and the stability of the receiver. It should be ensured that variations in the output pulse do not substantially exceed ± 5%.

VII. COMMENTS ON COMPARABILITY

The various methods for measuring particulate matter in ambient air samples use very different measuring effects. Thus, the question of comparability of measurement results arises. How thoroughly the result of one measurement method can be converted into that of another can theoretically be estimated in very few cases. Atmospheric dust and aerosols are differently shaped particles of uncertain composition. Statistical size distribution of these particles need not be constant. Thus, the parameters or parameter variations necessary for theoretical evaluation are not known at all or are known only approximately. The attempt to convert results from different measurement methods into one another by means of an empirical, generally valid relationship presupposes a certain stability of the particle parameters. For this purpose, it is assumed that composition of the particles in ambient air far from its source is independent of special emittents due to thorough mixing of the atmosphere during residence time. Thus, parameter fluctuations can more or less be neglected in the statistical mean. It is obvious that these types of conversion relationships between the results of various methods can only be derived from a comparison of a collective of measured values. When applying statistically averaged conversion factors to individual measured values, relatively large errors can occur. Determination of average conversion factors and evaluation of their range of validity is made more difficult by the fact that not only the measurement method itself but also technical design and location of the measuring

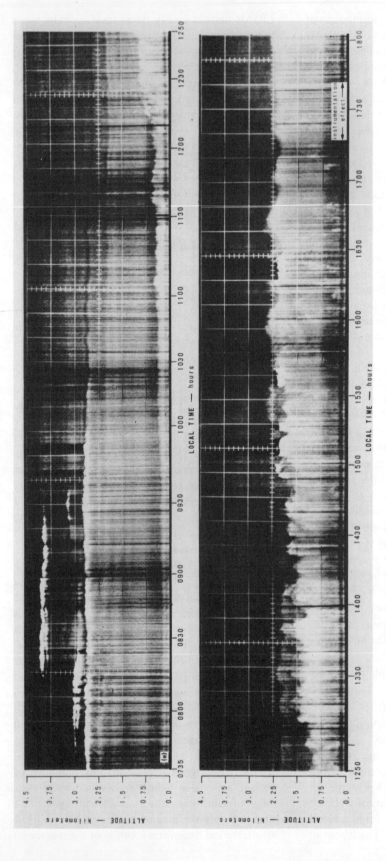

FIGURE 6.   Height/time cross section of the aerosol structure over St . Louis, Missouri. On August 13, 1971 as observed by SRI/EPA Mark VIII LIDAR system. (Reprinted with permission from the American Institute of Aeronautics and Astronautics, Uthe, E. E., *Joint Conference on Sensing of Environmental Pollutants*, American Institute of Aeronautics and Astronautics, New York, 1971.)

device in each instance must also be included. If the measuring units customarily used are compared, it must be taken into account that various particle size fractions with differing moisture components of the particles will be detected by differing measuring effects and by differing filter functions[1,3,5] of the sampling units. As a result, derivation of a conversion function with a nearly general validity is made more difficult if not impossible. Frequently the principle differences between the measurement methods become insignificant compared to those caused by the unit and the location. The values given in the literature, when the question of their general validity is raised, are to be viewed with this reservation.

Above all, factors or functions used for converting various results into mass concentration specifications are of special interest. The high-volume sampler is the reference unit. In conclusion, several results of comparative measurements of various methods with those of the high-volume sampler are quoted as examples. For a more detailed discussion of these conversion factors or conversion functions, the original publications should be consulted.

Comparative measurements within mass concentration measurement methods (to be more precise between the high-volume sampler, a dispersion photometer, and a unit using a β-radiation absorption basis) were performed in the heavily loaded area of Mannheim, West Germany.[60] The relatively small collectives of measured values resulted in the following regression line for the β-radiation absorption unit.

$$y = 0.91x + 18.2$$

and for the photometer with a straight sample line:

$$y = 0.96x - 37.7$$

where x is the measured value of the high-volume sampler, and the units of x and y are $\mu g/m^3$. A range up to ca. 350 $\mu g/m^3$ was covered with these measurements. Tests that were performed using the photometer showed a relatively strong dependence of the regression line on the manner of taking the sample and on the sample gas conduction.

In the aforementioned comparative measurement,[60] a particle counter, using the ion accumulation as a basis, was also included. The following regression line resulted for this device:

$$y = 0.18x + 35.8$$

Using the particle counter, particles greater than 10^{-2} μm were detected. Conversion of particle counts to mass concentrations depends primarily on the assumed average particle shape, on the particle size or a defined size distribution, and on an assumed average specific weight. The closer these parameters are to actual conditions, the better the correlation to mass concentration.[26] Favorable conditions for the conversion into mass concentration are particle counts per size ranges, so that the conversion can be performed with different parameters for different size ranges.

Comparative measurements between the high-volume sampler and the determination of the COH were performed in the New York area.[31] According to these measurements, a mass concentration of 75 $\mu g/m^3$ corresponds to a 0.53 COH/1000 ft. It should be noted that with the values for conversion quoted as examples, results are only valid for the location at which the measurements were performed. Especially in the case of measurement methods where great significance is attached to color of the dust, extrapolation of results to other locations is not possible without further work.

REFERENCES

1. VDI-Richtlinie 2463, *Messen von Partikeln in Aussenluft,* VDI-Verlag, Düsseldorf, Germany, 1974.
2. Junge, C. and Scheich, G., Studien zur Bestimmung des Säuregehaltes von Aerosolteilchen, *Atmos. Environ.,* 3, 423, 1969.
3. Davies, C. N., The entry of aerosol into sampling tubes and heads, *J. Phys. D.,* Ser. 2, 1, 921, 1968.
4. Böhlen, B., Über spezielle Probleme der messtechnischen Kennzeichnung von Schwebestoffen, *Chimia,* 23,(1), 1969.
5. Davies, C. N., Zur Frage der Probenahme von Aerosolen. Der Eintritt von Teilchen in Probenahmerohre und -köpfe, *Staub Reinhalt. Luft,* 28, 1968. Heft 6, S. 219.
6. Clements, H. A., McMullen, T. B., Thompson, R. J., and Akland, G. G., Reproducibility of the HI-VOL sampling method under field conditions, *J. Air Pollut. Control Assoc.,* 22(12), 955, 1972.
7. Jutze, G. A. and Forster, K. E., Recommended standard method for atmospheric sampling of fine particulate matter by filter media — High Volume Sampler. *J. Air Pollut. Control Assoc.,* 17(1), 17, 1967.
8. McKee, H. C., Childers, R. E., Saenez, O., Stanley, T. W., and Margeson, J. H., Collaboratory testing of methods to measure air pollutants. I. High-volume method for suspended particulate, *J. Air Pollut. Control Assoc.,* 22(5), 342, 1972.
9. Aurand, K. and Bosch, J., Gerät zur kontinuierlichen Bestimmung der Konzentration staubförmiger Luftverunreinigungen, *Staub Reinhalt. Luft,* 27(10), 445, 1967.
10. Nader, J. S., A versatile, high flowrate tape sampler, *J. Air Pollution Control Assoc.,* 9, 59, 1959.
11. Fischotter, P. and Dresia, H., Kontinuierliches Messen des Staubgehaltes in Luft und Abgasen mit β-Strahlen, *VDI-Z.,* 106, 1191, 1964.
12. Dresia, H. and Mucha, R., Registrierendes, radiometrisches Messgerät zur kombinierten Messung der Immissionen von Staub und Radioaktivität in Luft, *Staub Reinhalt. Luft,* 34, 125, 1974.
13. Gerätebeschreibungen zu den Staubmonitoren FH 62 A und FH 62 I, Frieseke & Hoepfner, Erlangen-Bruck, Germany.
14. Robock, K., Grundlagen der optischen Staubmessung, *Staub,* 22(3), 80, 1962.
15. Olaf, J. and Robock, K., Zur Theorie der Lichstreuung an Kohle und Bergepartikeln, *Staub,* 21(11), 495, 1961.
16. Mie, G., Beiträge zur Optik trüber Medien, speziell kolloidaler Metallösungen, *Ann. Phys.* Leipzig, 4, 25, 377, 1908.
17. Sigrist, W., Contribution to the Conference in connection with the Int. Air Pollut. Control and Noise Abatement Exhibition, Jönköping, Sweden, September 1–6, 1971.
18. Sigrist, W., Grundlagen und Technologie der Sigrist Photometer, personal communication, 1975.
19. Pinnick, R. G., Rosen, J. M., and Hofmann, D. J., Measured Light-Scattering Properties of Individual Aerosol Particles Compared to Mie Scattering Theory, *Appl. Opt.,* 12(1), 37, 1973.
20. HIAC — Automatic Particle Counter, High Accuracy Products Corp., Claremont, Cal.
21. Aerosol Particle Counting System, Bausch & Lomb, Rochester, New York.
22. A Controlled Environment Monitor, CI-208, Climet Instruments, Redlands, Cal.
23. Pisani, J. F. and Thomson, G. H., Coincidence errors in automatic particle counters, *J. Phys. E,* 4, 359, 1971.
24. Siegmann, H. C., Eine neue Methode der Staubmessung, *Tech. Rundsch.,* No. 2, 3, 1963.
25. Coenen, W., Staubmonitor zur betrieblichen Staubüberwachung, *Staub,* 23(2), 119, 1963.
26. Coenen, W., Registrierende Staubmessung nach der Methode der Kleinionenanlagerung, *Staub,* 24(9), 1964, S. 350.
27. Bricard, J., L'équilibre ionique de la basse atmosphère, *J. Geophys. Res.,* 54(1), 39, 1949.
28. Bricard, J., La fixation des petits ions athomsphériques sur les aerosols ultra-fins, *Geofis. Pura Appl.,* 51, 237, 1962.
29. Staubmessgerät DLM-101 und Staubmessgerät DLM-102, Forschungsanstalt für Strahlenmesstechnik, Vaduz, Liechtenstein.
30. Shaw, W. N. and Owens, J. S., *The Smoke Problem of Great Cities,* Constable, London, 1925.
31. Pedace, E. A. and Sansone, E. B., The Relationship between "Soiling Index" and Suspended Particulate Matter Concentrations, *J. of Air Pollut. Control Assoc.,* 22(5), 348, 1972.
32. Ingram, W. T. and Golden, J., Smoke Curve Calibration, *J. of Air Pollut. Control Assoc.,* 23(2), 110, 1973.
33. Hemeon, W. C. L., Haines, G. F., and Ide, H. M., Determination of haze and smoke concentrations by filter paper samplers, *J. of Air Pollut. Control Assoc.,* 3(1), 22, 1953.
34. Gruber, C. W. and Alpaugh, E. L., The automatic filter paper sampler in an air pollution measuring program, *J. of Air Pollut. Control Assoc.,* 4(3), 143, 1954.
35. Working Party on Methods of Measuring Air Pollution, chap. 2, Smoke, Publication No. 17 913, Organisation for Economic Co-operation and Development (OECD), Paris, 1964.
36. VDI-Richtlinie 2119, *Messung partikelförmiger Niederschläge,* VDI-Verlag, Düsseldorf, Germany.
37. British Standards Institution 1747, Methods for Measurement of Air Pollution. I. Deposite gauges, National Society for Clean Air, Brighton, England, 1969.
38. Bergerhoff, H., Staubpegelzonen nach Sedimentationsmessungen der Landesanstalt für Bodennutzungsschutz des Landes NW, Eigenverlag, Bochum, Germany, 1956.
39. Ost, K. and Mirisch, G., Über Messungen von Staubniederschlägen in der Umgebung eines grösseren Kraftwerks auf Steinkohlebasis, *Mitt. Ver. Grosskesselbesitzer,* 37, 689, 1955.

40. **Effenberger, E.**, Vergleichbarkeit der Messergebnisse der wichtigsten Geräte zur Messung partikelförmiger Niederschläge, *Staub Reinhalt. Luft,* 31(12), 496, 1971.

41. **Diem, M. and Jurksch, G.**, Vergleichsmessungen des Staubgehaltes der Luft nach Niederschlags- und Konzentrationsmethoden, *Staub,* 21(8), 345, 1961.

42. **Kolar, J.**, Die Umrechnung von Staubniederschlägen nach dem Diemschen Haftflächenverfahren in Messwerte des Bergerhoff-Gerätes, *Wärme,* 71(4), 142, 1965.

43. **Köhler, A.**, Über den Regeneinfluss bei Staubniederschlagsmessungen mit Haftflächen und Topfsammelverfahren, *Beitr. Phys. Atmos.,* 36(1/2), 148, 1963.

44. **Effenberger, E.**, Messfehler der gebräuchlichen Staubniederschlags-Messgeräte, *Staub Reinhalt. Luft,* 31(7), 273, 1971.

45. **Lucas, D. H. and Moore, D. J.**, The measurement in the field of pollution by dust, *Int. J. Air Water Pollut.,* 8, 441, 1964.

46. **Esmen, N. A.**, A direct measurement method for dustfall, *J. of Air Pollut. Control Assoc.,* 23(1), 35, 1973.

47. **Birkle, M.**, Lidar — Messverfahren im Umweltschutz, *Messtechnik,* 9, 280, 1973.

48. **Melngailis, J.**, The use of lasers in pollution monitoring, *IEEE Trans.,* GE-10, No. 1, Institute of Electrical and Electronics Engineers, New York, 1972, 7.

49. **Kildal, H. and Byer, R. L.**, Comparison of laser methods for the remote detection of atmospheric pollutants, *Proc. IEEE,* 59(12), 1644, 1971.

50. **Collis, R. T. H.**, Lidar a new atmospheric probe, *Q. J. R. Meteorol. Soc.,* 92, 220, 1966.

51. **Clemesha, B. R., Kent, G. S., and Wright, R. W.**, A laser radar for atmospheric studies, *J. Appl. Meteorol.,* 6, 386, 1976.

52. **Bringworth, B. J.**, Calculation of attenuation and back-scattering in cloud and fog, *Atmos. Environ.,* 5, 605, 1971.

53. **Bryant, H. C. and Cox, J. A.**, Mie theory and the glory, *J. Opt. Soc. Am.,* 56, 1529, 1966.

54. **Van de Hulst, H. C.**, *Light Scattering by Small Particles,* John Wiley & Sons, New York, 1957.

55. **Uthe, E. E.**, Study of laser backscatter by particulates in stack emission, Paper 71-1087, *Joint Conference on Sensing of Environmental Pollutants,* American Institute of Aeronautics and Astronautics, New York, 1971.

56. **Borchardt, H. and Rössler, J.**, Erfahrungen und Überlegungen mit Lidar am meteorologischen Observatorium Aachen, *Ber. des Dtsch. Wetterdienstes,* 16(125), 1971.

57. **Uthe, E. E.**, Lidar observations of particulate distributions over extended areas, Paper 71-1055, *Joint Conference on Sensing of Environmental Pollutants,* American Institute of Aeronautics and Astronautics, New York, 1971.

58. **Hamilton, P. M.**, The application of a pulsed-light-rangefinder (lidar) to the study of chimney plumes, *Philos. Trans. R. Soc. London Ser. A,* 265, 153, 1969.

59. **Allen, R. J. and Evans, W. E.**, Laser Radar (LIDAR) for Mapping Aerosol Structure, *Rev. Sci. Instrum.,* 43(10), 1422, 1972.

60. **Köhler, A. and Birkle, M.**, Streulichtphotometrische und gravimetrische Vergleichsmessungen atmosphärischen Staubgehaltes, *Staub Reinhalt. Luft,* 35(1), 1, 1975.

INDEX

Background reduction by anticoincidence, radioactivity
measured by, 66
Backscattering
electrons, effect of, 143—150
light, measurement with, 277—278
Band characteristics, infrared spectrometry, 211,
213—215, 217, 219, 225
Behavior, particulate
Brownian motion and diffusion, 11
coagulation, 15—16
condensation and evaporation, 15
determination of, 2
electrical mobility, see Electrical mobility, particulate
equivalent diameters, see Equivalent diameters,
particulate
optical properties, see Optical properties, particulate
resistance to gaseous medium, 8—11
sampling procedures, see Sampling, particulate
test aerosols, 23—24
Benzene-extractable compounds, analysis of, 228—234
Berghoff collector, dustfall measurement with, 17,
276—277
Beta filters, use in X-ray diffraction analysis, 182—183
Beta particles, measurement of in air, 67, 68
Beta-ray absorption instruments, use in monitoring,
267—268, 281
Beta rays, detection of, 95
Binding energies, electron, in electron spectroscopy,
192—197, 200, 202, 203
Biological tissues, asbestos-containing, analysis of, 171
Black and white episodes, described, 4—5
Blackening, spark source mass spectrometry,
determination of, 85—86
background blackening, effect of, 85—86
fractional, 85
Blackening of filter spot, measurement by in air
monitoring, 275
Blank/background factor, filters, in X-ray fluorescence
analysis, 34
Boltzmann equilibrium, aerosol particles, 11, 12
Boron, engine exhaust fumes containing, 36
Bragg's law, diffraction measurement, 32, 180
Bricard's theory, accumulation constant, 273
British Standard Smoke Calibration Curve, use in air
monitoring, 275
Brownian motion, in particulate analysis, 3, 11, 12, 14, 19
BSE images, use in electron probe microanalysis, 135, 136
Bubble-chamber method, enclosed, aerosol production for
X-ray fluorescence analysis, 34

C

Cadmium, morphology of, 138, 140
Calcite, analysis of, 185, 213, 214, 223
infrared spectrum, 213
Calcium energy loss spectra, 174
Calcium sulfate, analysis of, 249—251
Calibration curves, X-ray intensity measurements with,
149, 222, 223
Calibration methods in trace and microanalysis, see also
Standard reference materials, 255—260
Calorimetric analysis, described, 239

Camera, X-ray film, used in diffraction analysis,
described, 182
thermal applications, 246
Capacitor, use in particle counting, described, 273—274
Carbon dioxide analyzers, described, 249, 251
Carbon, presence in airborne particles, analysis of,
200—201, 205, 223—227
binding energies, 194—195, 200, 202
hydrocarbons, see Hydrocarbons
peak, 200—202
soot-like volatile, 200—201
transmission spectra of, 223—226
Carbonates, presence in airborne particles, analysis of,
214—217, 223, 225, 244, 247—249
Cascade impactor
atomic spectroscopy, use in, 46
described, 20
infrared spectrometry, use in, 210, 215—217, 220
neutron activation analysis, use in, 109—110
X-ray diffraction analysis, 183, 185
X-ray fluorescence analysis, use in, 28
Cathode lamps, hollow, light source in atomic
spectroscopy, 40
Cellulose filters, use in particulate analysis, 44—46,
63—65, 68, 109, 222, 265
Centrifuges, particulate sampling with, 19—20, 130
Certified natural standards, described, 256
Characterization of individual particles, methods of
analytical methods, survey of, 126—128
chromatography, 228—234
definition of analytical goal, 125—126
electron microscopy, see also Electron microscopy,
156—174
electron probe microanalysis, see also Electron probe
microanalysis, 134—154
electron spectroscopy, see also Electron spectroscopy
for chemical analysis, 194—205
ion probe microanalysis, see also Ion probe
microanalysis, 154—157
light microscopy, see also Light microscopy, 131—134
sampling, 128—131
Charge, resulting from aerosol production, 12
Charged particles, see also Electrical mobility, particulate;
Excitation energy
behavior of, 12
X-ray fluorescence analysis, use in, 32—33, 35
Charge limit, electron and ion, 12
Charging, field, aerosol particles, 12
Chemical analysis, electron spectroscopy for, see Electron
spectroscopy for chemical analysis
Chemical identification of submicron particles, 126, 134,
163—174
Chemical separation and enrichment procedures, atomic
spectroscopy, 46
Chemical shift, electron spectroscopy, 192—194, 198
Chemical states, elements, revealed by electron
spectroscopy, 195—201
Chromatography
gas, use in chemical characterization of particles,
228—234
liquid, use in chemical characterization of particles, 234
Chromium, flame profile of, 48

G

X

Y